MUSICAL MACHINES

Creating Technology for Music

Series Editor

V. J. Manzo, Professor of Music Technology and Cognition
at Worcester Polytechnic Institute

Supercollider for the Creative Musician: A Practical Guide
Eli Fieldsteel

Musical Machines: History, Theory, and Techniques for Makers and Musicians
Scott Barton

MUSICAL MACHINES

History, Theory, and Techniques for Makers and Musicians

Scott Barton

OXFORD
UNIVERSITY PRESS

OXFORD
UNIVERSITY PRESS

Oxford University Press is a department of the University of Oxford.
It furthers the University's objective of excellence in research, scholarship,
and education by publishing worldwide. Oxford is a registered trade mark of
Oxford University Press in the UK and in certain other countries.

Published in the United States of America by Oxford University Press
198 Madison Avenue, New York, NY 10016, United States of America.

Library of Congress Cataloging-in-Publication Data
Names: Barton, Scott author
Title: Musical machines : history, theory, and techniques for makers and musicians / Scott Barton.
Description: [1]. | New York : Oxford University Press, 2026. |
Series: Creating technology for music |
Includes bibliographical references and index. |
Identifiers: LCCN 2025051467 (print) | LCCN 2025051468 (ebook) |
ISBN 9780197599907 paperback | ISBN 9780197599891 hardback |
ISBN 9780197599938 | ISBN 9780197599921 epub
Subjects: LCSH: Mechanical musical instruments—Construction |
Mechanical musical instrument music—Philosophy and aesthetics
Classification: LCC ML1050 .B39 2026 (print) | LCC ML1050 (ebook) |
DDC 786.6/1923—dc23/eng/20260313
LC record available at https://lccn.loc.gov/2025051467
LC ebook record available at https://lccn.loc.gov/2025051468

DOI: 10.1093/oso/9780197599891.001.0001

The manufacturer's authorized representative in the EU for product safety is
Oxford University Press España S.A. of Parque Empresarial San Fernando de Henares,
Avenida de Castilla, 2 – 28830 Madrid (www.oup.es/en or product.safety@oup.com).
OUP España S.A. also acts as importer into Spain of products made by the manufacturer.

To my family, who are my center.

To my students, who are collaborators, teachers, and inspiration.

TABLE OF CONTENTS

FIGURES

TABLES

ACKNOWLEDGMENTS

Everything that I have accomplished has grown out of the boundless love and support of my family. My parents set the foundation; my mom was my first music teacher, led me in philosophical ponderings, and helped me become a better writer. My dad showed me the benefits of dedicated work, thorough evaluation, and a meticulous eye for detail. My wife and children are the lights of my life. They love and believe in me, and I in them. I thank the Good Lord for this pack that brings joy to all that I do.

My journey into the world of musical machines began during my master's work at the Brooklyn College Conservatory of Music, where I studied composition with Amnon Wolman. Amnon suggested that I compose for the Disklavier, a MIDI-controlled acoustic piano, which I did for my thesis *Etchings in Ice*. The experience opened my eyes to the musical potential of pairing computer control with acoustic sound production. It was a moment (of many) when a teacher had a profoundly positive influence on my life.

My involvement in the field of machine music took off exponentially when I began my doctoral studies at the University of Virginia in the Composition and Computer Technologies program. An engine of this journey was powered by my friends and collaborators, Troy Rogers and Steven Kemper. We formed EMMI (Expressive Machines Musical Instruments) during this period, which presented learning experiences around every corner in engineering, music, performance, collaboration, and fun. I learned a great deal from both of them and deeply appreciate the work and time we spent together. My teachers were equally impactful during this period in music, technology, and all points in-between; I am fortunate to have studied and collaborated with Judith Shatin, Matthew Burtner, and Ted Coffey.

I started as an Assistant Professor of Music at Worcester Polytechnic Institute (WPI) in 2012. I founded the Music, Perception, and Robotics Lab and established a suite of courses and project opportunities in musical robotics. These experiences introduced me to extraordinary students who have impressed me with their intelligence, knowledge, creativity, and dedication. The projects that are featured in this book, on the web, in publications, and in performance are the result of my collaborations with them. They have been instrumental in helping me realize my ideas and continue to inspire and educate me. This group includes (but is not limited to) Rayce Stipanovich, Nathan Hughes, Tim Murcko, Paulo Carvalho, Anurag Kesari, Mihir Acharya, Yihao Zhou, Ethan Prihar, Andrew Walter, Karl Sundburg, Tanuj Sane, Linda Baker, Alex O'Brien, Peter Mitrano, Michael Sidler, Matthew Bisson, Jordan Grotz, Sean O'Neil, Ian Gelman, Benjamin Mattiuzzi, Tommy Maloney, Kyle Reese, Hitanshu Shah, Kyle Mitchell, James Obermaier, Kieth Ruenheck, Tair Kareneyev, Chris Bove, and Beckett Peale. There are many more not listed here; I appreciate all of you.

I am grateful for the opportunities provided by WPI and the Departments of Humanities and Arts, Robotics Engineering, and Computer Science. Thanks to Kris Boudreau, Kate Moncrief, Mike Gennert, Jean King, and Fred Bianchi for supporting my ideas and career. Thanks to my colleagues and collaborators—Jim Cocola, Craig Putnam, Josh Rosenstock, Charlie Roberts, David Ibbett, Jake Whitehill, Dan VanHassel, Keith Kirchoff, Matt

Sharrock, and Ramon Castillo—who have inspired and educated me, and helped me voice and develop my work. A special shoutout to Nate Tucker, who has been a partner in exploring all kinds of interesting musical territory. Part of the reason I love my job is because I have such wonderful colleagues in the music discipline, and more widely across theater, art, and the humanities. Special thanks to V. J. Manzo for his collegiality, support, and friendship.

I would like to thank all the people who willingly and graciously share their knowledge in public and accessible ways. Their wisdom comes in many forms: academic publications, community forums, websites, online tutorials, YouTube videos, and open-source repositories. I have learned so much from these sources; they are an excellent complement to traditional educational channels (which remain worthwhile). This work is not always lauded, but it is valuable. It is a validation of technology's promise to do good.

Chapter 8—Inventor Perspectives—contains first-person reflections from prominent artists who have shaped the landscape of musical machines. Contributors include Nicolas Anatol Baginsky, Andy Cavatorta, Koji Shibuya, Sergi Jordà, Ajay Kapur, Steven Kemper, Troy Rogers, Eric Singer, Jim Murphy, and Mo H. Zareei. Thanks to all of you for thoughtful consideration of the questions posed, and for your contributions to the field.

ABOUT THE COMPANION WEBSITE

www.oup.com/us/MusicalMachines

Oxford has created a website to accompany *Musical Machines: History, Theory, and Techniques for Makers and Musicians*. Material that cannot be made available in a book, namely, technical descriptions and projects, is provided here. The reader is encouraged to consult this resource in conjunction with the chapters. Examples available online are indicated in the text with Oxford's symbol ⦿.

CHAPTER 1

THE BIG PICTURE

Scope and Purpose

Technology provides artists with tools to cultivate creative expression. Robotics—a synthesis of electrical engineering, mechanical engineering, and computer science—is fertile ground for such expressive exploration, given the computational power, physical capabilities, and opportunities for interaction afforded by modern materials and methods. Disciplinary convergence is at the core of these efforts, resulting in a perspectival and methodological sharing that is technically necessary and fundamentally humanistic. The results of such collaborations are machines that are increasingly able and autonomous yet remain expressions of what is deeply human. Adopting this as a central theme, this book aims to help those interested in exploring the expansive creative potential of musical machines. Doing so requires knowledge of the past, a vision for progress, the technical abilities to realize that vision, and the sensibility to proceed responsibly.

Musical machines are (electro)mechanical devices that produce musical sequences "on their own" using acoustic sonic objects. These systems hold unique expressive promise by combining the power of computer control, the sonic richness of acoustic sound production, and the visual communication of kinetic sculpture. Mechanical predecessors were ubiquitous before the twentieth century but were largely forgotten with the arrival of sonic recording and playback devices. The current interest in *musical machines* recaptures our ancient mechanical fascinations and fuses them with the affordances of the computer age.

Chapter 1 begins by asking questions: Why build musical machines? What motivations drive these efforts? What are the implications of introducing autonomous machinic agents into domains of human expressivity? Key concepts, definitions, and categories relating to musical machines are introduced in Chapter 2, providing context for historical interpretations in Chapter 3. Chapter 4 provides a template for the design of new machines that is applied to the three primary categories of musical machines: percussion (Chapter 5), strings (Chapter 6), and aerophones (Chapter 7). Technical sections are located on the Companion Website ⓟ, providing explanations and projects to help you build skills and knowledge in requisite areas: *Mechanisms*; *Actuators, Circuits, and Code*; and *Sensing, Communications, and Control.* These pages provide a reference to keep at hand as you work through the design process. They are located on the Companion Website to keep current with technological updates, maintain web links to components and content, provide video animations, and facilitate copying and pasting code examples. Chapter 8 is a collection of first-person reflections on motivations and methods from some of the most prominent builders in the field. Shifting from development to usage, Chapter 9 focuses on musical matters, exploring how to compose for musical machines. These myriad topics are synthesized in the concluding Chapter (10), which directs experienced eyes back to the big picture and forward to future possibilities.

Musical Machines. Scott Barton, Oxford University Press. © Oxford University Press 2026. DOI: 10.1093/oso/9780197599891.003.0001

Why Make Musical Machines?

Human musicians are imaginative, sensitive, and technically proficient. They express a range of feelings, working in brush strokes that are broadly ambiguous and microscopically detailed. They are flexible and communicative, adapting to a wide variety of musical scenarios. What do machines add to this already rich environment?

Machines' ability to act autonomously has captivated human attention for hundreds of years. They evoke wonder and a sense of magic as what seems impossible is tangibly realized. This sense of magic remains despite our increasing collective scientific and technological sophistication. Autonomous machines appear as agents in the physical world, yet we know they are not conscious: this irreconcilable contradiction sparks intellectual and visceral intrigue. They motivate associations with the familiar through physical gestures and sound production while simultaneously presenting something unique and alien. They inspire curiosity that seeks to understand how they work. We marvel at their complexity, appreciating the sophistication and diligence of human effort. We esteem designers and builders of such devices, endowing them with a status that encourages others to perpetuate the cause.

Machines are capable in ways we are not. They are fast, powerful, precise, accurate, predictable, and require little to no rest (qualities we aspire to and work toward as musicians, athletes, and workers). They are not constrained by human physiology, such as the number of fingers or the distance between them. They obviate conflicts of calendar and opinion (the latter has pros and cons). They are adept at complex calculations and can complete myriad tasks simultaneously. They can interpret incomprehensible volumes of information, offering new perspectives on past accomplishments and future possibilities.

Machines allow us to focus on what we value. By taking on tedious or undesirable tasks, they free us to devote attention to matters better suited to our talents, more enjoyable, or a combination of the two. Machines can give us more of what is supremely precious: *time*. They provide composers with a vehicle for their works to be performed. It is hard to find people who will dedicate the time and energy required to learn a new work, particularly when the latter is technically demanding. Developing musical machines offers challenges that yield feelings of accomplishment, triumph, and power when overcome. In making them, we replicate our physiological and psychological functionality, allowing us to understand ourselves better.

Musical machines open up possibilities for new art. The synthesis of computational control and kinetic acoustic bodies unlocks expansive realms of timbre, rhythm, harmony, relationships, form, texture, and gesture. As artists converse with these expressive elements, new musical works and styles are born. History has shown technology as a catalyst for musical innovation, from the electric guitar in rock to the synthesizer in electronic dance music (EDM). What kinds of technologies will inspire the development of future music? How can we harness the possibilities of computational precision and complexity while maintaining the acoustic richness and visual stimulation of traditional musical performance? One answer to these questions is the focus of this book: *musical machines*.

A WORD OF CAUTION

The preceding descriptions skew optimistic, but the picture is not all rosy. We may feel liberated by eschewing tedious, difficult, or repetitive tasks, but such moves have consequences. Manual work (writing text, playing an instrument, building a circuit, coding, composing a melody) is essential to building fundamental skills that are the stepping stones to more advanced abilities: thinking creatively, making distinctions, finding relationships, comprehending abstract ideas, communicating with others, and, more simply, being able to do things. By delegating too many of these responsibilities to machines, we risk atrophy of our fundamental abilities, which are related to higher-level aptitudes. For younger learners who have not yet acquired fundamental skills, removing early steps may leave higher platforms of advanced thinking inaccessible. The proliferation of generative artificial intelligence (AI) has rendered these science-fiction concerns into a tangible reality, bringing us face-to-face with critical questions about how we should and will learn, work, and contribute to the world. These fears do not mean that we should halt entirely; rather, that we should proceed with eyes wide open.

WHY NOT MAKE IT IN SOFTWARE?

Designing and building a musical machine requires significant knowledge, skill, experience, time, and money. *Why go to all of this trouble when you can make music in software?* Positing that the capabilities of a musical machine could be replicated in software implies that the destination was known before the journey began. The process of making a physical instrument is an exploration. Some qualities of the instrument are predictable (e.g., the general sound of a drum strike or string pluck), but many are not (e.g., the timbral character of an air stream directed over a custom PVC flute), which become clear while on the journey. Novel instruments cannot be sampled or modeled without building them physically first. The reality and intrigue of instrument building is that divergent purposes and processes produce diverse results. The method depends on a host of factors, both human—motivations, expertise, cultural conventions, aesthetic preferences—and contextual—accessibility and malleability of materials, tools, and timeframes. Some of these factors are inherently physical, leading to sonic and musical destinations that software approaches can only visit after having been erected.

Another problem with the *software question* is that it presumes the software path is easier and shorter. This may or may not be the case, depending on the people in question and the scope of what needs to be researched, designed, and built. Experienced programmers proficient in physical modeling may realize a sound-producing device more quickly and easily with a software approach than attempting to solder circuits and weld metal frames. A mechanical engineer with fabrication experience, an electrical engineer, or a musician new to engineering will have different experiences. The project's scope will also affect how difficult or time-consuming it will be. It is easier to realize a physical musical machine that strikes a drum than one that plays the clarinet. Software environments differ in the kinds of sonic explorations they afford, complicating the question of which approach is the easiest or fastest. It cannot be assumed that one method is easier or faster than another in general. Define your goal and then figure out the different ways you can get there. The approach you take should depend on what you want to do, what you know how to do, and what you want to learn. Going with what you know can make development easier and shorter in some phases, but it may not be the best way to bring your vision to fruition.

Musical machines combine the benefits of computer control with the richness of acoustic sound production in ways that software instruments do not. The surfaces of an acoustic instrument radiate in complex ways depending on their properties, which, when combined with a room's qualities, produce vivid sonic nuance. Recordings of acoustic instruments used in software samplers reflect a limited number of these facets in a process mediated by microphones, amplifiers, and converters. The ability of a physical model to comprehensively reproduce acoustic complexities depends on the accuracy and thoroughness of experimental measurements, the mathematical equations used to model them, and the algorithms that implement those equations. Whether or not the results of physical models can be distinguished from the real thing is an empirical question, but what is clear is that the task can be exceedingly difficult. Marc-Pierre Verge, co-founder of Applied Acoustics Systems (a company that makes digital instruments based on physical models), reflects: "the real interest of physical modeling is not so much how realistic it can be but how natural and expressive it is." (*AAS-Tech Talk—Physical Modeling*, n.d.). Signals produced from samplers or physical models are mediated by digital-to-analog converters, amplifiers, and speakers, which impart their character to the sound.

Musical machines' visual forms and movements create interest and clarify the causality of sound production. Their embodiment invites human interaction in ways that computer keyboards and screens do not. Their forms evoke associations in the minds of musicians and audiences that invite reflection about the nature of and relationships between humans, machines, creativity, expression, and culture. They present us with dimensions within which we can see, feel, and listen. Loudspeakers are amazing instruments but are not particularly exciting to look at in live performance, nor do they reveal anything about the causality of sound production.

Physical musical machines are idiosyncratic in ways that software instruments are not. Nominally identical parts vary in size, response times, noise, movement, and friction, even when carefully designed and fabricated. This variation has musical consequences in pitch, rhythm, timbre, dynamics, and articulation. It contributes mechanical noise, which can help shape an instrument's sonic character (and distract if too conspicuous). Idiosyncratic characteristics are a significant factor in determining which ideas sound good on the machine and which do not. They change over multiple time scales (e.g., seconds, days, months) as the instrument is subjected to internal and external forces, making them difficult to capture, predict, and model. They must be learned and adapted to in a way similar to working with human musicians. The process and results are fundamentally different when working with software instruments, whose lack of variation is often a defining characteristic.

Human in the Machine

Musical machines are humanistic expressions. They actualize our desires and inspire our creativity. They realize our imagination in the physical world. They are familiar and foreign, inspiring amazement and curiosity about how they work. Leichtentritt (1934) describes eighteenth-century Europe's fascination with automata as a "longing for the irrational, supernatural" (p. 22) to counterbalance rationalistic dispositions. The calculated engineering woven into the fabric of our modern world finds similar relief in artistic domains. Musical machines become *instruments* as they facilitate expression and inspire imagination, essential to music-making. As we learn to use or play an instrument, the abilities and limitations of both player and object are revealed. The engagement is symbiotic: the instrument shapes our musical ideas, which, in turn, define the instrument's identity. These technological objects appear separate and distanced from us, yet they are profound reflections of human identities and values. Making such objects is not just a matter of fabrication techniques or learning to program a computer; it is about understanding and exploring aspects of the human experience we seek to express.

EXPRESSIVITY

Humans are expressive, fueled by the desire to communicate motivations, emotions, preferences, appraisals, and value judgments to others. Conveyances can be direct, powerful, and confrontational, or quiet, restrained, and delicate. Our expressions take root in music, language, and physical movement, shaped by particularities in content and performance. They are devices of a rhetorical strategy that seeks to convince the listener or viewer of an affective state(s) or proposition held by the composer or performer. A symbiotic relationship between performer and listener forms where the listener empowers the performer by entraining to actions (e.g., singing or playing along), validating their skill (e.g., "she is an agile pianist"), or showing that the performer's emotional state or message has resonated in their lives (e.g., crying). These gestures energize the performer's actions and inspire them to create new works that find meaning in audiences' lives.

Artificial objects "cannot be expressive . . . since machines do not have anything to express" (Jordà, 2005, p. 233). They do not feel emotion; they do not embrace cultural traditions; they do not ponder existential questions. They do not fall in love or have their hearts broken, feel part of a group or excluded from it, or champion social change. They feel no compulsion to share anything with other objects or agents. Accepting this, one might think that a machine would flounder in a fundamentally expressive context such as music. Indeed, such attitudes have surfaced: in *The Menace of Mechanical Music*, composer and conductor John Philip Sousa (1906) writes:

> *Let us not hamper [music] with a machine that tells the story day by day, without variation, without soul, barren of the joy, the passion, the ardor that is the inheritance of man alone.* (p. 7)

Interpreting an artificial object as an end in itself with no past or apparent connection to humans is misguided. Artificial objects reflect the goals, desires, values, and cultural conventions of the people who designed and built them. Human expressions inhere in objects and are revealed when the objects are seen, heard, and used. Musical machines are not inherently expressive; they enable people to be expressive (Dobrian & Koppelman, 2006). Some of my most genuine, powerful, human, and spiritual musical experiences have occurred from listening to recordings. Recorded music is mediated by technology through transduction, amplification, mixing, distribution, and playback, which is easy to forget or ignore, given how common and even *natural* the activity has become. We accept that acoustic instruments can authentically communicate human expressions, but these objects are also mechanically endowed. Does the capstan, damper, hammer, or wippen make a piano less capable of human expression than an instrument with fewer mechanisms, such as a violin? Does the amplification, distortion, and feedback of Jimi Hendrix's guitar impede his expression? On the contrary, such technologies facilitate human feeling, creativity, and connection.

It seems then that musical machines, as Jordà says, "can be used to transmit human expressiveness, and they can do that in many ways and with different degrees of success" (2005, p. 233). If we want them to be more successful in expressive endeavors, we need to better understand the circumstances when they fail. What factors contribute to these occlusions of expressive intent?

In the *Extended Lens Model* (Juslin & Laukka, 2004), composers and performers communicate emotions by *encoding* them using *cues* that manifest in musical aspects such as pitch and tempo. These cues do not map to emotions one-to-one; rather, they interact with each other and contribute in proportions depending on the context. Context in musical experience is rich and labyrinthine, influenced by factors such as genre, venue, presentation format, performer gestures, lighting, stage design, and instrumental interfaces (Marynowsky et al., 2023). Listeners *decode* cues and their interactions contextually to understand the meaning of the communication. As this process repeats, *schemas* are formed, which are abstract models of sequences of events that listeners use to navigate musical experiences (Snyder, 2000). They are templates we use to interpret and navigate unfamiliar situations to achieve our goals and communicate with others. Machines' inability to successfully convey emotion occurs because of disparities between a musical experience and the schema it evokes.

Schematic disagreements result in miscommunications. One type of miscommunication occurs when the builder or programmer of a musical machine attempts to convey emotions by encoding them in musical cues as a human performer does, but the listener sees only a machinic performer incapable of feeling anything that could be expressed. In this case, the schema does not provide a map to attribute communication to a person; the listener decodes no expressive cues and concludes no emotion was present. Dobrian & Koppelman (2006) touch on this idea when contemplating a gesture and its trace: "Is the line, then, expressive or does it seem expressive because it is a trace of what I perceive/read as an expressive human action?" (Murata, M. as quoted in Dobrian & Koppelman, 2006, p. 2). Musical machines make physical gestures when producing sound, but the conditions, processes, and apparatuses that generate them differ from those of human performers (to varying degrees). The sonic *traces* produced by human and machine gestures also can vary, but their similarity is independent of the similarity of gesture production (i.e., a non-anthropomorphic machine can produce a sonic trace that is more similar to a human performer's than an anthropomorphic machine). This creates schematic divergence. Breaking the gesture-trace schema creates an impasse for expressive communication.

Another flavor of schematic divergence stems from the relationship between inputs and outputs. Jordà (2005) defines the *efficiency* of an interface as the ratio of input to output complexity. Dobrian & Koppelman (2006) argue that systems that produce complex outputs from simple inputs are often seen as lacking expressivity. The transparency of how inputs are mapped to outputs can also affect expressivity. Clear mappings facilitate expression; ambiguous ones inhibit conveyance. It is no wonder that a musical robot that provides little evidence of its inputs would be seen as deficient in this context. The issue is not a lack of a relationship between inputs and outputs; rather, human actions and expressions (inputs) are temporally and spatially disjunct from their sonic consequences (outputs), obscuring causality and leading to missed or misattributed communications, given that most lack a schema that explains the relationship between the two.

Temporal and spatial distance—between composers, performers, and listeners of music—does not always inhibit expressive communication. Most of the music we hear was composed in a time and place different from the one we inhabit when we listen. Notions of time and place become even more nebulous as recordings are made during different periods, in different places, by different people, and are processed in ways that form virtual identities by obscuring historical realities. For some sounds, the instrument and playing method are apparent, while other sounds provide little indication of either. One might think this would make expressive communication difficult, but a wealth of personal listening experience has shown me the opposite is true. Some musical aspects—a soul-warming melody, propulsive rhythm, or passionate solo—transcend time, place, and media. The contexts differ, which likely means a listener at The Five Spot Café in 1957 would receive a Coltrane performance differently than one streaming a recording of the same show in a 2025 California apartment, but some essential elements persist, connecting us to those of different locales and eras. As musical machines produce such elements, they can link to this communicative web.

Another type of schematic misattribution occurs when a listener decodes expressive cues and attributes them to the machinic performer rather than to the person who made or programmed the machine. Human performers have fantastic emotional and expressive capabilities; a machine assuming a *performer*'s role may be viewed in a similar light (literally and figuratively). It is not that a listener or viewer thinks the machine actually feels the same emotions as a human, but we are disposed to project onto others, so attribution occurs on some level (Marynowsky et al., 2023). Watch the video of Boston Dynamics engineers trying to knock a bipedal robot over: How does it make you feel? (Media Example 1.1 ▶). Schematic associations prevail even when logic tells us otherwise.

Machines' emotional communication is limited as they cannot provide the auditory and visual cues that human performers convey. These cues are numerous and diverse, involving pitch, rhythm, dynamics, timbre, articulations, facial expressions, and body movements (see Juslin & Laukka, 2004, for a comprehensive list). The ability to produce such a range of cues is rooted in humans' extraordinary capabilities. These include speed and strength, but also more nuanced abilities: the delicate proportion of thumb flesh and pick in a guitar articulation, the modulation of breath in a flute passage, or the tension of the roughly twenty skeletal muscles in the human face. Arfib et al. (2005) define expressivity as the capacity to convey an emotion, sentiment, or message. Humans' aforementioned abilities increase expressive capacity. As machines are far more limited in these physical, cognitive, and behavioral domains, their capacity to convey expression is smaller.

How can musical machines better facilitate human expression? In cases of schematic incongruence, listeners and viewers must develop new schemas cognizant of the human efforts behind performative machines. Some of the responsibility falls on listeners to be curious about how things work, learn from those who know more, and seek varied experiences. Some of the responsibility falls on designers, builders, composers, and performers, who can teach audiences about unfamiliar creative artifacts and experiences they present. Doing so can redraw expressive boundaries in agreement with the characteristics and possibilities of the tools used. For example, Tara Rodgers (2003) identifies randomization, glitches, and an *aesthetics of failure* as part of expressivity in digital music practice, equally applicable to musical machines.

Limited parametric variation (e.g., dynamics, timing) contributes to judgments of low expressivity. Greater expressivity can be achieved through continuous parametric variation, extended ranges, higher resolution, and adaptability to different styles (Arfib et al., 2005). These capabilities allow for deviations from the score, another hallmark of expressivity (Lim et al., 2012). Great effort has been devoted to expanding the parametric palette of musical machines. Increased dynamic range, microtonality, portamento (Murphy et al., 2013), physical cues (Hoffman & Weinberg, 2011), timbre (Shibuya, 2011), multiple actuators (Kapur et al., 2011), and body movements (Li et al., 2019) seek to increase expressive potential.

This does not mean that parametric variation is monotonically related to expressive potential. A machine that can produce sounds with greater speed, more temporal precision, and a wider timbral and dynamic range is not necessarily more expressive. To facilitate expressivity, a machine must be capable of technical feats (playing fast, proper intonation) and *virtuosity,* as described by Dobrian & Koppelman (2006), where basic functions are no longer the primary focus, leaving space to devote attention to higher-level matters such as expression. In this way, musical machines, particularly **cooperative** ones, offer new possibilities for human artists.

Expressivity depends on the information a gesture conveys (Camurri & Volpe, 2011). Information emerges from the actor's form, gestures, and context, but it also depends on the *message* transmitted. Machines can communicate some messages better than others. A message may lack affect or valence, which might be interpreted as a lack of expression or a type of expression in itself. These messages are well-suited to musical machines. It can be challenging to clarify that this type of expression is the composer's desire and not the performer's deficiency.

The project of composers is figuring out which messages persist through machinic lines of communication.

Thus, we ask, what is *idiomatic* for machines? Perhaps the goal should not be to make machines better at expressing human sentiments, but to forge a new expressive language that suits machinic characteristics and conventions. The result may be a hybrid in which humans remain the impetus behind machinic performance, but some domain of the experience is reserved for aspects that are machinically idiomatic. Doing so builds a culture *between* humans, robots, and other agents (Samani et al., 2013). In some scenarios, humans suspend disbelief, allowing fictional transmedia narratives of machinic autonomy to thrive, as Davies & Crosby (2016) describe. Exemplified by the robotic band Compressorhead, origin stories are crafted with few public references to the machines' developers or technological constitution. Human audiences clap and whoop during performances. Machines become musicians, performers, celebrities, and producers of culture. The identity and possibilities of these cultures lie not only in those who design and build the machines, but also in those who (inter)act with them.

AESTHETICS

Our aesthetic preferences, shaped by individual and social factors, affect how we evaluate musical machines and their sonic outputs. During a radio show interview, the host asked me what I thought about the news of an AI-assisted work of visual art that had won a fine arts competition. A reporter on the show who covers the art and performance scene in Boston described the work as "exquisite" but also expressed trepidation about how it was made. The process involved providing prompts to an AI (Midjourney, https://www.midjourney.com/home/), evaluating hundreds of output images, and processing the most compelling results. The reporter was discontented that practice eschewed the "hand" of the artist, referring to the "direct" correspondence between generative actions and artistic results. Similar disdain was expressed through the internet's many social media tentacles (Harwell, 2022).

This work reveals a distinction between *artifact* and *process*. The artifact is valuable in itself. If you hear a beautiful melody, does it matter if it was written by someone of your time and place? Does it matter if it was created by singing or configuring note-blocks in a digital audio workstation (DAW)? Does it matter if it was made by a person who combined existing melodies "manually" or by an algorithm doing the same automatically? A work's shapes, sounds, colors, and textures are, *on their own*, communicative, forming a basis for evaluation.

At the same time, ignorance of context and process can prevent one from seeing the value in a musical work. Meaning in Nancarrow's *Study No. 14* is illuminated as one learns the generational algorithms, formal structure, tempo relationships, and piano-roll production central to its development. Ignorance of process can lead to dangerous misinterpretations, particularly when involving the art of different cultures. Engaging with a work's context and processes reveals how human motivations, values, and practices differ. It may be tempting to deem some of these inferior, particularly the ones that diverge from the conventions held dearest. Are pen and paper better than computer-based notation programs? Is recording your own bass line more legitimate than inserting a sample from a downloaded pack? Is composing your own chord progression more valuable than using one generated by a computer? One might be enticed to think that some approaches (often conventional ones) involve more work and require more expertise, and thus are preferable. The practice of audio sampling provides an arena to test this proposition.

Sampling appears to some as a technology-enabled shortcut for making music (Insanul, 2023). It does not require experience composing melodies, chord progressions, or playing and recording instruments; just find a sample and drop it into your DAW. From this orientation, one does not see the sampler's labor digging through crates of LPs in search of quotations to be recontextualized or the complementary music that must be imagined to make a new work. Samplers audition, juxtapose, and superimpose fragments according to their aesthetic preferences. They develop techniques through persistent engagement in their process, resulting in greater efficiency and better results. They try, get frustrated, and find joy in successful realizations. Composers' activities are qualitatively similar, though their attention is directed at different parts of the creative process. It does not follow that one of these approaches is inherently better or more creative: like painting and sculpture, these artists work with different materials.

The expanding generational power of AI means that artists will increasingly assume the sampler's role, curating, collecting, and assembling artifacts produced elsewhere. Musicians who use computers to generate ideas are caught in the same scrutinous gaze directed at samplists. The former designs algorithms, sets parameters, curates outputs, and organizes results. When viewed through the lens of traditional composition, essential parts of the craft are absent. Computer-aided practice may seem easier, lazier, and to require less skill. In some cases, this is true. Writing a one-sentence prompt into an AI music-generator is easier than writing a dense 10-minute orchestral work using pen and paper. Other cases are less clear. What if the prompt was revised 1,000 times, each iteration involving a thorough review of the output and dedicated modifications to the text? What if the orchestral work consisted of an established form populated by conventional chord progressions and derivative themes? What if the user of generative AI also designed and programmed the system? These complications make it challenging to issue universal value judgments about a generalized artistic process relative to another. New methods are accompanied by new kinds of work that may be unrecognizable to those steeped in existing practices. Preferences in these cases reveal less about the quality of a particular piece of art and more about an individual's familiarities, dispositions, and their cultural context.

Endowing an autonomous musical machine with the ability to utilize "its own" aesthetic preferences is a formidable problem. Should it play *riff A* or *riff B*? After how many repetitions of a pattern should something new happen? Which ideas should be developed further? Human musicians know the answers to these questions without extensive reflection. The ideas we like (and those we do not) are intuitively apparent. These preferences are defined in each of us as we navigate the world's food, music, art, clothes, and people, molding our personal identity. Recognizing that aesthetic preferences play an enormous role in the creative process, how can we bestow such characteristics on a machine?

A first step would be to better understand our own aesthetic preferences, which are mysterious despite their significant influence on our lives. Dobrian (1993) argues that a creative decision in the compositional process is made on the basis of one of three possibilities: aesthetic criteria, a system (e.g., an algorithm), or an arbitrary result. Rule-based systems may seem objective, but they were formed by earlier choices settled either by aesthetic criteria or arbitrary processes, leaving us with these two as primitives. When asked, "*Why do you like x*?" the matryoshka doll that is the rationale for aesthetic preferences eventually reveals the answer "because I do," which is difficult to explain further. Alternatively, the answer may be chosen arbitrarily among available options.

Painting a complete picture of the causality and rationale for aesthetic preferences is outside the scope of this book. Here, the interest is in the character of these aesthetic murals rather than a comprehensive explanation of how they came to be. In what domains can aesthetic preferences be defined to shape the identities of artificial agents? One might prefer meters in seven, "clean" timbres, stepwise melodic motion after a leap more than a major 3rd, or secondary dominants within chord progressions. Together, these smaller preferences, which differ depending on context, form a much larger picture that helps define the style of a musical agent. Designers could model such preferences after their own, someone else's, a group of people, or they could make something new. Insights about my own preferences came to light as I designed and programmed systems for robotic musical improvisers. I was confronted with situations in which a musical choice needed to be made or the outcome was not to my liking, requiring me to specify the aesthetic criteria to be used. This is a good way for musicians to learn about their own creative process, regardless of the "success" of the musical machine. These *preference maps* can be the basis for musical choices using rule-based, probabilistic, or random processes.

Machine Learning tools offer powerful ways to approach the aesthetics problem and have been implemented in diverse creative disciplines (Gao et al., 2019; Li et al., 2020). Developers focus on training data and model parameters rather than the myriad individual scenarios that require an aesthetic choice. While this may seem more "objective" than specifying a set of rules and probabilities, developers are still exercising aesthetic judgments when choosing training data and evaluating model outputs to determine if further modifications are necessary. While these choice-making responsibilities can be delegated to other algorithmic processes, such as using recurrent neural networks (RNNs) to judge fitness in genetic algorithms (Mitrano et al., 2017), humans remain in the loop, imparting their preferences on machinic outputs.

THE TURING TEST

Musical machines are inevitably compared to human counterparts. One type of comparison involves the machine performing works that are hallmarks of human virtuosity. Rimsky-Korsakov's *Flight of the Bumblebee* is a popular proving ground for these kinds of showdowns. The interest is if the machine can match the human in speed, accuracy, dynamics, articulation, and "feel" and if we can tell the difference between the two. Distinguishability is the central concern of the *Turing Test*, an idea introduced by Alan Turing (1950) that has exerted great influence on the development and evaluation of artificial agents. The test involves a human judge, a human conversant, and an artificial conversant. The judge interacts with both conversants and tries to determine which is which. If the judge cannot accurately determine which is human and which is the computer, then . . . well, the implications of that scenario have been debated hotly for the last 75 years. Sidestepping that rich history, let us consider a few related points germane to those designing and building musical machines.

First, ends and means should not be confused when making human–machine comparisons. If we consider only the artifacts or traces of a performance, rather than its motivations, intentions, affects, and methods, we risk missing the message. A prominent idea of twentieth-century art is that the concept behind a work is as (or more) important than the artifact. If we are to understand what another person is trying to say—if we are to experience a work of art fully—then the context that preceded the results must be considered. These contexts define human creative experience. This is not a binary situation where a receiver has either a complete set of objective information or none at all. Instead, cues are viewed differently depending on individual perspective. It is not so much a case of getting as many cues "right" as possible; it is about becoming aware of the character and possibilities of the context in which the work was realized. A gesture might have emerged as the expression of deep emotion or the output of an algorithmic process. Which scenario was the case matters; thus, we cannot equate motivations or methods based on the similarity of their results alone. We are reminded of Searle's Chinese Room argument, which warns of the dangers of confusing ends with psychological realities. The tricky part is that motivations and methods are not always apparent, leading to speculation, misunderstanding, inquiry, and discovery, which are inherent parts of the creative experience. Making context clear in the case of machinic musical performers is a philosophical and technical challenge.

Second, human–machine comparisons activate a network of associations, qualities, and judgments that may or may not be appropriate, depending on the machine's purpose. In cases where the designer aims to replicate human performance, direct human–machine comparisons are appropriate. Often, these examples purport to research human physiology, motor control, performance practice, acoustics, or seek to tackle an engineering problem. For those who see musical machines as tools of artistic expression, direct comparisons with human performers may be less relevant. Musical machines are capable in ways that humans are not, so why not explore those possibilities rather than trying to force them into a mode of performance or evaluation that is not idiomatic to them? We do not criticize the accelerating sequence of snare hits in an EDM riser as lacking because its rigidity fails to match the temporal malleability of human performance. The sequence explicitly avoids human emulation and, in doing so, gains its own expressive identity that is characteristically *machinic*. We then need not worry about the results of the Turing Test (or others like it) because, in these cases, the purpose of the test is incongruent with the purpose of the expression.

AUTHENTICITY AND MASKS

Can we connect to machine performers as we do to human performers? As fans, we feel we know the people we admire on stage, but what do we really know about them? To what extent are these identities fabricated out of imagery, mores, and communications that correspond only in part to the people to which they are connected? In listening to an artist sing or play, we feel like the object of their expression, that the artist is speaking directly to us. This connection is real in some ways: an artist conveys something that we understand or feel ourselves. The ideas and spirit of the communication are preserved despite the voyage through intermediary layers (producers, audio engineers, stylists, marketing campaigns, friends' opinions). As an artist's ideas and feelings are authentic, we come to know that person in a meaningful way.

Conversely, the relationship between audience and artist is also illusory. Communication is largely unidirectional (from artist to audience). The expressions of an artist may make a listener feel that they know the artist intimately and comprehensively when, in actuality, what is known is superficial and narrow. The intermediary levels between artist and audience may also be more influential than many listeners think, as sonic artifacts pass through and are shaped by the hands mentioned earlier.

Often, a performer plays the role of a *character*, expressing feelings, ideas, positions, and behaviors that are not their own. An actor on stage can exude calmness through their body language, facial expressions, and vocal intonation, even when they feel nervous or excited. A musical performer may flamboyantly and confidently jaunt around the stage, while in "real" life is humble, insecure, and introverted. These performers are not frauds; rather, they realize that part of performance is to signal qualities—confidence, proficiency, passion—valued by an audience, who gain satisfaction from seeing them actualized and embodied.

Incongruities between a person and their performance identity create a *persona*. How close does the person have to be to the persona for us to find value in a performance? What if the performer is not the person at all, but an *impersonator*: someone intimately familiar with the persona and can produce it to the extent that the audience accepts the rendition as a substitute for the real thing? These questions are increasingly relevant as technologies shape performance identities. Modern performances involve representations of human performers (e.g., holographic projections) where only disembodied traces of the persona are presented. Hybrid cases abound where a person and their representation are present simultaneously, such as in lip-synced performances. What then do we make of a machine that assumes the role typically inhabited by a human performer? How might we further adapt our conception of what a *performer* is?

Representational disjunctions are not necessarily problematic. On the contrary, they are part of the shared understanding that composers, writers, producers, performers, and audiences enter into when making and experiencing art. As viewers, we know a fictional work is not real and that the actors bringing it to life differ from the characters they portray. *Characters* are a fundamental vehicle for human expression, bringing internal imagination to life through external personas that audiences engage with. In performance, we are not focused on the thoughts, motivations, and experiences of the actor (this is less of an issue for a machinic performer devoid of such things); we are interested in those of the character that the actor represents, which are expressions of other humans. Machinic performers are capable of being *characters*, which may be a fruitful direction to pursue when presenting musical machines. Of course, in human performance, the actor is one of the humans who contributes to the expression along with the writer, director, and so on, which is not necessarily the case with machines.

A QUEST TO BECOME MECHANICAL . . . AND THE ART THAT EMERGES WHEN WE DO

We strive to become mechanical in many domains. High-level tennis players and golfers practice their strokes thousands (in some cases millions) of times to become faster, more precise, more efficient, and more reliable. They review the details of their mechanics, tweaking and refining to make gains in any relevant area. Musicians practice similar behavior. Pianists and guitarists play the same finger exercises again and again to build speed, reduce mistakes, and increase fluidity. Many of the attributes that humans seek in such activities—precision, speed, stamina, evenness, repeatability—are the characteristics that define machines.

One reason athletes and musicians develop mechanical abilities is that they can devote attention to other aspects of their performance. Instead of focusing on each part of a tennis swing or the position of each finger in an arpeggio, such actions happen "automatically" without conscious thought. This allows the tennis player to strategically visualize the next sequence of shots or the musician to connect to the emotions they seek to express. The automatization of technique, particularly of low-level coordinated physical actions, frees a person to engage with unequivocally human worlds, including emotion, artistry, and expression. *Becoming mechanical facilitates higher levels of human potential.* What kinds of performances are possible if a machine handles the low-level tasks that it excels at, freeing the human to operate in higher-level domains?

We aspire to the mechanical to gain control. In one model of musical communication prominent in European Art Music, composition and performance are separate processes. The composer specifies instructions in a score, such as instrumentation, pitches, and rhythm, and leaves other elements up to the performer to shape, such as phrasing, dynamics, articulation, and even tempo. Sometimes, performer interpretations diverge from idealized visions, frustrating those composers who seek "to control fully the interpretive aspect of the musical art" (Fowler, 1967, p. 49). One way to achieve such control is to provide more specification in the score, exemplified in the work of composers such as Edgard Varèse. Others seek to make performative choices using technological resources such as digital sequencers, computer algorithms, and musical machines. Assuming a high degree of control has a price: by circumventing human performers to gain responsibility for some aspects of interpretation, you take on all of them, which can be tedious and overwhelming to manage.

THE RESISTANCE

We seek to become more machinic; machines become more like us. Talk of human–machine cooperation agitates fears that machines and humans will become indistinguishable. Some are unnerved as they see an increasing list of machinic abilities accompanied by a decreasing list of ways humans are unique. There are economic consequences of this development when machines can complete tasks more quickly, precisely, or cheaply than humans. The resistance to musical machines is even more pronounced because it engages in artistic realms that have exemplified and defined what it means to be human. Replacement concerns are not new to the musical world, accelerating with the development and proliferation of digital instruments starting in the 1980s. From the Virtual Orchestra (Bianchi & Campbell, 2000) to wedding DJs, there are many scenarios in which machines have assumed performative roles previously held by humans.

Replacement fears catalyze technological aversion, but extremes of this position ignore technology's essential role in human creativity. Mediating objects—*technologies*—have inspired human music-making for thousands of years. Music of the good old days was not made by bands of naked Luddites who eschewed anything artificial; modern music has not cut humans out of the loop altogether. Devices such as the piano, electric guitar, synthesizer, and sampler made new kinds of expression possible and, in doing so, led to the emergence of new works, styles, and identities (rock, hip-hop, techno, guitar heroes, DJs) imbued with the characteristics of the technologies used to make them.

As the capabilities of objects develop, people use them in new ways, altering the roles of both humans and objects in the process. This sometimes means fewer human jobs for musicians (party bands, session players) and those in the music industry (copyists, tape machine technicians, CD manufacturers, and distributors). Some changes to human roles are additions: instrument builders, programmers, sound designers, and engineers. In some cases, technologies offer opportunities for artistic expression, such as a musical theater production using a virtual orchestra that could not otherwise afford an ensemble of human musicians. As a performing musician, I do not want to see anyone lose their job, and I am unwavering in my belief in the magic of human musical performance. At the same time, creators and practitioners of art must accept that the context in which their work is received changes according to a host of social, technological, legal, and economic factors. If audiences find the holographic image or robot DJ more compelling than a live person, then human performers must come to grips with this reality and adapt their practice if they wish to compete. There will always be a place for humans and machines in the theater of musical performance, and the pendulum will swing between the two depending on what is "mainstream" and what is "new" or "innovative." These worlds need not be separate; their synthesis has great potential.

CONTROL

The desire for and fear of machinic abilities return us to the topic of *control*. While technological innovations have been driven by the question of what machines *can* control, an equally important question is: What *should* machines control? Answers depend on human motivations and values. There is great joy and accomplishment in the *ergotic* experience of playing a musical instrument, which involves forces, displacements, and exchanges of energy (Cadoz, 2009). Feeling the energy flow from your core through your arms, hands, and fingers to form a symbiotic physical connection with a dynamic object such as a musical instrument can be euphoric. These moments are accompanied by a powerful sense of control that would be a shame to dilute through machinic re-assignment. At the same time, a machinic collaborator may provide music that inspires your playing, which would otherwise not be realized. In these cases, you may seek a machine that can *control* compositional or improvisatory processes, producing music you can play or play with. In other instances, musical ideas might be there, but instrumental proficiency may not; machines offer a way to control the sound production process while the human controls *what* the machine will play. One of these possibilities is not inherently more valuable than the others; it depends on which ones yield results and feelings the human desires.

In pondering the possibilities of machinic control, remember that machines facilitate human expression. A machine can press keys, hit drums, generate sequences of notes, and determine statistical regularities in a musical corpus, but it does not have inspirations, volition, or feelings; thus, the ways in which it exerts *control* are limited. The highest level of control, which contributes to music's purpose, is born out of our awareness, ability to feel, and need to connect to others. This level of control compels us to stretch conventional boundaries while remaining connected to heritage. It moves us to refine techniques that will capture the attention of audiences. These aspects remain characteristically human despite increasing machinic involvement in domains such as music. How that essential human control manifests varies depending on the technological intermediaries present, but these fundamental aspects persist.

SYMBIOSIS AND PRESERVATION

As we make and use tools, we realize what we find useful, interesting, definitive, or fun. As our tools can do things we cannot, they express our goals and dreams beyond our cognitive and physical limitations. Machinic abilities may result in unique creative expressions, but they are not entirely distinct from our own. They are an extension of ours, germinating from our ideas and values. Technological development is an expression of human creativity. At the same time, machinic agents lack consciousness, empathy, and feeling—the foundations of understanding. The latter will remain a central and differentiating factor for animals, being, artistic expression, and life.

Summary

New types of expression develop as technological tools come into the hands of artists. Robotics is fertile ground for such expressive exploration, given the computational power, physical capabilities, and opportunities for interaction afforded by self-acting machines. The interest here is in the creative possibilities of *musical machines*, which are (electro)mechanical devices that produce musical sequences "on their own" using acoustic sonic objects. These objects hold unique expressive promise as they combine the possibilities of computer control, sonic richness from acoustic sound production, and visual communication enabled by kinetic sculpture.

We start by asking questions: *Why build musical machines? Why build physical machines when you could design sounds in software?* In doing so, we think about the human motivations and implications of such endeavors and the pros, cons, and alternatives to given approaches.

The exploration of these questions reveals that, despite the seeming independence and autonomy of such machines, they are expressions of what is deeply human. This position is supported by deeper contemplations of expressivity, aesthetic preferences, similarities and differences, authenticity, desires, control, resistance, and synthesis within human–machinic contexts.

References

AAS- Tech Talk—Physical Modeling. (n.d.). Applied Acoustics Systems. Retrieved August 14, 2020, from https://www.applied-acoustics.com/techtalk/physicalmodeling/

Arfib, D., Couturier, J.-M., & Kessous, L. (2005). Expressiveness and digital musical instrument design. *Journal of New Music Research*, *34*(1), 125–136.

Bianchi, F. W., & Campbell, R. H. (2000). THE VIRTUAL ORCHESTRA: TECHNICAL AND CREATIVE ISSUES. *Journal of Sound and Vibration*, *232*(1), 275–279. https://doi.org/10.1006/jsvi.1999.2699

Cadoz, C. (2009). Supra-instrumental interactions and gestures. *Journal of New Music Research*, *38*(3), 215–230.

Camurri, A., & Volpe, G. (2011). Multimodal analysis of expressive gesture in music performance. In J. Solis & K. Ng (Eds.), *Musical robots and interactive multimodal systems* (Vol. 74., pp. 47–66). Springer. https://doi.org/10.1007/978-3-642-22291-7_4 ().

Davies, A., & Crosby, A. (2016). Compressorhead: The robot band and its transmedia storyworld. In J. T. K. V. Koh, B. J. Dunstan, D. Silvera-Tawil, & M. Velonaki (Eds.), *Cultural robotics* (pp. 175–189). Springer International Publishing. https://doi.org/10.1007/978-3-319-42945-8_14

Dobrian, C. (1993). *Music and artificial intelligence.* http://music.arts.uci.edu/dobrian/CD.music.ai.htm

Dobrian, C., & Koppelman, D. (2006). The 'E' in NIME: Musical expression with new computer interfaces. *Proceedings of the 2006 Conference on New Interfaces for Musical Expression*, 277–282.

Fowler, C. B. (1967). The museum of music: A history of mechanical instruments. *Music Educators Journal*, *54*(2), 45–49. https://doi.org/10.2307/3391092

Gao, X., Zhou, C., Chao, F., Yang, L., Lin, C.-M., & Shang, C. (2019). A robotic writing framework–learning human aesthetic preferences via human–machine interactions. *IEEE Access*, *7*, 144043–144053. https://doi.org/10.1109/ACCESS.2019.2944912

Harwell, D. (2022, September 2). He used AI to win a fine-arts competition. Was it cheating? *Washington Post.* https://www.washingtonpost.com/technology/2022/09/02/midjourney-artificial-intelligence-state-fair-colorado/

Hoffman, G., & Weinberg, G. (2011). Interactive improvisation with a robotic marimba player. *Autonomous Robots*, *31*(2–3), 133–153.

Insanul, Ahmed. (2023, November 13). *Is hip-hop sampling lazy? The history of sampling is the same arguments over and over again. Complex.* https://www.complex.com/music/a/insanulahmed/hip-hop-sampling-2023-interview

Jordà, S. (2005). *Digital lutherie: Crafting musical computers for new musics' performance and improvisation* (Dissertation, Universitat Pompeu Fabra). https://www.tdx.cat/handle/10803/575372#page=1

Juslin, P. N., & Laukka, P. (2004). Expression, perception, and induction of musical emotions: A review and a questionnaire study of everyday listening. *Journal of New Music Research*, *33*(3), 217–238. https://doi.org/10.1080/0929821042000317813

Kapur, A., Darling, M., Murphy, J. W., Hochenbaum, J., Diakopoulos, D., & Trimpin. (2011). The KarmetiK NotomotoN: A new breed of musical robot for teaching and performance. *NIME*, 228–231.

Leichtentritt, H. (1934). Mechanical music in olden times. *The Musical Quarterly, 20*(1), 15–26.

Li, J., Hu, T., Zhang, S., & Mi, H. (2019). Designing a musical robot for Chinese bamboo flute performance. *Proceedings of the Seventh International Symposium of Chinese CHI*, 117–120. https://dl.acm.org/doi/abs/10.1145/3332169.3332264

Li, J., Peng, H., Hu, H., Luo, Z., & Tang, C. (2020). Multimodal information fusion for automatic aesthetics evaluation of robotic dance poses. *International Journal of Social Robotics, 12*(1), 5–20. https://doi.org/10.1007/s12369-019-00535-w

Lim, A., Ogata, T., & Okuno, H. G. (2012). Towards expressive musical robots: A cross-modal framework for emotional gesture, voice and music. *EURASIP Journal on Audio, Speech, and Music Processing, 2012*(1), 3.

Marynowsky, W., Knowles, J., Bown, O., & Ferguson, S. (2023). Sonic robotics: Musical genres as platforms for understanding robotic performance as cultural events. In B. J. Dunstan, J. T. K. V. Koh, D. Turnbull Tillman, & S. A. Brown (Eds.), *Cultural robotics: Social robots and their emergent cultural ecologies* (pp. 219–235). Springer International Publishing. https://doi.org/10.1007/978-3-031-28138-9_14

Mitrano, P., Lockman, A., Honicker, J., & Barton, S. (2017). Using recurrent neural networks to judge fitness in musical genetic algorithms. *Proceedings of the 5th International Workshop on Musical Metacreation (MUME) at the 8th International Conference on Computational Creativity (ICCC), 6.*

Murphy, J., McVay, J., Kapur, A., & Carnegie, D. (2013). Designing and building expressive robotic guitars. *Proceedings of the 2013 Conference on New Interfaces for Musical Expression (NIME), Daejeon, Korea.* http://www.nime.org/2013/program/papers/day4/paper12/36/36_Paper.pdf

Rodgers, T. (2003). On the process and aesthetics of sampling in electronic music production. *Organised Sound, 8*(03), 313–320.

Samani, H., Saadatian, E., Pang, N., Polydorou, D., Fernando, O. N. N., Nakatsu, R., & Koh, J. T. K. V. (2013). Cultural robotics: The culture of robotics and robotics in culture. *International Journal of Advanced Robotic Systems, 10*(12), 400. https://doi.org/10.5772/57260

Shibuya, K. (2011). Violin playing robot and kansei. In J. Solis, & K. Ng (Eds.), *Musical robots and interactive multimodal systems* (pp. 179–193). Springer Berlin Heidelberg. http://link.springer.com.ezproxy.wpi.edu/chapter/10.1007/978-3-642-22291-7_11

Sousa, J. P. (1906). The menace of mechanical music. *Appleton's Magazine, 8*(3), 278–284.

Snyder, B. (2000). *Music and memory: An introduction.* MIT Press.

Turing, A. M. (1950). Computing machinery and intelligence. *Mind, LIX*(236), 433–460. https://doi.org/10.1093/mind/LIX.236.433

CHAPTER 2

KEY CONCEPTS

Definitions

Many objects can be considered *musical machines*, including pianos, drum machines, synthesizers, computers, and robots. Determining the commonalities and differences between these examples helps us better understand their capabilities, usage, and expressive potential. This chapter starts by defining terms, recognizing that dictionaries and encyclopedias represent particular perspectives at particular times. The meaning of concepts and the words that describe them change in the face of artistic, cultural, and technological innovation.

Machines, Mechanisms, and Instruments

MACHINES

The *Oxford English Dictionary* defines a machine as

> A complex device, consisting of a number of interrelated parts, each having a definite function, together applying, using, or generating mechanical or (later) electrical power to perform a certain kind of work (often specified by a preceding verbal noun). ("Machine, n.," n.d., 6.b.)
>
> A living being considered to move or act automatically or mechanically, rather than of its own volition; *esp.* a person who acts mechanically or unthinkingly, as from habit or obedience; a person who acts with mechanical precision or efficiency. Also: a system or organization of an impersonal or inflexible character. ("Machine, n.," n.d., 8.b.)

and by *Merriam-Webster* as

> an assemblage . . . of parts that transmit forces, motion, and energy one to another in a predetermined manner. ("Definition of MACHINE," n.d., 1.d.1)

A machine is a complex entity constituted by simpler parts that uses energy to perform work automatically. Insensitive ("mindless," perhaps) automaticity has been a fundamental aspect of how machines have been characterized: press *start* and the machine proceeds unknowingly, attempting to lift a load that is too heavy, providing heat when we are already warm, or dispensing candy from an empty vending machine row. This aspect of machinic identity resulted from an inability to sense, process, and react to information from the external world. These machines could be considered *automatic* because they operate with some degree of self-sufficiency, but their lack of awareness (of self or others) contrasts sharply with the autonomy that conscious beings embody. Recent technological developments have endowed machines with the ability to sense, interpret, and react to the world around them. As a result,

Musical Machines. Scott Barton, Oxford University Press. © Oxford University Press 2026. DOI: 10.1093/oso/9780197599891.003.0002

machinic autonomy has expanded, though we should be careful not to equate such abilities with either awareness or consciousness. In any case, our understanding of what a machine is and what it does, and reflexively what a human is and what a human does, must adapt to these changing capabilities.

The forms and characteristics of machines vary widely. Some are enormous, powerful amalgams of metal. Some are ancient, basic tools made of stone or wood. Some use electricity; some are controlled by computer programs. Some are microscopic and biological, such as the protein-synthesizing molecular machine *ribosome*. Some exist physically, some exist virtually, and some fall somewhere in-between. The category of *machine* is flexible.

The type of machines that are the focus of this book convert energy into mechanical forces that move components (e.g., **mechanisms)** when prompted by a control (e.g., an electrical signal). Differentiations are made based on energy source (e.g., internal vs. external, electric vs. fuel-based combustion) and type of motion produced (i.e., rotational vs. linear). The devices that effect motion, including **actuators**, **motors**, and **engines**, are essential to mechanism design.

MECHANISMS

Mechanisms are assemblages of parts that transform input forces and movements into output forces and movements. Mechanisms are a fundamental part of human-playable musical instruments. The *Oxford English Dictionary* illustrates musical connections in its definition of the word "mechanism" ("Mechanism, n.," n.d.):

> 1849 *Brit. Patent 12,378* (1857) 2 My invention consists of so arranging the mechanism of a flute that the closing of the C sharp and the B natural holes may be simultaneous by the action of the second finger.
>
> 1876 J. Stainer & W. A. Barrett *Dict. Musical Terms* 283/2 *Mechanism*, that part of an instrument which forms the connection between the player and the sound-producing portion.

The definition provided in *A Dictionary of Musical Terms* from 1876 (Stainer & Barrett) provides a profound statement that contradicts the opinion that machines separate us from sound and music. Mechanisms connect humans to sound-making objects. *Machines offer access to sound worlds otherwise inaccessible, facilitating human musical expression.*

A mechanism's identity is defined by the creative actions it enables. A piano contains over 8,000 parts, the most complex of which constitute the *action* responsible for coordinating the pressing of 88 keys with the striking and damping of 88 strings. To apprehend the reason for and design of these mechanisms, imagine a time before the piano's invention. Before you sits a large zither (a board with numerous strings tuned to different pitches). You pick up a rod and strike the strings percussively. Desirous of variation, you wrap the end of the rod in felt to make the sound softer and rounder. You make another one of these rods so that you can play with one in each hand. This is fun for a while, but you yearn for the rich harmonies you know exist beyond the dyadic capabilities of your bespoke system. Playing rapid scales and arpeggios is also difficult because you must cross your hands over each other and move your entire arm with each strike. The inability to damp strings occasionally leads to sonic soup and inhibits your ability to articulate specific phrases.

The thought experiment gives insight into the kinds of musical expression the piano's mechanisms make possible. The keyboard interface takes advantage of our numerous and dexterous fingers, allowing for complex chords, rapid scales, and sweeping arpeggios. The mechanical connections between key and hammer allow the velocity of one to translate to the other, allowing for an expansive dynamic range (when coupled with a sizable soundboard). The diminutive relative forces and motions of fingers are amplified into an enormous sound world. The automatic damping system quiets strings when keys are released or left untouched. It is easy to take such ingenuity for granted when mechanisms are seen as a cold intermediary between humans and music, but the opposite is true: mechanisms are both a result and facilitator of human expressivity.

Acoustic instruments misleadingly appear free of technological mediation. Guitars and violins contain tuning machines, pegs, saddles, bridges, and nuts that apply forces that ensure uniform position and tension of strings and the transmission of strings' vibrations to the instrument's body (the "work" of the device). Drums contain mechanical devices to tension a membrane uniformly and reliably. Flutes and clarinets are equipped with key mechanisms that open and close tone holes on the instrument's bore. As these mechanisms accommodate human physiology, they are interfaces that transform the actions of one domain (human) into those required by another (tone hole configuration of an instrument).

INSTRUMENTS

> A tool, implement, or utensil used to execute a piece of work; (now) *esp.* one used in delicate, precise, or skilled work, or for artistic, medical, or scientific purposes. ("Instrument, n.," n.d., 1.b.)

The objects mentioned previously (piano, guitar, violin, drum, flute, etc.) contain mechanisms, so are they *musical machines*? Few designate them as such, opting instead for the term *instruments*. Even complex devices containing many mechanisms, such as a pipe organ, are considered *instruments*, not *machines*. Conversely, (electro)mechanical devices that produce music via prerecorded media (records, CDs, mp3 files) are *machines*, not *instruments*. What if that record-playing machine appears in the hands of DJ Qbert, who uses it to weave stuttering, rhythmic lines of notes and noise with a virtuosity previously unseen and unheard? Is the *machine* then an *instrument*?

The distinction between *machine* and *instrument* is not based on an object's physical characteristics or mechanisms' number, type, or complexity. The difference resides in how the object is *used* or, in the case of music, how it is *played*. Cadoz (2009) argues that musical instruments involve **ergotic** interactions between a player and a material object involving forces, displacements, and exchanges of energy. The extent to which an interaction is ergotic can help distinguish what we mean when we use the term *play*, when talking about interacting with both a musical instrument ("I am playing the piano") and a musical machine ("I am playing a record").

The human *plays* the instrument by physically interacting with it, exchanging energy, experimenting, repeating actions, incrementally improving, and having fun. Developing proficiency requires thousands of hours of practice as player and instrument symbiotically mold to each other's physicality. Forearm muscles strengthen through the piano action's resistance; calluses develop on the tips of guitarists' fingers in response to the tension of steel strings. The development of physical strength, coordination, and technique is mirrored by the sounds produced by the instrument, which are increasingly swift, fluid, and expressive.

Playing a record (or CD, album, song, etc.) also involves musical choices ("Which record shall I play?") that require physical execution (taking the record off-the-shelf and out of its jacket, putting it on the record player, starting the record player, dropping the needle, etc.), but they do not meaningfully affect the essence of the music produced. The fundamental character of John Coltrane's *Giant Steps* persists whether I put it on the record player or you do. One could argue that the work's fundamental character persists whether it was played on (by?) a digital system through expensive monitors in a recording studio or an old record player in a living room (the work would not sound exactly the same in both scenarios, but its essence remains intact). This kind of *play* involves fewer choices and actions that require minimal practice, expertise, and physical exchange between human and object. A human makes a choice, configures the machine accordingly, and starts the process ("press play"); the machine takes care of the rest.

The difference between *machines* and *instruments* is, then, less a matter of technological presence or sophistication and more of human effects on musical outputs. An object becomes an *instrument* when it helps humans express musical ideas, desires, and feelings. The products of these interactions may occur on *low* levels, such as shaping the timbral character of an individual note, or *higher* ones, such as crafting the dynamic contour of a musical phrase. One intriguing possibility of musical machines is that they could perform lower-level tasks, such as generating pitch and rhythmic sequences on acoustic instruments. At the same time, a human musician could tend to higher-level matters, such as tempo, dynamic contour, and orchestration, through computer-based interfaces.

The entrance of digital and robotic technologies into the domain of acoustic sound production invites us to reconsider the previous implication that instrumental playing requires ergotic interactions. Consider someone with physical limitations using an eye-tracking interface to generate musical sequences: are they not playing an instrument? Programming environments and DAWs may require little in terms of forces, displacements, and exchanges of energy, yet can generate a panoply of musical sounds, gestures, and works. If an album of music was made exclusively using such tools, was no instrument played? These scenarios form the character of modern music-making and make definitions of instrumentality rooted in a particular kind of physical relationship problematic. We revise the definition: when humans make choices and perform actions *with* an object whose identities inhere in the music produced, the object is an *instrument*. Choices, actions, materials, and configurations coalesce with their sonic results; we *hear* a player's choices and interactions with an instrument in a musical trace. This opens the door for *musical machines* to be instruments, even though the human inputs are distant or invisible. Whether or not the object is an *instrument* (like a guitar) or a *machine* (like a CD player) depends on the nature of the human input and its reflection in the musical output.

Like any other definition, exceptions and ambiguities abound; the preceding offers a starting point. The relationship between *players* and *instruments* helps us understand extant musical machines and imagine ones yet to be realized. It is a central idea in the later sections of this book.

Acoustic / Electric

The extent to which a technology is acoustic or electric has been a significant factor in defining the identities of instruments throughout music history. The sonic, visual, and associative differences between a Stratocaster and an acoustic guitar or a Yamaha DX7 and a harpsichord are dramatic and have shaped the character of musical genres. Bob Dylan "went electric" at the Newport Folk Festival in 1965 to the consternation of a folk audience's acoustic identity. As music historian Elijah Wald reflects: "what happened at Newport was much more complicated than just Bob Dylan plugging in an electric guitar." (NPR Staff, 2015). The schism did not arise solely because of timbral preferences; instead, *electric* and *acoustic* had become cultural identifiers in a tumultuous political and social environment. They were markers of (in)authenticity that engendered trust, fear, inclusion, and exclusion (see Wald, 2015 for more). While environments, signifiers, and signifieds have changed, the essence of this division, linked to the *nature / technology* duality, remains in play today (regardless of when *today* is).

Accepting that the *acoustic / electric* division is important, how do the instruments in these categories differ? One crucial area of difference involves sound production. Generating sound requires an *action* that **excites** an object into **oscillation**, which is **amplified**. For acoustic instruments, excitation is usually provided by a human physical gesture such that the force, velocity, and range of that gesture has an analogous effect on the sound produced. It is the difference between striking a drum *pianissimo* versus *fortissimo*. Excitation in electronic instruments is more variable, depending on the system's design and mapping. Lightly pressing "p" on a computer keyboard could unleash a 100 dB gong strike, while a homerun-hitting swing against a sensor-equipped plate could produce the sound of two leaves rustling in a light breeze. The lack of definitive correspondences rooted in physical experience is a primary differentiator between *electric* and *acoustic* instruments.

Acoustic and electric instruments differ in how their oscillating components are experienced. In acoustic instruments, a physical object (e.g., a string) oscillates at a frequency and amplitude discernible by our perceptual systems. While the individual oscillations of acoustic elements are invisibly rapid, the holistic activity is visually and haptically apparent. In many electric instruments, the oscillating component is a signal generated by an electronic circuit. The activity of a typical electronic oscillator has no physical presence at all; it is invisible and concealed within an enclosure (for safety and operational reasons). We have a visceral bond with acoustic oscillators that does not mirror in electronic counterparts.

How a sound is amplified, the third phase of sound production, influences how the instrument or machine is perceived and categorized. All instruments generate sound via oscillatory movements that create fluctuations in air pressure (e.g., the soundboard of a piano; the voice coil-cone unit of a loudspeaker); *acoustic* and *electric* instruments create these movements in distinct ways. Acoustic and electric guitars contain tensioned strings whose vibrations are transmitted via bridges coupled to larger resonating objects. The *exciters* and *oscillators* on the two are the same; the difference is that the vibrations of the electric guitar are *amplified* via electromagnetic transduction, amplifier circuits, and speakers, while the vibrations of an acoustic guitar are *amplified* via a physical body (soundboard) that acts as a resonator. In both cases, amplification shapes the amplitude, timbre, and identity of the sound; differences in the amplification method define instrumental identities.

The extent to which amplification methods are **embodied** in an instrument affects the latter's identity. If a microphone amplifies an acoustic guitar, the guitar does not become electric. If a piezo disc is attached to an acoustic guitar's body to generate a signal that can be amplified, the instrument becomes *acoustic / electric. Electric* instruments, therefore, require technological elements to be embodied, attached, or woven into the fabric of the instrument rather than being external to it. Embodiment in the acoustic / electric distinction also plays a role when considering *excitation*. If a human and an electric solenoid-based mechanism produce similar sonic results when pressing a piano's keys, the instrument is still *acoustic*. In this case, the key-pressing mechanism is external to the piano as the microphone is external to the acoustic guitar; the instrument's acoustic nature is preserved.

While associations between **interfaces** and music-making practices exist (e.g., grid-based MIDI controllers and electronic music), an interface does not dictate whether an instrument is acoustic or electric. The musical keyboard was developed as a mechanical device to play acoustic instruments (e.g., piano), but it persisted as the model in the design of interfaces for electric instruments (e.g., analog synthesizers) and digital instruments (e.g., keyboard MIDI controllers that could be mapped to control a panoply of sampled or digitally synthesized sounds). The physical constitutions of these iterations differ, but the fundamental design and role as a conduit between human intention and sound-production system, as an *interface*, does not. Some interfaces are more idiomatic to some musical practices than others (e.g., a grid of touch-sensitive pads suits percussion-playing more than a computer keyboard), which instrument builders should consider as they envision how their device will be used.

Tracing the formation of acoustic / electric identities leads us to the phonograph: a pivotal invention in musical history that drew lines along which the *acoustic / electric* divide formed. A phonograph comprises a rotating disc whose surface is etched with grooves representing a musical performance. A stylus contacts the surface, gliding along the grooves on the disc. The stylus is connected to a diaphragm that creates corresponding fluctuations in air pressure, amplified by an acoustic horn. The device has the primary components of a musical instrument: the disc is the *exciter*, the stylus-diaphragm the *oscillator*, and the horn the *amplifier*. As stated previously, whether the excitation process or interface involves human or mechanical action or not does not solely define an instrument's *acoustic* or *electric* identity. Here, mechanisms are involved in sound production, but electricity is not, so one could argue the phonograph is an *acoustic* device. Later, with the development of the *record player*, the oscillations of the tone arm produce an analogous electrical signal that is amplified through electrical means and projected through loudspeakers. As the transduction and amplification process affect the sound produced, the record player is an *electric* device, technically and musically speaking. Note the sound wouldn't be meaningfully affected by the device rotating the turntable (e.g., manual crank, spring, electric motor), so that wouldn't factor in the distinction. Whether or not the phonograph or record player is a *machine* or an *instrument* will then depend on how it is used, as previously discussed.

The extent to which sound is *acoustic* or *electric* affects our sense of time and place. The experience of sitting in a room listening to, watching, and feeling a musician play a piece of music on a piano is characteristically *acoustic*. In contrast, some recorded *electronic* music leaves little indication of the time, place, people, or processes involved in its realization. Instead, it creates a virtual world whose qualities are endowed through the processing of sonic novelty, association, and the interactions between the two. The relationship between composers, performers, and listeners is changed, which can be metaphorically understood as *distance* or *obfuscation*. Distancing or obfuscating causal aspects does not inhibit electronic music's impact or ability to communicate with listeners, though the nature of that communication differs from that experienced in acoustic realms. At the same time, electronically produced sound can be immediate and visceral in its own ways (e.g., the grain of an overdriven amp or the edge of a sawtooth synth). Spatiotemporal identities depend on the sounds, performance, physical context, and the musical and technical experience of the listeners.

The *acoustic / electric* divide is framed as a binary pair, but the boundaries of these categories are nebulous. If an electric mechanism that presses the keys of a piano is replaced by a pneumatic or hydraulic one, is the instrument fundamentally different? There is an essential distinction between *technical* descriptions and *musical* ones. If we are describing the technical system, then yes, the means used to power actuators is an important detail. If those means have negligible *musical* effects, they do not require distinctions; thus, *electric*, *pneumatic*, and *hydraulic* might be conflated in a category such as *machinic*. If those means affect the music produced, such as in articulation or timbre, they deserve to be acknowledged in musical categories. These are details to attend to in analysis and design (the possibility of different musical results from diverging technical implementations is intriguing).

We should be careful not to consider all examples within a category such as *acoustic* or *electric* as the same, nor should we see *acoustic / electric* identity as a force that always separates. In some pockets of the rock world, an electric guitar is effectively of the same instrumental type as an acoustic drum set, while a synthesizer inhabits another group. The synth and the guitar both rely on electrical technologies and may even be fed through the same amplification chain but are categorically distinguished because of a host of factors, including physical interface, timbral manipulability, attitudes toward technological innovation, and stylistic convention. In other parts of the rock world, such boundaries have eroded or were never erected in the first place. Rigid categorical boundaries dilute explanatory power, which isn't a reason to avoid categorization; it is a reminder that context matters.

Distinctions blur as worlds coalesce. Electronic and acoustic sounds combine in **electroacoustic** musical practice. Mechanical, electrical, and control systems fuse in **mechatronic** devices. Electroacoustic and mechatronic capabilities synthesize to make new kinds of musical instruments. Categories can provide a foundation from which to learn and emulate but moving across their boundaries and combining them can be an exciting source of creative potential.

Control

Early acoustic instruments were controlled directly by human musicians who used their arms, hands, fingers, feet, and mouths to produce sound. Early *electric* instruments such as the *clavecin électrique* (1759), *Telharmonium* (1897), and *Theremin* (1928) were also controlled manually. Both branches of history saw developments that changed the types of human actions required to produce musical sound. These developments resulted in autonomous systems that could control themselves and algorithmic systems and digital instruments that enhanced how humans could control musical outputs.

AUTONOMOUS SYSTEMS

In both *acoustic* and *electric* lineages, *autonomous* systems emerged that could play music "on their own" without direct human control. The latter statement is misleading, for the designing, building, and programming these devices (e.g., arranging the pegs on the barrel) was still the work of humans who wished to hear musical ideas. Humans maintained significant control of the music in these cases, but in ways that were *distanced* in time and place from musical performances. The nature and degree of how *musical machines* function "on their own" vary. Some instruments proceed as instructed, oblivious to the world surrounding them, while others can sense, interpret, and respond to their environment. The systems that control machinic actions differ, defining categories of autonomous musical machines that include *deterministic*, *generative*, *robotic*, *mechatronic*, and *cooperative*.

Deterministic

Deterministic musical machines play musical sequences as they were encoded, whether on pegged barrels, paper rolls, or computer memory. They do not make interpretive choices or generate new sequences. They do not sense or react to their environment and are, therefore, incapable of interacting with others. Most *musical machines* before the twentieth century were deterministic. Relatedly, an **automaton** (from the Greek for "self-mover"), a term used to describe many early musical machines, is "a mechanical device which . . . executes a function on its own and in a completely determined way" (Steinbuch, 1965 as quoted in Koetsier, 2001). Modern mechatronic musical machines can also have this character. Connecting a mechatronic musical machine to sensing and generative systems can change the extent to which it is deterministic, so it is usually more accurate to describe most modern musical machines as operating in a deterministic mode, as opposed to saying they are inherently deterministic.

Generative

Some machines can **generate** musical sequences that are not predetermined. These systems are **algorithmic** as they formalize processes for generating music (Rowe, 2001, *Machine Musicianship*). Algorithmic systems enable musical machines to make some of the choices that human composers and performers typically make, thus expanding their functionality and potential for musical expression. For example, algorithms can determine a sequence of pitches to be played, the temporal intervals that separate those pitches, dynamic contours, and timbral changes.

Robotic

Technological developments have enabled machines to sense internal states and external environments. As information from these sensory capabilities is processed and used to guide (inter)action, a *robot* is formed. Internal sensing devices can indicate components' positions, vibration frequencies, velocities, and acceleration. Internal sensing capabilities improve the efficiency and precision of movements, compensate for physical changes in materials, and protect against dangerous conditions for the robot and those around it. External sensors can interpret sound, physical spaces, expressive gestures made by co-performers, audience reactions, and environmental conditions. For example, a robotic drummer may interpret a melody played by a human guitarist and generate an accompanying rhythmic pattern in response. External sensory abilities enable musical and artistic possibilities through interactions with performers, audiences, and physical environments. A word of caution: "robotic" is often used broadly to indicate a range of mechatronic devices that might be deterministic, generative, or robotic (as defined here). Given the distinctions between these groups are meaningful, the term *robotic* is reserved according to the above definitions.

Mechatronic

Mechatronic instruments consist of (electro)mechanical actuation systems and electronic control systems. For example, a percussion machine that plays a drum via a lever mechanism actuated by an electric solenoid controlled by a musical sequence stored on a microcontroller is *mechatronic*. Mechatronic instruments could be deterministic, generative, or robotic, depending on the software used to control them, so some confusion abounds. Often, systems are distinguished as *robotic* or *mechatronic* depending on whether they have sensing, feedback, and planning capabilities (*robotic* systems can sense, plan, and react accordingly, while *mechatronic* systems are less able in these areas). The distinction between the two partially depends on *automatic* versus *autonomous* action. Like most categories, there is a gray area between these domains, so flexibility in definitions is necessary (see Indri & Oboe, 2020 for further discussion of these concepts).

Cooperative

A **cooperative** musical machine requires a human and a machine acting together to produce music (Barton et al., 2017). The parties involved are responsible for different complementary tasks needed to produce the desired sounds; neither party can do the job alone. The difference between musical machines that are cooperative and those that are not involves the relationship between inputs and outputs and the *level* of musical organizations that are produced. *Non-cooperative* (which is different than *uncooperative*!) instruments have a one-to-one correspondence between input and output. Press one key of a piano and one note sounds. The instrument is always reactive; its musical abilities are *low level* (e.g., individual notes), and it cannot autonomously create *higher-level* organizations such as rhythms or melodies. In contrast, *cooperative* machines can function autonomously (e.g., producing musical sequences), but their outputs are incomplete. They require coordinated human actions (also, by themselves, incomplete) to produce a musical result that is whole and *whose identity is unrealizable without such cooperation*. An example of a cooperative musical machine is *Cyther* (Barton et al., 2017), a zither whose strings are hammered and damped by computer-controlled solenoids from the underside while plucked, struck, stopped, and damped by a human from the top. The two interact to create musical gestures that are only made possible through coordinated action. As with other categories of musical machines, the defining factors are not just about components, technical design, or even capabilities; they are about how the machine is used in the context of the music produced.

EXPANDED CONTROL POSSIBILITIES

Advances in computer technologies since the mid-twentieth century enabled autonomous music-making as their mechanical predecessors did, except this time with vastly expanded control possibilities. Inputs are no longer limited to the predetermined sequences that governed musical automata of the past. Modern computational tools offer a vast array of control possibilities, enabling users to manipulate characteristics of an individual sound, generate a set of complete works, or anywhere in-between. The history of these generative computer algorithms is rich, stretching back to the 1950s and accelerating in recent years with the development of machine learning and AI methods (see Ames, 1987; Fernández & Vico, 2013; Wang et al., 2024 for a review). **Digital musical instruments (DMIs)** provide physical interfaces for humans to access these extended possibilities. These devices have given humans increasingly precise control of musical and sonic details while at the same time offering ways to engage with *higher-level* musical dimensions (e.g., "energy") via *macro*-controls that affect multiple *lower-level* parameters (e.g., note pitch, volume) simultaneously. Combining algorithmic generators, digital interfaces, and robotic acoustic instruments enables humans to exert control of music-making processes in new and exciting ways.

Summary

Machines are complex entities constituted by parts that transform energy, forces, and motion to perform work. They comprise **actuators**, **motors**, and **engines**, which convert energy into mechanical forces that move mechanisms when prompted by a control (e.g., an electrical signal). The mechanisms of machines transform input forces and motion into output forces and motion. Mechanisms can only move in constrained ways and are connected either directly (cam, gear) or via an intermediary (linkage, chain, belt). Mechanisms connect players to instruments and are central to realizing musical ideas, gestures, and expressions.

The distinction between a **machine** and an **instrument** is based on how the object is used rather than on its physical characteristics or the number, type, or complexity of its constituent mechanisms. *Instruments* are objects that help humans express ideas, desires, and feelings in musical forms. Such expression has historically required **ergotic** interactions between *players* and acoustic instruments, but technological developments have changed this relationship, opening the door for instruments that are **musical machines**. An instrument is identified by how it produces sound in terms of **excitation**, **oscillation**, and **amplification**, as well as its human **interface**. Instrument components can be **acoustic**, **electric**, or some combination of the two. How acoustic and electric elements are **embodied** influences the instrument's identity. The value of these categorizations extends beyond taxonomic interest; the extent to which an instrument is acoustic or electric defines how it will be used. (Electro)(Acoustic) identity conveys time and place and shapes musical genres and individual styles. The synthesis between computer control and acoustic sound production blurs the lines between acoustic and electric in ways that can lead to new instruments and music. Thinking about these distinctions helps us learn more about the types of instruments we use and make, but also inspires ideas about substitutions and transformations that could lead to compelling hybrids.

The marriage of acoustic instruments with electronic systems has resulted in autonomous systems that *control* themselves to varying degrees. **Deterministic** musical machines produce encoded musical sequences without interpretation or purposeful variation. **Generative** machines can produce musical sequences that are not predetermined. **Mechatronic** instruments consist of (electro)mechanical actuation systems and electronic control systems. **Robotic** musical instruments can sense their internal state and external environment, generating feedback to generate responses to contextual inputs. Some instruments are **cooperative**, which require coordinated human and machine action to make music. These may be mechanical (e.g., the pianola), mechatronic, or robotic. Examples in all of these categories can be **musical machines**, the defining characteristic being the ability to produce musical sequences "on their own" using acoustic sonic objects. *Musical machines* are differentiated from instruments that solely use electrical circuits to generate and amplify musical signals, such as synthesizers, and **DMIs**, which involve a physical interface and a digital sound-producing engine. The development of computer-based algorithmic tools has expanded musical control possibilities, automating processes and enabling new ways to translate input actions to musical outputs.

References

Ames, C. (1987). Automated composition in retrospect: 1956–1986. *Leonardo*, *20*(2), 169–185.

Barton, S., Prihar, E., & Carvalho, P. (2017). Cyther: A human-playable, self-tuning robotic zither. *Proceedings of the International Conference on New Interfaces for Musical Expression*, 319–324. https://doi.org/10.5281/zenodo.1176266

Cadoz, C. (2009). Supra-instrumental interactions and gestures. *Journal of New Music Research*, *38*(3), 215–230.

Definition of MACHINE. (n.d.). In *Merriam Webster*. Retrieved August 27, 2021, from https://www.merriam-webster.com/dictionary/machine

Fernández, J. D., & Vico, F. (2013). AI methods in algorithmic composition: A comprehensive survey. *Journal of Artificial Intelligence Research*, 48, 513–582.

Indri, M., & Oboe, R. (2020). Mechatronics versus robotics. In *Mechatronics and Robotics* (pp. 1–14). CRC Press.

Instrument, n. (n.d.). In *OED Online*. Oxford University Press. Retrieved July 21, 2022, from http://www.oed.com/view/Entry/97158

Koetsier, T. (2001). On the prehistory of programmable machines: Musical automata, looms, calculators. *Mechanism and Machine Theory*, *36*(5), 589–603.

Machine, n. (n.d.). In *OED Online*. Oxford University Press. Retrieved February 5, 2019, from http://www.oed.com/view/Entry/111850

Mechanism, n. (n.d.). In *OED Online*. Oxford University Press. Retrieved February 7, 2019, from http://www.oed.com/view/Entry/115557

NPR Staff. (2015, July 25). 50 years ago, Bob Dylan electrified a decade with one concert. *NPR*. https://www.npr.org/2015/07/25/424977355/50-years-ago-bob-dylan-electrified-a-decade-with-one-concert

Rowe, R. (2001). *Machine musicianship*. MIT Press.

Steinbuch, K. (1965). Automat und mensch. In K. Steinbuch (Ed.), *Automat und Mensch: Kybernetische Tatsachen und Hypothesen* (pp. 390–411). Springer. https://doi.org/10.1007/978-3-662-00604-7_22

Wald, E. (2015). *Dylan goes electric!: Newport, Seeger, Dylan, and the night that split the sixties*. HarperCollins.

Wang, L., Zhao, Z., Liu, H., Pang, J., Qin, Y., & Wu, Q. (2024). A review of intelligent music generation systems. *Neural Computing and Applications*, *36*(12), 6381–6401. https://doi.org/10.1007/s00521-024-09418-2

CHAPTER 3

A BRIEF HISTORY OF MUSICAL MACHINES

Organizing History

the art of music has always courted mechanical aids (Fowler, 1967, p. 49)

The history of autonomous musical objects is partitioned near the beginning of the twentieth century, with everything before it grouped under the umbrella of the golden age of mechanisms and automata, often referred to as the *mechanical* age, and everything after it, the *modern* age. Musical machines before the twentieth century were abundant and technically advanced. The Banū Mūsā brothers invented an automatic water organ in the year 850 in Baghdad. Mechanical carillons were present in the Netherlands in the 1200s. Hydraulic and pneumatic technologies developed by the Greeks were revived by Renaissance scholars and builders, who implemented them in water organs, musical scenes, and instrument-playing automata. Music boxes and clocks became popular in the seventeenth and eighteenth centuries. Piano players and player pianos proliferated in the late nineteenth and early twentieth centuries. There have been many volumes written about this history (Engramelle, 1775; Leichtentritt, 1934; Bedini, 1964; Fowler, 1967; Ord-Hume, 1983; Reblitz, 2001), and institutions are dedicated to preserving its artifacts, such as the Museum Speelklok (the Netherlands, https://www.museumspeelklok.nl), MIM (Musical Instrument Museum, Arizona, USA, http://mim.org), and Morris Museum (New Jersey, USA, https://morrismuseum.org). These fantastic resources offer a more complete picture of this amazing history.

The *modern age* ushered in by the twentieth century saw two defining periods: the first powered by electric actuation, the second shaped by the possibilities of computer control. The history of these periods has been the focus of other works (Kapur, 2005; Robinson et al., 2023; Solis & Takanishi, 2007), and examples are detailed in later chapters. Some of these examples will be mentioned here to illustrate the similarities and differences between machines and the lineage and innovation of their development.

Reviewing this history reveals aspects that have changed substantially, such as the interpretive and generational power of computers, while others have persisted across time, geography, and culture, such as the magic of autonomous action and the desire to augment human abilities. Our fascination with what is new is a product of what is old. Wintergatan's *Marble Machine* (Media Example 3.1 ▶) straddles historical divisions, which is one of the reasons it has captivated audiences so expansively. In this chapter, historical investigation serves neither to paint a comprehensive picture of all things machinic nor to trace a definitive path of linear development. Instead, the goal is to identify conceptual, artistic, and technical categories that define the essential character and components of musical machines, illustrated by examples across place and time. Categories that emerge include *purpose, energy source, sound-producing mechanisms, sonic objects, encoding devices*, and *musical value*.

Musical Machines. Scott Barton, Oxford University Press. © Oxford University Press 2026. DOI: 10.1093/oso/9780197599891.003.0003

Identity, Purpose, and Usage

The twentieth century brought a multitude of technologies—phonographs, record players, magnetic tape-playing devices, CDs, iPods, streaming music services—that changed how people recorded and listened to music. Before such inventions, the primary way to hear music was to attend a live performance or to play music yourself. Few people could play keyboard instruments in the eighteenth century (Fowler, 1967), so machines such as music boxes, home barrel organs, pianolas, and player pianos allowed people to hear and enjoy music they otherwise would not. Comprehending this reality is challenging in a modern day where music is near-ubiquitous. It is in our cars as we drive, stores as we shop, and gyms while we exercise. The development and proliferation of transportable music players and wireless headphones have shown the extent to which our attachment to music shows no bounds. While the functionality of some of these early machines seems modest in comparison, the appreciation they received at the time was not, for they provided special experiences.

The motivations behind and meaning of pre-twentieth-century musical machines extended beyond sound. They were works of visual art that exhibited expert craft and codified cultural values. As musicologist Hugo Leichtentritt expresses, "these earliest mechanical instruments show that inventive genius, skill of handicraft, taste, intelligence, and imagination, combined to fashion a thing of beauty and a work of art, which cannot always be said of their modern successors" (1934, p. 26). *The Musician* by Jacquet-Droz in 1774 exemplifies such vision, skill, and dedication. The automaton plays five unique musical tunes by pressing the keys of an organ with its fingers. The machine replicates human action with a chest that rises and falls in a breathing rhythm, a torso that moves, and eyes that align and shift with the hands to represent focus (see Media Example 3.2 ▶).

Representing human forms and capabilities in **anthropomorphic** objects is a thread woven through historical periods. In addition to *The Musician*, prominent early examples include *The Flute Player* by Jacques de Vaucanson (1737), which contained artificial lungs, trachea, mouth, moveable lips, and tongue to play the flute "as a human would." Cabinetmaker David Roentgen, watchmaker Peter Kintzing, and artisans from a variety of trades created an Automaton of Queen Marie Antoinette, *The Dulcimer Player* (*La Joueuse de Tympanon*, circa 1780), which can play eight different musical tunes by striking a dulcimer with hammers while moving its arms, head, and torso in human-like ways (see Figure 3.1).

FIGURE 3.1. **Left: the automata of de Vaucanson, right: La Joueuse de Tympanon (image by Jean-Pierre Dalbéra from Paris, France—La joueuse de tympanon (musée des arts et métiers, Paris), CC BY 2.0, https://commons.wikimedia.org/w/index.php?curid=37213558).**

Anthropomorphic emulation continues to motivate designers and builders in the present day, from machinic bands such as *Z-Machines* to Waseda's robotic flutist and saxophonist.

The motivations behind such representations were introduced in Chapter 1—The Big Picture. One of the primary ways we make "new" things is by **emulating** what already exists. The ubiquity of human musical practice provides a bountiful source to copy. At the same time, and paradoxically, how we physically and mentally accomplish our musical feats is somewhat of a mystery to us; thus, some who make anthropomorphic machines seek to learn more about human processes by replicating them. We emulate ourselves to inspire interest. Seeing and hearing something recognizable evokes a host of psychological associations (e.g., related works and artists, personal experiences). When foreign elements are also present, typical in artistic and machinic contexts, our curiosity is sparked, seeking to learn more about similarities and differences relative to what we know.

Some musical machines do not emulate human forms, serving as **attachments** to existing instruments or manifesting as works of visual art in the form of kinetic musical sculptures. Orchestrions of the eighteenth and nineteenth centuries combined multiple instruments in a single structure that clarified instrumental identity and showed attention to form, proportion, and decoration. Unlike anthropomorphic examples, many of these works made the mechanical means of sound production visible, presenting a viewer with a challenge to figure out *how it works*, another source of interest (Figure 3.2).

FIGURE 3.2. Orchestrion (image By Miloš Jurišić—Museum of Science and Technology Belgrade, CC BY-SA 3.0, https://commons.wikimedia.org/w/index.php?curid=77502917).

The extent to which the means of action are **transparent** is part of the purpose of a musical machine and is a factor that differentiates them. *Joueuse de Tympanon*, the *Marble Machine*, and a record player are transparent in different ways. *Joueuse de Tympanon* does not reveal internal mechanisms, but we see the automata excite the strings. The invisibility of internal mechanisms creates a sense of mystery and magic, while the transparency of the sound-production process grounds the system in an accessible reality. The *Marble Machine* makes the object's inner workings visible, inviting viewers to investigate it and ponder how it works. At the same time, mystery and magic persist due to the challenge of figuring out how the components function, the amazement at the technical achievement, and the uniqueness of the presentation. A record player makes some parts of the sound production process visible via a stylus that follows a rotating, etched disc, but the transformation of that physical interaction into sound that emerges from speakers is invisible.

Sound generated by a computer is less transparent, and these devices conjure less interest in how they work despite their sophistication and complexity. Part of the reason is their ubiquity; they are a pervasive part of modern life to which we have become habituated. Another reason is that they generally are not designed to engender interest in how they work. Circuit boards, power supplies, and connectors are housed in sleek, minimalistic enclosures, and code is invisibly stored in computer memory. As a result, causal processes are inaccessible and external forms are nondescript: they are purposely **not** transparent. A third reason is that the ways that computers work are not comprehensible to those without specific technical knowledge. A human can understand the causal chains that make a basic machine act because of shared roots in the physical world. Displaying an electrical circuit or a computer program is not as universally revealing, for both are governed by conventions of human encoding discernible by relatively few. Learning them is analogous to learning a new language. With that said, the right presentation of technological inner workings can potentially facilitate audience understanding and curiosity; this is a design challenge for builders.

Transparency is a meaningful quality of musical machines that reflects their purpose and affects how they are perceived. Transparency can refer to the visibility of components and processes or can be interpreted metaphorically to indicate the extent to which aspects of a machine are accessible to viewers, for example, involving stylistic or technical knowledge. Transparency of causality, actuation, and movement facilitates understanding. The **invisibility** of components and processes can inspire curiosity and interest but, to its extremes, can obstruct human engagement. A single musical machine may be transparent in some ways and invisible in others in varying proportions.

Technical Design

Musical machines can be differentiated based on technical design. The energy that powers the machine is one essential element, as are the mechanical systems implemented to produce sound. This categorization approach may be of greater interest to designers and builders of musical machines than their audiences: use as you see fit.

ENERGY SOURCE

Musical machines are powered by water, air, weights, springs, and electricity. Some of the earliest musical machines were powered by water, including the water organs of the Banū Mūsā brothers of Baghdad (850) and the Villa D'Este in Tivoli, Italy (circa 1570). In a water organ, water is routed into a pipe such that a pressure source (such as gravity or a pump) directs the water into the instrument. Air is introduced into the stream of water, and the two are separated in an internal chamber where the compressed air excites the organ's pipes. The water drives a wheel, rotating a pegged barrel containing the musical sequences to be played.

Air is another source of energy for musical machines. The *Automatic Flute Player* by the Banū Mūsā brothers of Baghdad in the ninth century generated airflow by filling a reservoir with water that forced air into the playing system. Pneumatic mechanisms were also used in instruments such as the player piano in the late nineteenth and early twentieth centuries. Actuation is catalyzed via air pressure differential, which occurs as a human pumps foot treadles to create negative pressure in a chamber. As a perforated disc or roll rotates, holes in the surface allow air to be sucked into the system, effecting a musical action (e.g., vibrating a reed, pressing a key). Air remains an energy source in more recent *pneumatic* systems such as the *Z-Machines* robotic band or the MPR Lab's *Percussive Aerophone* (Sundberg et al., 2018). While modern pneumatic systems integrate electric components (e.g., motors for compressing air, solenoids for opening valves), air ultimately drives the actuators. Pneumatic systems can exert considerable forces at great speeds.

Weights and springs supply energy to initiate musical actions. Springs store potential energy, driving mechanisms to perform musical actions such as plucking a string or striking a drum. Weights and springs have been featured in many early instruments, such as musical clocks and music boxes. They continue to be used in recent examples, providing counterbalancing forces for robotic arms and returning solenoid armatures to initial positions. As with the treadle-based pneumatic instruments, a source of energy that performs the work of tensioning the spring mechanism is required, which was historically the responsibility of a human.

Electricity transformed the landscape of musical machine design. Most modern efforts use electric **motors** (which produce rotational motion) and **actuators** (which produce linear motion) to excite sonic objects. Electricity has advantages. It is available in effectively limitless quantities and can power a vast array of mechanisms. It is fast, quiet, reliable, and capable of driving actuators with significant speed and force. It can be transformed and modulated (e.g., via pulse-width modulation [PWM]) according to the needs of a system. It requires no human work, such as treadle-pumping or spring winding. Its downsides include physical danger, cost, and ecological implications. While some musical instruments used electricity before the twentieth century, such as the *clavecin électrique* (a carillon invented by Jesuit priest Jean-Baptiste Thillaie Delaborde in 1759), they were few and far between. Conversely, musical machines that do *not* rely on electricity are rare today; a notable example is the *Marble Machine*. The revolutionary impact of electricity is one of the primary planks in the wall erected at the dawn of the twentieth century, separating the "old" from the "new."

SOUND-PRODUCING MECHANISMS

How a musical machine produces sound is critical to its identity. A drum sounds when struck as a string sounds when plucked, but the methods for doing so vary. The mechanisms responsible for such actions have been essential throughout the history of musical machines. Levers, linkages, gears, wheels, axles, pulleys, screws, and cams were used before the twentieth century to (un)cover tone holes and press keys to realize autonomous music performance. The advent of electric motors and actuators changed this landscape. In some cases, electric actuators reduce the need for mechanical complexity. For example, the intricate coordination between the hammer and damper in a piano's action could be simplified using solenoids or servomotors that directly excite and attenuate a string's vibrations. That said, mechanisms remain a fundamental part of recent mechatronic and robotic musical instruments, particularly those that seek to make their movements more visually apparent and engaging.

Because mechanisms have been present throughout the history of musical machines, *mechanical* is not a particularly descriptive term to delineate some musical machines from others. Instead, the methods and mechanisms used to produce sound can be the basis for categorical inclusion. For example, the pitch of a string can be changed by an array of actuators located at different positions (EMMI's *PAM*), a moving bridge that travels from one end of the string to the other (League of Electronic Musical Urban Robots' [LEMUR] *GuitarBot*), or by changing the tension of the string (MPR Lab's *Cyther v3*). These distinctions, which can be further specified by actuator and mechanism type and configuration, help us identify similarities and differences between individual devices, regardless of the period or geographic location they inhabit.

Sonic Objects

Musical instruments or the sonic objects they sound provide another basis for distinction, which can be grouped into three categories: *percussion*, *strings*, and *aerophones*. The later chapters of this book are structured according to these groupings and illustrate examples from the *modern* period. In pre-twentieth-century examples, the percussion category includes mechanical carillons and assorted percussion instruments such as xylophones, bells, drums, and cymbals found in orchestrions.

The *strings* group includes percussive string instruments such as the dulcimer of *La Joueuse de Tympanon* and bowed string instruments such as violins. The *Hupfeld Phonoliszt-Violina* (1907) was an exceptional string-playing machine, sounding three violins (each with one string) using 49 mechanical "fingers," a circular bow, and a piano. The design is unique in that the rotating bow is stationary while mechanisms move the violins to contact the bow to modulate bowing pressure and create different musical effects (Figure 3.3). The ingenuity of the machine was so considerable that it was featured at the World's Exhibition of 1910 in Brussels.

FIGURE 3.3. The Hupfeld Phonoliszt-Violina (image By Jorge Royan, CC BY-SA 3.0, https://commons.wikimedia.org/w/index.php?curid=23403774).

Piano players and *player pianos* were among the most prominent musical machines of the mechanical age. The *piano player* was a device that would *push-up* (another name by which they were known) to a regular piano and play it via felt-covered "fingers." While the components of a self-playing piano were disparately present by the 1870s, American Edwin S. Votey assembled them into a complete system in 1897 called the *Pianola*. Human-pumped foot treadles created suction in a chamber, driving a motor to spin a paper roll on which music was encoded, activating pneumatic mechanisms when perforations in the roll passed over a tracker bar. For those who already owned a piano, an attachment such as the piano player was convenient as one did not need to buy another instrument, and the attachment could be removed if one wished to play the piano manually. Eventually, the piano player's mechanisms were integrated into the instrument's structure, forming the *player piano*. Player pianos became big business as the instruments proliferated in the early part of the twentieth century: the Aeolian Company, a leading manufacturer of the instruments, constructed a building as its headquarters on 42nd Street in Manhattan, New York City. Popularity peaked in the 1920s and declined due to the rise of electronic recording and broadcasting, industry dynamics, and the stock market crash of 1929 (*The Pianola Institute—History of the Pianola—An Overview*, n.d.).

The *Aeolian Harp* is unique: it is an acoustic string instrument that produces sound without direct involvement from humans or machinic actuators. It "acts on its own," so it belongs in a discussion of *musical machines*. Its 8–12 strings are of different thicknesses and tuned to the same pitch. They are sounded by air currents through a phenomenon known as *vortex shedding*. As air flows, vortices form on alternating sides of the string. The resulting oscillations excite the strings' modes of vibration depending on the speed of the airflow and the string's tuning, often emphasizing upper harmonics. The sonic result is ethereal and awe-inspiring, which needs to be heard for yourself (Media Example 3.3 ⓟ).

Many early instruments were *aerophonic*, including water organs, barrel organs, flute-playing machines, and flute clocks. Water organs are among the earliest musical machines, including examples by the Banū Mūsā brothers of Baghdad (850) and the Villa D'Este in Tivoli, Italy (circa 1570). The oldest known playable barrel organ was built in 1502 (Ord-Hume, 1983), and the instrument proliferated in the subsequent centuries. Johann Strasser made a notable mechanical organ in 1802, which contained pipes separated into two orchestras capable of playing symphonies, overtures, and piano concertos (Leichtentritt, 1934). Flute-playing machines also have had a long history, exemplified by the steam-powered *Automatic Flute Player* by the Banū Mūsā brothers of Baghdad in the ninth century and *The Flute Player* by de Vaucanson. Flute clocks were small organs that were actuated by clockwork mechanisms. Given their relatively small stature, they could be featured flexibly, such as in restaurants, to facilitate wider enjoyment of music.

Some musical machines combine multiple instruments. Jacques de Vaucanson made a shepherd automaton that could play a pipe and bang a drum with its other hand. The *orchestrion* (late eighteenth century) assembled multiple instruments from the orchestra (e.g., pipes, percussion) into a coordinated ensemble. Johann Maelzel's *panharmonicon* combined winds and strings (1805). The Cremona *Theatre Orchestra* was an autonomous 20-piece band that played during motion pictures from 1914 to 1924. Such developments reveal our thirst for sonic diversity in timbre, pitch range, dynamics, and articulations. One response is to add more instruments to the ensemble, as the preceding examples pursued, but doing so introduces the complexity of coordinating them. Another possibility is to enhance the instrument's functionality to produce a wider array of sounds. Which path is taken depends on technical and economic factors, but it is also a fundamental question in the compositional process: *What can I do with what I already have? Do I need more?*

Music Encoding

Musical machines can be classified based on how they encode, store, and recall information. Early machines' sound-producing actuating systems could perform sundry musical works, but their encoding systems were limited. The primary ways musical instructions have been stored and recalled have involved pegged barrels, perforated discs or rolls, and computer memory.

PEGS AND BARRELS

Imagine the (not so distant) time before electricity was harnessed and computers were spitting out 1s and 0s: How could an instrument play music by itself in such conditions? How could it "remember" a piece of music that could be recalled accurately at a specified time? One answer to this question was a barrel (cylinder, drum) with pegs (pins) that extrude from its surface, as seen in Figure 3.4.

FIGURE 3.4. **Pegged cylinder of a mechanical carillon at the Belfry, Ghent, Belgium (By simonly—https://www.flickr.com/photos/simononly/7875515678/sizes/l/, CC BY 2.0, https://commons.wikimedia.org/w/index.php?curid=20929659).**

As the barrel rotates, pegs contact mechanisms that initiate the sound-making process. Typically, each peg corresponds to a specific pitch. The ability to produce many pitches is realized by distributing numerous pegs across the cylinder's circumference. All pegs in a row (parallel to the axle) are played simultaneously, while peg columns (perpendicular to the axle) correspond to different pitches. The distance between pegs and the barrel's rotational speed determines the time interval between notes. Cylinders can be made from assorted materials, wood and metal being the most common.

The pegged barrel was one of the most important innovations in musical machine technology and was featured in manifold instruments over hundreds of years. The Banū Mūsā brothers described a water-powered organ that used interchangeable pegged cylinders in their *Book of Ingenious Devices* in 850, identified as one of the earliest known musical machines (Fowler, 1967). Pegged cylinders were used in the mechanical carillons—pitched keyboard percussion instruments consisting of chromatically tuned cast bronze bells—of the Low Countries (Belgium, the Netherlands, Luxembourg, and parts of France and Germany) at the beginning of the thirteenth century. They were also used in organs in English churches, smaller barrel organs intended for home use, musical clocks, watches, and music boxes from the seventeenth to nineteenth centuries.

The design of pegged barrel systems depends on the instrument being played and the builder's musical goals. In some cases, pegs contact mechanisms, generating movements that excite sonic objects. For example, in a mechanical carillon, pegs contact levers as the barrel spins, raising and releasing hammers that strike the bells. In others, there are fewer intermediaries between the pegs and the sonic objects. In a music box, the pins pluck the comb's tines directly. Pegs can be fixed on a cylinder, determining the sequence of pitches and their rhythmic proportions (the only variable is the speed at which the barrel is rotated). Practically, this means one song or piece per cylinder, so the ability to swap cylinders becomes necessary. In the case of larger orchestrions with large cylinders, systems were designed to automatically swap out barrels so that different musical pieces could be heard. An innovation to the pinned barrel featured holes along the surface of the cylinder in which the pegs can be configured, enabling multiple musical sequences to be *programmed* and played. The *Automatic Flute Player* by the Banū Mūsā brothers of Baghdad in the ninth century is the earliest known example featuring a programmable pinned barrel (Koetsier, 2001).

The pegged barrel system was a laudable invention but has musical limitations. Some pinned barrels could not be reconfigured. The ones with such capabilities required significant time and effort to translate or realize new musical works; thus, pieces were often not changed as a result (Fowler, 1967; Leichtentritt, 1934). The length of a piece and the rhythms that can be performed are limited to the characteristics and configuration of the cylinder. Playing a new piece requires replacing the cylinder, which can be demanding depending on its size and materials.

PERFORATED DISCS AND ROLLS

The advent of pneumatic energy sources enabled a new form of musical encoding: perforated discs and rolls. The pneumatic mechanism of a player piano illustrates how this recording medium works. A **tracker bar** has a row of holes, each corresponding to a pitch. As a hole in the paper passes over the tracker bar, air is admitted into a **channel** that forces a **pouch** to rise, pushing a **valve** up. As the valve rises, it opens a passage that connects a **pneumatic** to a source of suction generated by continuously pumping bellows, forcing the pneumatic to collapse. The pneumatic is coupled to a piano action via a linkage, forcing the hammer forward to strike the string (see Figure 3.5).

The size of holes in a roll or disc corresponds to note length. A motor controls how fast the spool of paper unwinds; speeding or slowing the rotation affects the produced music's tempo. The **metrostyle** was introduced by the Aeolian Company in 1903, allowing a player to vary the tempo of the music via a lever. Music rolls were printed with a vertical wavering line on the score that, when followed with the metrostyle, would produce tempo changes that

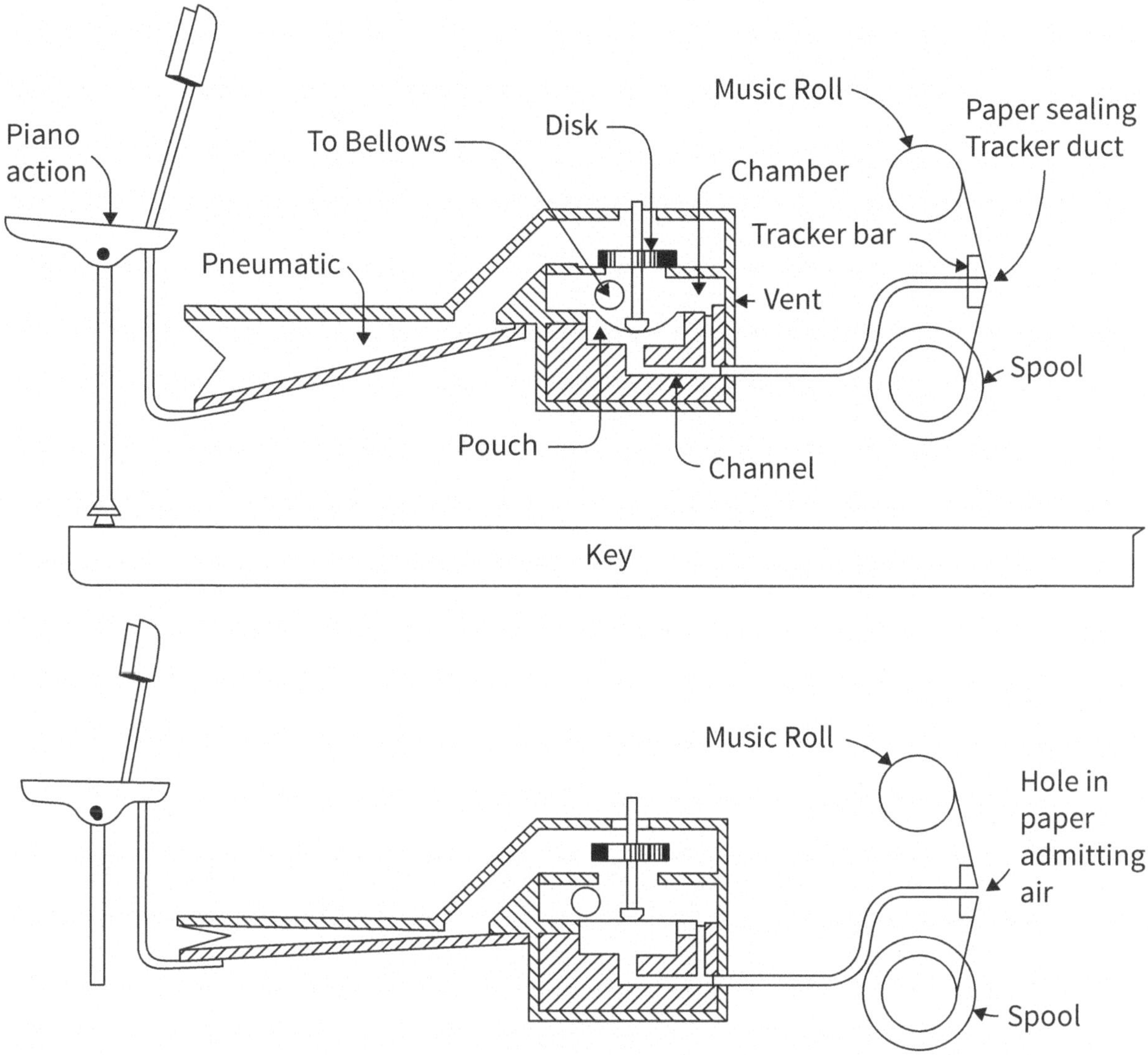

FIGURE 3.5. **Pneumatic mechanism of a player piano (image from White, 1925).**

complemented a piece. The **themodist** was introduced in 1906, allowing a player to highlight a melodic voice by referencing perforations on the sides of the rolls and using the instrument's subduing levers to attenuate the accompaniment parts.

Machines were invented to automate the encoding of musical scores. An early example was a music typewriter that could notate whatever was played at the piano, envisioned by Father Engramelle in the eighteenth century. This idea was further developed by Unger in Brunswick and realized by Hohlfeld in Berlin as a clockwork-powered mechanism consisting of paper rolls and pencils, which drew lines and dots when the piano keyboard was played (Leichtentritt, 1934). The idea developed in inventions such as the *Q-R-S marking piano* in the early twentieth century. The 88 keys of the device were pneumatically connected to a stylus positioned over a roll of paper that passed over a carbon cylinder. When a key is pressed, the pneumatic collapses, bringing the stylus in contact with the paper and the carbon cylinder, leaving a mark corresponding to the note's pitch and duration (*Q-R-S Marking Piano*, n.d.). These machines allowed performances by famous pianists to be recorded, which could then be played back in people's homes via acoustic player pianos such as the Aeolian *Duo-Art*, American Piano Company's *Ampico*, and the *Welte-Mignon*.

Musical works could also be encoded on perforated discs. The *Ehrlichs Organ* was such an example, employing flat paper records that contained classics, marches, hymns, and love songs, primarily of European origin. The *Polyphon* (1870) was a disc-playing music box made by the Polyphon Musikwerke in Leipzig, Germany. Its punched metal discs contacted ratchet wheels (ratchets allow movement in only one direction) that plucked the tines of a comb as it rotated, powered by a hand-wound clockwork motor. The *Regina*, whose method of sound production was similar to the Polyphon's, contained a self-changing attachment that could hold twelve 27″ discs, invented in 1897 by Gustave Brachhausen.

Perforated discs and rolls had advantages over the pegged barrel system. Making holes in a roll of paper was far easier than configuring pegs in a cylinder (given the right tools); rolls and discs were small, light, and transportable; musical works of variable lengths could be encoded and played back; rolls and discs could be changed to play different musical works; and performance nuances in timing, duration, tempo, and dynamics could be preserved. The visualization of notes on a piano roll is clear and intuitive: it is no wonder that the model is preserved in today's DAW "piano roll" MIDI sequencers. Paper rolls and discs also have limitations. They are ultimately static forms of musical representation. Once a hole is made in a roll or disc, that is effectively it: changing the sequence of pitches or rhythmic timing requires making a new version. Their materiality leads to issues over time, such as ripping (a notable con of paper that accompanies its advantages in weight and space) and warping.

COMPUTER MEMORY

Instead of orienting pegs in a barrel or punching holes in a paper roll, musical sequences can be stored in computer memory. The capabilities of computer memory address many of the shortcomings of previous media. The size of these stores is unfathomably large, easily reconfigured, and can be stored and accessed at fantastic speeds. This means storing and recalling massive libraries of musical works and filtering, analyzing, transforming, and generating information in real time, opening up new worlds of musical creativity and expression. The coupling of such information-processing abilities with physical actuation systems unlocks the artistic promise of musical machines and is, therefore, the core of modern design efforts.

Autonomy

The extent to which musical machines are *autonomous*, as described in Chapter 2—Definitions, can be used to distinguish their history. The vast majority of musical machines before the twentieth century were deterministic. Some recent examples of modern instruments, such as the *Marble Machine*, are also deterministic (a function of using sound-production systems prevalent in the *mechanical* age). Computer-based musical machines can also reproduce a predetermined score as notated. These instructions may be embedded in the instrument (e.g., on a microcontroller) or on an external device (e.g., communicated by a separate computer connected to it).

Modern *generative* systems typically use computer-executed algorithms. These can be defined as rules, probabilistic models such as Markov chains, random processes, machine learning methods, and neural networks. Few generative musical machines existed before the twentieth century; the *Componium* is an exceptional example. The machine, invented in 1821 by Dietrich Nicholas Winkel and currently residing in the Brussels Museum of

Musical Instruments, "composed" music automatically. The machine is an automatic organ that contains two barrels that rotate simultaneously. The barrels take turns performing two measures of music. The non-sounding barrel slides horizontally and uses a roulette-like flywheel to randomly select the next variation that will be played (*Componium | MIM*, n.d.). The permutations afforded by the system result in "novel" sequences, representing an early example of a compositional technique that would gain significant popularity in the computer age.

The abilities of *robotic* systems are afforded by modern technological developments, including sensors, microcontrollers, motors, and platforms for disseminating technical information. The technical details, musical potential, and past realizations of these machines are the focus of Chapters 5–7.

Robotic implementations make *cooperative* instruments possible, but the latter's history dates back before the twentieth century. The *piano player* is a cooperative instrument from the *machine age*. A human pumps the foot treadles to create an air pressure differential, while the machine produces pitches and rhythms according to the perforations on the music roll. Freed from the obligation to produce specific pitches at specific times, humans can attend to higher-level matters, such as varying the tempo or subduing the accompaniment part to bring out the melody. The relationship between pianola and player has been analogously compared to conductor and ensemble (*History of the Pianola—Pianolists*, n.d.). The ensemble and the pianola can play a piece "on their own," but the realization of the music is enhanced by someone whose job is to focus solely on interpreting collective and emergent aspects, such as tempo and balance, which the conductor and pianola player do. In the case of the pianola, human and machine *cooperate* in the process of making music, each responsible for different roles. This mode of interaction motivated the invention of the human-playable robotic zither *Cyther* (Barton et al., 2017) and works such as my *Experiment in Augmentation 2*.

Musical Value

Technical and contextual descriptions of early musical machines valuably illustrate historical lineage, but how did they sound? In describing the barrel organ of 1810, Burney states, "recent improvements of some English Artists have rendered the barrel capable of an effect equal to the fingers of the first-rate performers" (Fowler, 1967, p. 46). Strasser's mechanical organ of 1802 was capable of crescendo, sforzando, and rubato and apparently delineated voices in ways that were preferred to the actual organ (Leichtentritt, 1934). The violin and piano-playing *Hupfeld Phonoliszt-Violina* could produce different dynamic levels, crescendos, decrescendos, and accents. The instrument's sound quality is impressive, which can be experienced today thanks to preservation and restoration efforts.

At the same time, many early machines were musically limited. Actuation systems of most instruments could not produce variation in timing, dynamics, and timbre, which limited sonic nuances essential to musical performance. They were confined to predetermined scores fixed on pegged barrels or perforated rolls or discs. This latter point is a matter of perspective, as Fowler (1967) points out: "the objectivity of the perforated Pianola roll or the magnetic tape is as much a creative challenge on the one hand, as aleatoric or "chance" music is on the other" (p. 49). Indeed, composition occurs in a system of constraints, one of which is the encoding process.

Prominent composers wrote works for musical machines despite the instruments' limitations. Handel composed for machinic chimes, carillon, and flute-playing clock, while Mozart did the same for pinned-cylinder instruments. A barrel organ made in 1819 by Diederick Nickolaas Winkel contains a piece entitled *Tiroler Lied gevarieerd door Mozart*, speculated as a lost work of W. A. Mozart (Ord-Hume, 1983). Beethoven composed "Wellington's Victory" for a version of Maelzel's *panharmonicon*. Debussy, Ravel, and Mahler journeyed to the recording studios of their day to have performances captured by player-piano systems. Some composers specialized in writing for carillons in the eighteenth century, such as Pothoff in Amsterdam, van den Gheyn in Louvain, and J. A. H. Wagenaar in Utrecht. Songs, chorales, and even polyphonic motets could be realized on the instruments (Leichtentritt, 1934). Pieces composed for fixed media, such as pinned barrels, are historically important because they preserve musical works as they were originally heard and, thus, arguably as they were *intended* to be heard. A transcription of a Haydn piece as performed by a flute clock (made by Joseph Niemecz) shows differences with the notated score, providing evidence of how styles of musical notation and interpretation have changed (Ord-Hume, 1983). It would be irresponsible to assume the activities of these composers indicated a view that musical machines were equally capable as human performers, for they could have been motivated by a host of reasons, including curiosity, novelty, economics, and personal relationships. Nonetheless, such usage showed that experts viewed these machines as more than niche technological curiosities and were interested in exploring their creative potential.

The value of musical machines may have to do less with their inherent capabilities and more with how they are presented, distributed, and used. Ord-Hume (1983) argues that before the nineteenth century, when musical machines were a relative rarity, they were closely connected to the composers who wrote for them and served to document those works *as they were heard at the time* in ways that paper scores could not capture. As these instruments were produced in larger quantities in the nineteenth century to reach larger markets, their musical value and importance diminished. Early musical machines that "lived side by side with other musical instruments and were viewed less as interpretational novelties and more as musical equals" devolved into entertainment machines that "masked the true ability of these unique musical time-machines" (Ord-Hume, 1983, p. 171). As we contemplate these perspectives from our current position, we see how these attitudes persist today. Current musical machines are sometimes denigrated as contrivances and novelties. This view is formed for a host of reasons, including the composition, performance, visual aesthetics of the machine, and the context of the experience. These aspects are under builders' control (some more than others, depending on the situation) and should be kept in mind when designing objects and experiences intended to communicate and connect with others.

Summary

The history of musical machines is separated at the beginning of the twentieth century, demarcating the *mechanical* from the *modern* age. The chronological recounting of this history has been captured in other volumes, so here, the intent is to identify qualities and categories that help make connections between historical periods. The **purpose** of musical machine builders has been to facilitate broader performance and enjoyment of music, demonstrate high levels of artistic and technical craft in sonic and visual domains, and explore *human* and *machinic* identities via anthropomorphic and non-anthropomorphic representation. **Transparency** of motivations, interpretations, and means of action is a central aspect of the design and experience of musical machines. Technical aspects of machines can also be used to differentiate between them. **Energy sources** are one such aspect, coming from water, air, weights, springs, or electricity. The **mechanical systems** that produce sound are another, involving levers, linkages, gears, wheels, axles, pulleys, screws, and cams. Myriad approaches can realize the same musical goal (e.g., producing different pitches on a string instrument), so categorization via mechanical design is essential to designers and builders. **Sonic objects** are a musically-focused way to distinguish between machines, comprising the categories of *percussion, strings*, and *aerophones*. How music is **encoded** impacts performance capabilities. Methods for doing so have ranged from pegged barrels to perforated discs and rolls to computer memory. The extent to which musical machines act "on their own," to which they are **autonomous**, is a primary distinguishing aspect. History reveals four primary modes of machinic autonomy: **deterministic**, **generative**, **robotic**, and **cooperative**. In all cases, the musical value of these machines is of primary importance. Such value is derived partly from the machines' capabilities and partly from how they are presented, distributed, and used.

References

Barton, S., Prihar, E., & Carvalho, P. (2017). Cyther: A human-playable, self-tuning robotic zither. *Proceedings of the International Conference on New Interfaces for Musical Expression*, 319–324. https://doi.org/10.5281/zenodo.1176266

Bedini, S. A. (1964). The role of automata in the history of technology. *Technology and Culture*, *5*(1), 24–42.

Componium | MIM. (n.d.). Retrieved July 16, 2022, from https://www.mim.be/en/collection-piece/componium

Engramelle, M. D. J. (1775). *La tonotechnie: Ou L'art de noter sur les cylindres et tout ce qui est susceptible de notage dans les instruments de concerts méchaniques*. Minkoff reprints. https://books.google.com/books?hl=en&lr=&id=4MoXAQAAIAAJ&oi=fnd&pg=PR1&dq=Tonotechnie&ots=QrdpeUQD2k&sig=tLg2gC3pXNKbjTug_cmttOco6JI

Fowler, C. B. (1967). The museum of music: A history of mechanical instruments. *Music Educators Journal*, *54*(2), 45–49. https://doi.org/10.2307/3391092

History of the Pianola—Pianolists. (n.d.). The Pianola Institute. Retrieved July 20, 2022, from http://pianola.org/history/history_pianolists.cfm

Kapur, A. (2005). A history of robotic musical instruments. *In Proceedings of the International Computer Music Conference*. International Computer Music Conference, Barcelona, Spain. http://citeseerx.ist.psu.edu/viewdoc/download?doi=10.1.1.88.4599&rep=rep1&type=pdf

Koetsier, T. (2001). On the prehistory of programmable machines: Musical automata, looms, calculators. *Mechanism and Machine Theory*, *36*(5), 589–603.

Leichtentritt, H. (1934). Mechanical music in olden times. *The Musical Quarterly*, *20*(1), 15–26.

Ord-Hume, A. W. (1983). Cogs and crotchets: A view of mechanical music. *Early Music*, *11*, 167.

Q-R-S Marking Piano. (n.d.). Retrieved July 15, 2022, from https://www.asme.org/About-ASME/Engineering-History/Landmarks/161-Q-R-S-Marking-Piano

Reblitz, A. A. (2001). *The golden age of automatic musical instruments: Remarkable music machines and their stories* (Q. D. Bowers, Ed.). Mechanical Music Press. https://www.mechanicalmusicpress.com/rbook/goldage.htm

Robinson, F. A., Bown, O., & Velonaki, M. (2023). The robot soundscape. In B. J. Dunstan, J. T. K. V. Koh, D. Turnbull Tillman, & S. A. Brown (Eds.), *Cultural robotics: Social robots and their emergent cultural ecologies* (pp. 35–65). Springer. https://doi.org/10.1007/978-3-031-28138-9_3

Solis, J., & Takanishi, A. (2007). An overview of the research approaches on musical performance robots. *International Computer Music Conference Proceedings, 2007*. http://hdl.handle.net/2027/spo.bbp2372.2007.179

Sundberg, K., Barton, S., Walter, A., Sane, T., Baker, L., & O'Brien, A. (2018). A robotic percussive aerophone. *Proceedings of the 18th International Conference on New Interfaces for Musical Expression*, 409.

The Pianola Institute—History of the Pianola—An overview. (n.d.). Retrieved January 17, 2017, from http://pianola.org/history/history.cfm

White, W. B. (1925). *Piano playing mechanisms*. EL Bin. https://www.mmdigest.com/Smythe/Piano_Playing_Mechanisms.pdf

CHAPTER 4

THE PROCESS OF MAKING A MUSICAL MACHINE

Part 1: Context and Ideation

Context and Ideation

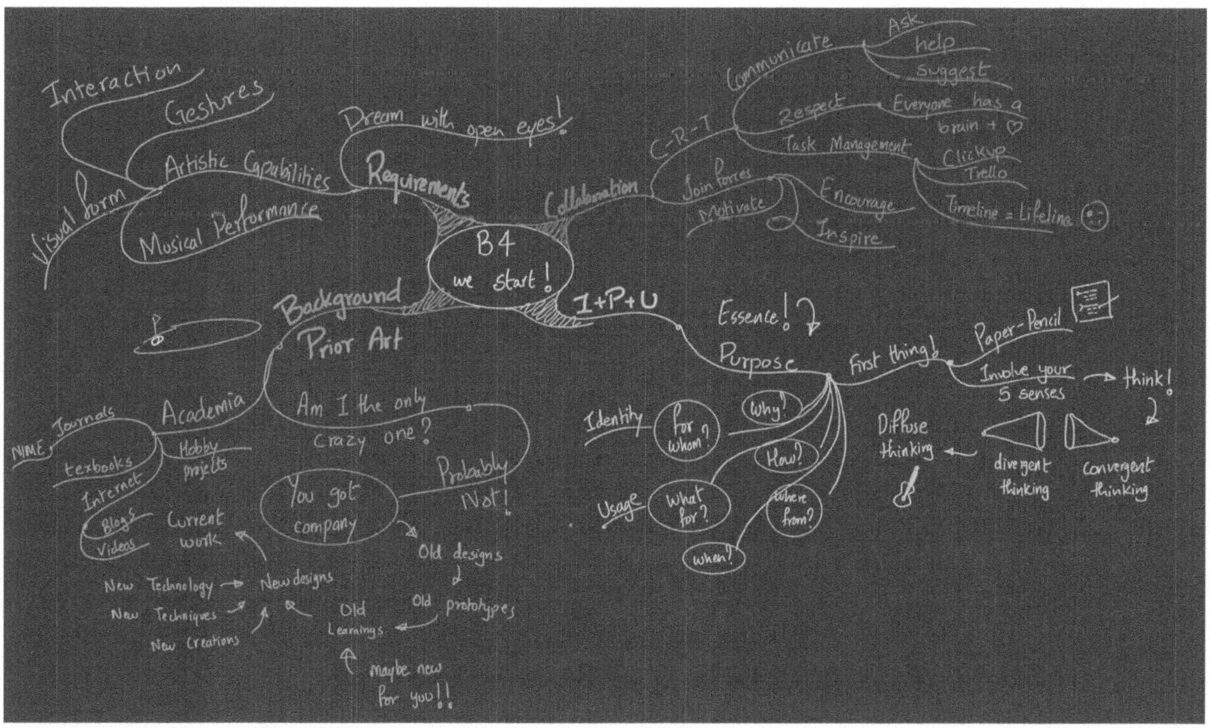

FIGURE 4.1.1. **Mindmap by Yash Garje.**

Having engaged with philosophical quandaries, key concepts, and historical examples, you are prepared to embark on the journey of creating your own musical machine. How does one go about this process? Musical machines are inherently physical, but their essence is sparked in human imagination. You may have an idea and are tempted to start building right away. While hands-on experimentation is necessary early in the design process, proceeding impetuously can lead to problems that are not realized until after many hours of development and building time. Conceptualizing and designing a musical machine requires addressing musical, visual, and technical questions from the beginning. This chapter provides a template for the ideation and design process, offering concepts, techniques, and questions to analyze

Musical Machines. Scott Barton, Oxford University Press. © Oxford University Press 2026. DOI: 10.1093/oso/9780197599891.003.0004

existing examples and shape ideas for new mechanical instruments. This template will structure the instrument chapters that follow.

Collaboration

An initial decision in the ideation process is whether you will go about it alone or with others. Making a musical machine involves programming, electrical engineering, mechanical engineering, visual design, and music. While it is possible to be proficient in each of these areas, it is often better to seek collaborators who can contribute to the project in the domains in which you lack expertise. Groups with diverse skills, experiences, and knowledge are smarter and more able. Dialogue with others can inspire ideas that would have otherwise laid dormant. Most importantly, collaborations facilitate human connections. They allow you to share your excitement and interests with others who are similarly minded. They can lead to new friendships. They can be great fun. Robots may appear cold and alienating, but they are actually the product of fundamental human interactions. A group's chemistry needs to be right for such collective augmentation to occur, but you can have an active role in creating that chemistry through the way you act and communicate with others.

The benefits of teamwork depend on the organization and relationships of the team members. Working with a group of people who have varied talents and who are organized, diligent, and communicative can not only help you produce higher-quality results in a smaller amount of time but also provide you (and the rest of the team) with creative inspiration and an opportunity to learn. Conversely, poor communication and organization can produce results that are less than what a single diligent and capable individual could accomplish.

> *Years of advising group projects has taught me that teams' biggest hurdles are rarely technical; they usually involve teamwork.*

How do you ensure that your team is successful? Good teams typically excel in **communication**, **respect**, and **task management**. Teammates must communicate *fluidly* to ensure tasks are done correctly and on time. Tips on being a better communicator and teammate:

1. Reach out to others when you need help, have a question, or need another perspective.
2. Respond to messages in a timely and thoughtful way.
3. Respect your teammates as competent, sensitive, and capable. Sometimes, this requires being understanding when life comes up. Sometimes, this requires forgiving and forgetting. You do not have to be best friends with the people that you work with, but you do need to respect each other if you are going to realize your goals and feel good about the experience.

Good teams manage tasks proficiently. Identify larger categories (e.g., electrical system, mechanical system, picking mechanism, frequency detection, etc.) and tasks within each of these categories (e.g., "select motor"). Task management software can be helpful in this effort, though a spreadsheet works just fine. Agree on a method that works for the team, create tasks, assign tasks to team members, and create due dates for each task. View the entire timeline for the project as a Gantt chart. This process makes you think about what you need to accomplish in the time allocated to the project and shows the project schedule in a single view. It also helps keep the proportions of effort correctly balanced. Too often, the early part of the project takes too long because the team thinks they have all the time in the world, which takes away from later tasks such as testing and evaluation. As a result, some (important) tasks are done hastily (or not at all).

Stick to your timeline. Treat the due dates you set seriously; consider them as you would a test or a presentation you are scheduled to deliver. You will get a task to a point where the results are acceptable but could be improved. The period after this point often has diminishing returns relative to the time and effort expended. If you are at that point and the schedule says you need to advance to the next task, pause, consider the larger project timeline, and move on as appropriate. You can always come back to revise and refine part of the project later (and time away might bring a beneficial perspective). It can be hard to accurately estimate how much time a task will take, especially if you are new to the process. Stay focused on the small scale, which you can control more easily. Allocate blocks of your daily and weekly schedule to the task and treat them as meetings you must attend. This will make revisions to the longer-term schedule more realistic. As aspects of the project take longer than anticipated, either (1) drop some of the tasks, (2) move the due date of the project, or (3) enlist more help. If more people need to be involved in a project, the question is when to bring them on. There is no "right" time, though having additional perspectives at the beginning is often helpful.

Toward a Successful Collaboration

- communicate with others in a timely and thoughtful way
- respect your teammates
- create, allocate, and schedule tasks
- stick to your timeline

With your team assembled (this might just be you, which is OK), you are ready to enter the ideation process.

Identity, Purpose, and Usage

What is the purpose of your machine? What kind of music will it make? Who is it designed for?

What is the essence of your idea? What kinds of sounds will the machine produce? What will the visual experience of the machine be like? How will it play or be played?

To answer these questions, define the purpose of the machine. It could be a muse for your compositional inspiration, a means to explore new musical expression, or a collaborator and partner in improvisation. It could bring a physical dimension to the performance of computer music. It could be a tool for learning about music and technology. It could serve as a research platform for studying acoustics or human performance. It could enable those with impairments or provide therapy to those in need (see Solis & Takanishi, 2007, for further discussion of different purposes).

Who is going to use your machine? Is it made for programmers or the tech-adverse? Acoustic instrumentalists or DAW-based DJs? Tone-row manipulators or "intuitive" artists?

What is required to transport the device? What maintenance is needed, and can it be performed by the user? One way to address these questions is to create **personas**: fictional characters that represent the people who might use your device. Having a better sense of the people using the machine can guide your design.

Spend enough time in the ideation process so that inspiration percolates when waking in the morning or walking between appointments. These are powerful creative periods that are accessed through dispersed rumination. *Inscribe* your thoughts in text and images. Discuss ideas with collaborators. During this stage, resist making too many specific design choices. You are trying to clarify your motivation and sketch a high-level idea of the machine that can be realized in multiple ways, which you will articulate later. Keep your mind open so that you can be influenced by the work of others, which is the next step in the process.

Background and Prior Art

> What concepts and prior works are relevant to your idea? How were these works envisioned and realized? What aspects will you emulate? Where can you innovate?

With a clearer sense of your motivation, the next step is to assume that others have already traveled the path you are on. Your charge is to discover their work to build knowledge, make interpretations, and find inspiration. Investigate machines, musical works, people (builders, composers, performers, audiences), cultural contexts, technical concepts, and practical applications. This research process is invaluable, yet it is sometimes treated cursorily by those eager to get their hands dirty. Becoming familiar with related work reveals a network of achievements and potentials that define improvements and innovations. It prevents you from reinventing the wheel, repeating past mistakes, or falsely signaling innovation. It exposes you to the technical knowledge needed to understand and realize your idea. Some of the information encountered in research can be learned through firsthand experience, but this approach requires a great deal of time and effort. Education through background research is a shortcut that exponentially accelerates becoming smarter, more able, and more valuable to a team.

Practitioners in the field of musical machines are diverse; thus, background research should be conducted across diverse arenas. Peruse academic publications in journals and conference proceedings (The International Conference on New Interfaces for Musical Expression—NIME; The International Computer Music Conference—ICMC; *The Computer Music Journal*), review textbooks that explain technical concepts, explore online videos made by artists and technologists, and venture into the community resources of the maker world (e.g., Arduino). Become acquainted with best practices, approaches worthy of emulating, and opportunities to be creative.

How should the prior art of musical machines be classified? Categorization by builder is one option. This method identifies the styles of particular people or groups, but it is less useful for showing features or approaches common among different designs. An alternative is to categorize by instrument type, either by using the familiar Western system of Greek origin, which includes winds, strings, and percussion; the Hornbostel–Sachs taxonomy, which distinguishes instruments by the means and materials of sound production (e.g., chordophones); or some other variant. These approaches are useful as they align with our sense of timbral categorization, but focus primarily on sonic objects rather than the systems that play them. The designer of a musical machine must consider both the object and the playing system; both must be a part of a categorical structure. Categorization systems become problematic when requiring an example to be placed in one bin, even though it may belong equally well in others.

To address these issues, the approach here is to identify *properties and relations* that define musical machines. These can be located along continua rather than being confined to singular attributes, which allows the strength or degree of a characteristic to be described. Multiple properties and relations can be applied to the same musical machine. The following section will explore properties and relations in more detail.

Properties and Relations

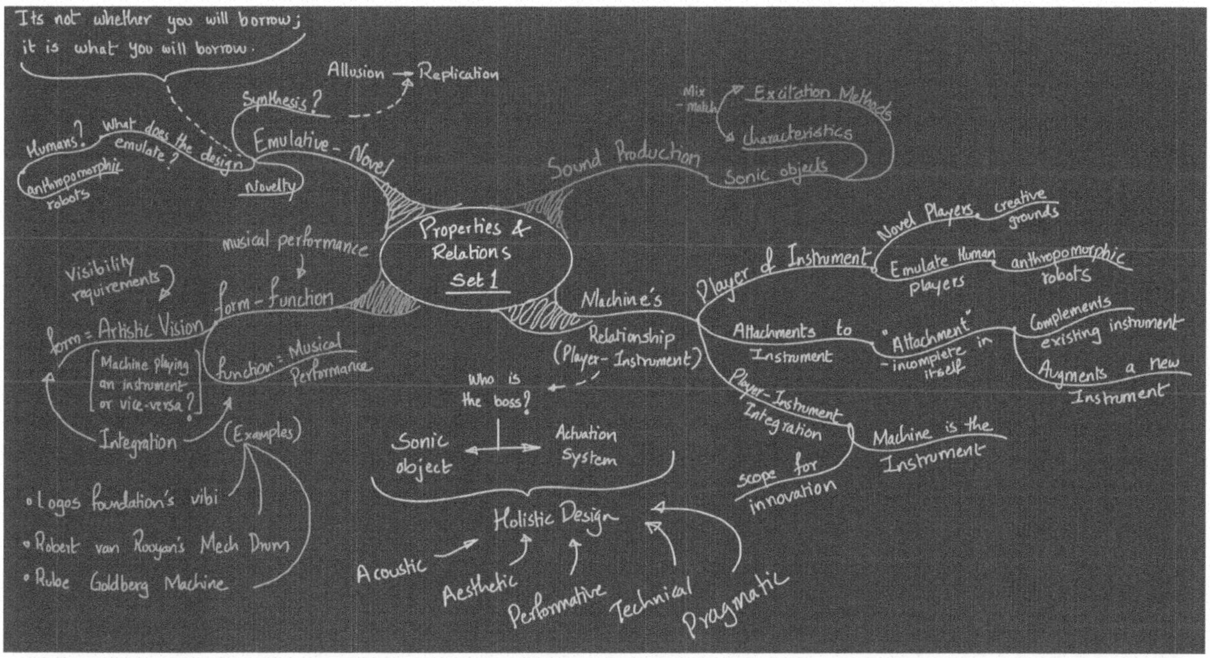

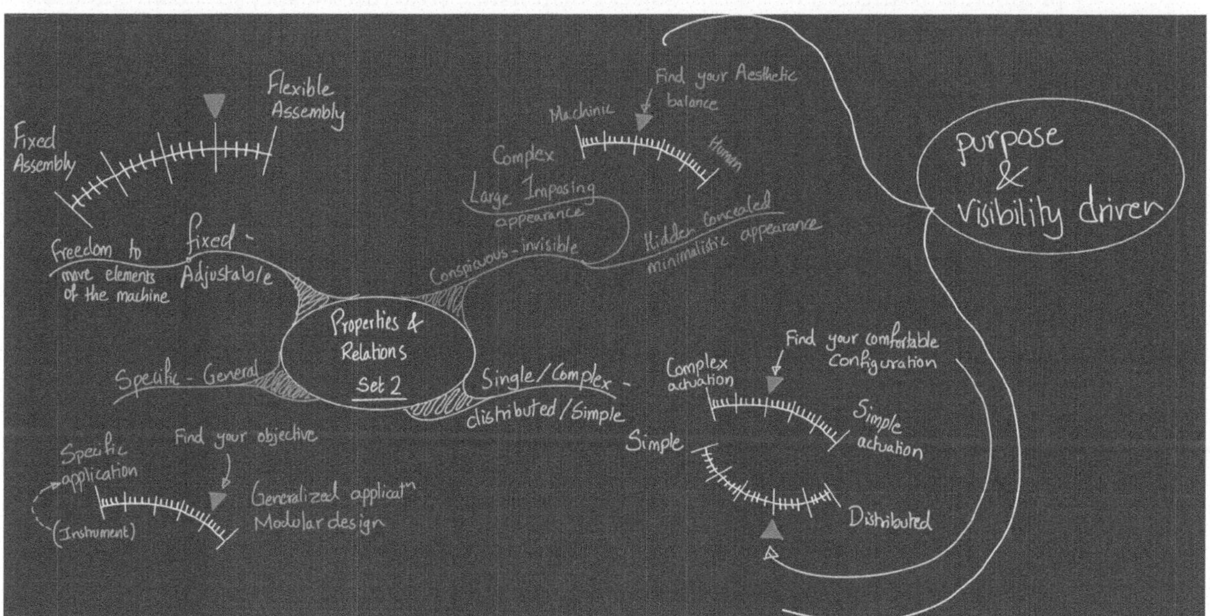

FIGURES 4.1.2 (TOP) AND 4.1.3 (BOTTOM). **Mindmaps by Yash Garje.**

A musical machine's identity materializes in a set of *properties* and *relations* represented by bipolar dimensions defined by opposites (see Bianchi et al., 2011; Bianchi & Savardi, 2008, for an exploration of the psychological importance of contrariety), including:

- *sound production*
- *players* ⇔ *instruments*
- *form* ⇔ *function*
- *emulative* ⇔ *novel*
- *human* ⇔ *machine*
- *conspicuous* ⇔ *invisible*
- *single / complex* ⇔ *distributed / simple*
- *specific* ⇔ *general*
- *fixed* ⇔ *adjustable*

This set is useful for both analyzing existing works and creating new musical machines. They apply to all phases of the process. They can be used to organize and compare examples of prior art, providing direction about where to imitate and where to innovate. They can help shape an initial concept and set priorities. For example, a primary consideration may be the extent to which the machine's form is *emulative* or its actuating components *invisible*. They are equally as important in designing an excitation system (e.g., will it be *fixed* or *adjustable*?) or determining the machine's structure (e.g., will it lean toward *form* or *function*?). Keep these *properties and relations* in mind from the start and apply them universally. Some of these dimensions may be more relevant than others depending on your situation; do not feel compelled to address all of them in the same proportions. Entertain how each of them may be relevant and focus on those that will inspire and direct you.

SOUND PRODUCTION

How musical objects produce sound is a primary way to distinguish among them. Sound production is affected by a sonic object's material(s), structure (shape, size, configuration), and excitation methods. For example, the characteristics of a non-free aerophone's structure (e.g., tube length and diameter) will affect the sound produced when the column of air within it is excited. At the same time, the air column can be excited in varied ways, including directing air over an edge, setting a reed into vibration, "buzzing" lips into a mouthpiece, or vibrating a membrane coupled to an electromagnetic driver. Sound production thus involves different *roles* for different components. Thinking in this way allows one to see similarities and differences between musical machines. New instrumental possibilities emerge as different types of sonic objects and excitation methods are mixed and matched.

PLAYERS ⇔ INSTRUMENTS

The relationship between a *player* and an *instrument* is primary in defining the character of a musical machine. Within this umbrella are three categories: *players* of instruments, *attachments* to existing instruments, and systems in which a player and an instrument are *integrated*.

Machines that are *players* of instruments are distinct, complete, and whole, independent of the instrument they play. The instrument is also a distinct, complete whole with a sonic and cultural identity that different players in different musical scenarios can use. Examples of this category are the general-purpose anthropomorphic robots that play human instruments, such as a guitar, violin, drums, or trumpet (Figure 4.1.4). An advantage of *emulating* an extant player is that it provides a model to guide the design. A disadvantage is that complexity may be introduced that is not required to achieve the desired task (i.e., a robotic arm may not be necessary when a properly positioned solenoid can also do the job). Emulative machinic players evoke associations with human characteristics, techniques, idiosyncrasies, and styles. In some cases, such associations are welcomed; in others, they may limit or even detract from the perceived value of the machine or what it can do. Instruments designed to be played by humans (e.g., guitar) increase the chance that the actuating system will be perceived as a *player*, given our associations between the object and human performers. Sonic objects that are not typically played by humans (e.g., a glass bottle) can reduce such associations and still be seen as distinct from the player. Novel (i.e., non-emulative) machinic players are also possible, which could assume innumerable shapes and sizes and move in unfamiliar but intriguing ways. These possibilities fuel the imagination (some of them are realized in the creations of Animusic), presenting fertile territory for future builders to explore.

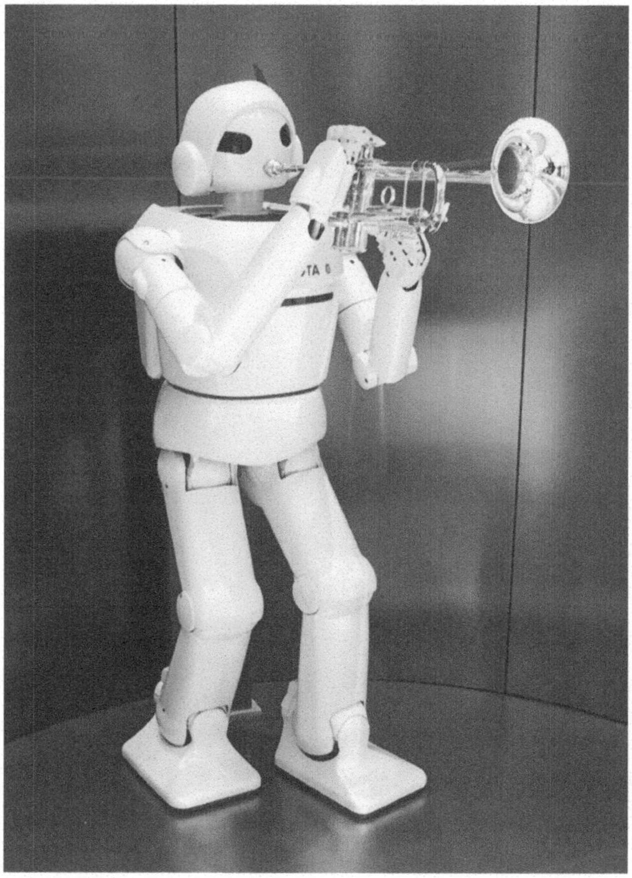

FIGURE 4.1.4. **Toyota Robot playing a trumpet (image By Chris 73 / Wikimedia Commons, CC BY-SA 3.0, https://commons.wikimedia.org/w/index.php?curid=73899).**

The second category includes machines designed around and *attached to* existing instruments. The distinction between this kind of machine and the *player* described previously is that, here, the machine is not a separate whole unto itself. Instead, it complements an instrument's form, features, affordances, and constraints. Using an existing instrument is advantageous because it provides a reference. Its physical characteristics afford and constrain what sounds are possible and how they are produced, which help guide the choices involved in designing an actuating system. A disadvantage of this approach is that instruments designed for human players may not accommodate machinic actuation. Familiar instruments are also associated with a voluminous library of music and techniques, which are imparted upon the minds of designers, composers, performers, and audiences. These may or may not be desired, depending on the builder's goals.

The third category consists of sonic objects and machinic actuating systems developed together as *integrated* wholes. Sonic objects are configured according to the capabilities of actuating systems. The form and structure of the machine take acoustic, aesthetic, performative, technical, and pragmatic factors into account. Instead of a machine that *plays* an instrument, the machine and the instrument are a single whole. A developer of an *integrated* system is not constrained by extant instruments, so they have more starting points (e.g., from an actuator first, from the string first) and paths to follow. As the idea comes to fruition, the sonic object can be adjusted to accommodate developments in the actuation system and vice versa (this is not so easy when working with a piano). The quantity of choices makes this a challenging category to work in, but the rewards are new sounds, gestures, and music. An example is the MPR Lab's *Parthenope*, a robotic siren that produces sound by releasing compressed air through perforated spinning discs, discussed in more detail in Chapter 7—Aerophones (Figure 4.1.5). Synthesis of actuation and sound production systems offers exciting ways to create new audiovisual art.

FIGURE 4.1.5. **Parthenope by WPI's MPR Lab.**

As with any categorization system, there are gray areas between groups, such as LEMUR's GuitarBot or EMMI's PAM (Figure 4.1.6). Actuating systems and sonic bodies are integrated in these instruments, but the two are distinct visibly and physically in some ways (e.g., the strings) compared to *Parthenope*.

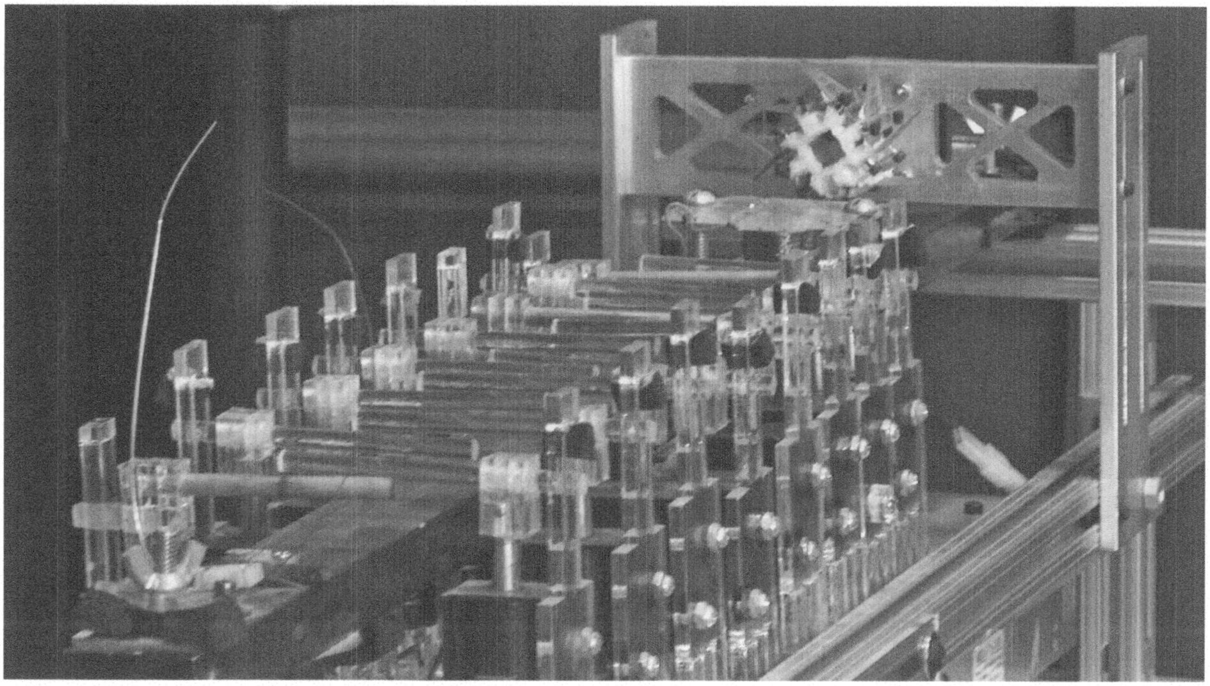

FIGURE 4.1.6. **PAM by EMMI and WPI's MPR Lab.**

The relation between *players* and *instruments* involves who controls what. In the case of a human player and an acoustic instrument, this relation is dictated by the physical and cognitive abilities of the human and the affordances and constraints of the instrument relative to those capacities. While it may be tempting, we cannot simply say that a human unidirectionally controls an instrument, for the characteristics of an instrument (thickness and tension of strings, size and spacing of keys, transformation of input forces and velocities of mechanisms, etc.) play a role in such interactions. Machines are affected similarly by the acoustic instruments they play. More complex situations involve *cooperative* instruments that allow human and machinic players to interact through the same instrumental medium, such as the MPR Lab's *Cyther* (Barton et al., 2017). In these examples, designs accommodate the unique strengths of humans and machines (or not), and the interaction between the two can result in new musical communications. For example, a machine may handle "low"-level choices, such as determining specific pitches and onset timing, while the human focuses on "higher" matters, such as the contour and punctuation of phrases or the summation of individual voices into more complex textures.

FORM ⇔ FUNCTION

A musical machine's mechanical and structural aspects may be utilitarian, depending primarily on *function*. *Function* refers to musical performance and interaction requirements as well as pragmatic concerns. Positioning the actuating mechanisms, securing the circuit boards and power supplies, routing the wires, tensioning strings, making parts accessible for repair or adjustment, and ensuring transportability are functional matters. Conversely, there may be an artistic vision for the *form* of the machine that is not directly related to functional concerns. Such a vision may seek to emulate familiar bodies (e.g., humans) or may explore abstract shapes, sizes, surfaces, and angles. Emphasis on *form* is embodied in the world of dynamic sound sculpture (Figure 4.1.7).

Compromises are necessary as functional necessities are balanced against artistic visions voiced in visual and gestural characteristics.

FIGURE 4.1.7. Left: Conloninpurple by Trimpin (Image Courtesy of Tacoma Art Museum, photo credit: James Porter). Right: Gravity Harps by Andy Cavatorta (image from andycavatorta.com).

Functional approaches seek to maximize musical performance. These designs prioritize speed, temporal accuracy, dynamics, and pitch range. This approach requires minimizing superfluous parts and motion, often making the presence and movements of machinic components inconspicuous, if not invisible. Often, these machines are discreet *attachments* to a sonic object or are *integrated* with it. The result is an instrument that *sounds itself* rather than a machine that plays an instrument. The removal of the *player* in this process affects audiences' understanding of how sound is produced and musical performance practice. When experiencing such a machine performance, attention often turns to human performers interacting with the ensemble, the static visual qualities of the instruments, or the music itself (or some combination of the three). This approach posits instrumental identity in the qualities and configurations of sonic objects rather than in the technologies or gestures used to activate them. Lacking an egocentric celebration of technological accomplishment, these instruments invite audiences to contemplate the ideas and actions of the composers and human performers who bring the music to life with the help of these machines. Many of the Logos Foundation's instruments, such as <vibi> (Figure 4.1.8), illustrate this category.

FIGURE 4.1.8. <vibi> by the Logos foundation (image from https://logosfoundation.org/instrum_gwr/vibi.html).

Functional machines can also be *players* of instruments. The same prioritization of musical performance applies here, but there is no desire to integrate the actuation system and the sonic body. Often, the rationale of this approach is to make the machine *general* so that it can play a variety of instruments. Robert van Rooyan's *Mech Drum* exemplifies this category (Figure 4.1.9). A *general functional* machinic *player* could be anthropomorphic, but requiring no superfluous parts or movements engages the fascinating relation between anthropomorphic complexity (e.g., more degree of freedom [DOF]) and the ability to play diverse musical instruments. This relation is one for future builders to explore.

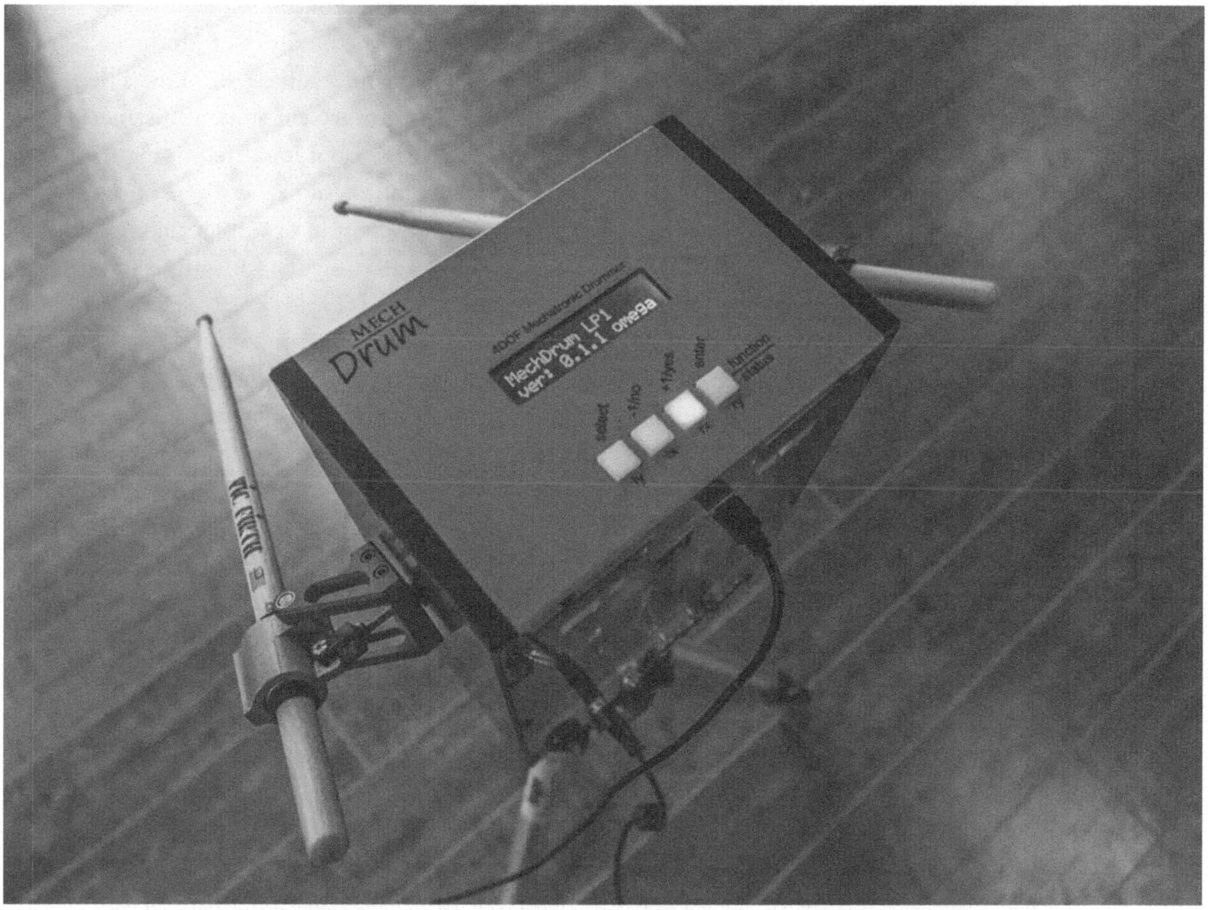

FIGURE 4.1.9. **Mechdrum by Robert van Rooyan (image from https://mechdrum.wordpress.com/).**

Designs that prioritize *form* look beyond musical performance requirements. In some of these works, sonic objects are subsumed by structural and machinic entities that assert visual and gestural identities (e.g., *Gravity Harps*, Figure 4.1.7; the work of Felix Thorn, Figures 4.1.14 and 5.1.11). These are dynamic sculptures; artistic forms that move and make sound. Imagine a Rube Goldberg machine sprinkled with mechatronically actuated sonic objects that speak as causal reactions unfold around them. Alternatively, designs prioritizing *form* could consist of artistic *players* of sonic objects (examples are featured in Chapters 5–7). These *players* may or may not bear likeness to known animals or structures, opening a fantastical world of possibilities.

EMULATIVE ⇔ NOVEL

What will your design emulate? What will it synthesize? What will it offer that is new?

When pondering the examples related to your idea, which aspects do you want to emulate and which do you not? It is naïve to think that what we make is wholly novel. We use the creations of others (musical scales, programming languages, physical tools) to make creations of our own. The question is not *whether* you will borrow but *what* you will borrow.

In the case of musical performance, a primary source of borrowable material comes from our own human form and function. Here, *function* refers to the qualities of produced sound, while *form* refers to the extent to which a machine looks or moves like a human. There may be some combination of both. **Anthropomorphism** is a primary categorical distinction for musical robots and robots in general (e.g., the Toyota Partner Robot, pictured in Figure 4.1.4). As designs become increasingly anthropomorphic, the connections between humans and machines are plainer to see, but such connections are still present in examples that are not anthropomorphic. Such is evident in the case of players and instruments. Our understanding of a guitarist is one who plays a guitar. A machine that plays a guitar leads one to see the machine as a guitarist, even if the machine is boxy, cylindrical, metallic, or wooden.

Emulating humans has advantages. It is easier to imitate that which is close, available, and controllable. We understand how a human plays an instrument (to varying degrees), so if we want a robot to do the same, we build a robot that functions like a human. The Cog team at MIT (Figure 4.1.10) gives two reasons for pursuing anthropomorphic form (*Cog Project Overview*, n.d.). First, our (humans) physical form affects how we mentally represent, understand, and interact with the world. If you want a machine to act in human-like ways, it must have a human-like form. Second, the development of intelligent machines requires interaction with humans, and humans feel more comfortable interacting with machines that appear in a familiar form. Other researchers have used anthropomorphic robots to better understand human performance (Solis & Takanishi, 2011).

FIGURE 4.1.10. MIT's Cog Project (image By Daderot—Own work, CC0, https://commons.wikimedia.org/w/index.php?curid=31244145).

Employing humanoid forms helps artists define musical and performative roles. These can include energetic leaders, virtuosic soloists, or deferential accompanists. As the designer of an anthropomorphic robot adheres to or challenges established conventions, they comment on human and machinic identities. Familiarity motivates audiences to make comparisons and generate questions about abilities and uniqueness. One of the ramifications of anthropomorphic emulation is fueling the *machine-replacing-human* trope common to robotic and technological narratives. The extent to which this interpretation is accurate depends on the machine's capabilities and the context defined by the designer.

Beyond human comparisons, likeness can extend to animals, objects, structures, and types of motion. Inspiration is everywhere: the movement of an antelope, the body of a millipede, the communications network in a city. One may emulate the unyielding resourcefulness of water as it negotiates the rocks in the bed of a stream, the oscillations of air around an edge, or the orbit of the planets around the sun. These examples provide structure to aspire to and to build from.

Ultimately, machines are not humans, antelopes, planets, or buildings. When they emulate, they do so metaphorically; their gestures fall between *allusion* and *replication*. Part of the design process involves determining how strongly your machine will connect to external references: To what extent will your machine evoke familiar notions of form, movement, and performance? If you seek to avoid such references, to what extent will you attempt to establish new realms of identity? Where will your instrument be located along a continuum whose poles are *emulative* and *novel*?

emulative ●————————● novel

FIGURE 4.1.10A. EMULATIVE–NOVEL DIMENSION

This dimension can be used to evaluate wholes and parts. For example, the player can be considered independently of the instrument, which can be considered separately from the timbre of the sound produced. There could be an emulative player and a novel instrument, a novel player and an emulative instrument, an emulative player and an emulative instrument, or a novel player and a novel instrument. In-between possibilities abound: An instrument could sound a lot like a guitar and a little like a trumpet. Its structure could be a sphere that alludes to a building.

Individual dimensions combine in composites that evoke the excitement of innovation and the comfort of familiarity. Often, the weight rests in the middle as emulation and novelty are balanced in individual aspects and the collective whole. Too emulative and the object is derivative. Too novel and the object is esoteric or crosses categorical lines to become something else. Presence in the middle encourages us to see the object as relevant *and* interesting. It allows us to carry traditions of human musical activity forward. With all this said, there are no absolute rights and wrongs concerning emulation and novelty. An anthropomorphic robot that plays a guitar has value, as does an odd-looking assemblage that produces sounds only the avant-garde would love. Where will your machine (and its components) be located on this continuum?

HUMAN ⇔ MACHINE

The extent to which an artificial object presents as *human*, *machine*, or somewhere in-between is a primary consideration in the design, use, and interpretation of musical machines and robots in general. These associations determine how humans connect to machines (or to other humans via machines), affecting the communication of narrative, personality, and affect. A machine's location on this dimension is influenced by physical form, materials, and movements, as well as musical timing, articulations, and dynamics. Emulation is a significant factor in evoking such associations, but *emulation ⇔novelty* and *human ⇔ machine* are not synonymous. First, it is possible for a machine to emulate a human, animal, object, or another machine, so *emulative* in and of itself does not imply *human* or *machinic* qualities. The same could be said for *novelty*. Second, contextual factors (e.g., the musical work or stage design) may make the same machine seem more *human* in some cases and more *machinic* in others. The machine could be located at one point on the *emulative ⇔ novel* dimension and at different points on the *human ⇔ machine* dimension, depending on the context. The preceding necessitates a separate *human ⇔ machine* dimension, even though overlap with the *emulative ⇔ novel* dimension will sometimes occur; combine and separate the two as needed.

CONSPICUOUS ⇔ INVISIBLE

A musical machine may be *conspicuous*: large, imposing, dramatic, obvious, or complex. Sometimes, these qualities emerge from functional needs, such as needing to secure many actuators in different positions. In others, conspicuous qualities are the expression of an artistic vision that sees value in scale. The other pole of the continuum is populated by electromechanical systems that are minimally apparent, purporting to get out of the sonic object's way. The aesthetic values of this approach de-emphasize machinic presence, reinforcing notions of conventional instrumental identity and human performance. At the same time, humans are conspicuously absent, evoking magical notions that fascinated audiences of nineteenth-century automata. Designing components or processes to be conspicuous or invisible is thus a matter of making causality transparent (or not), which defines accessibility and inspires curiosity.

SINGLE / COMPLEX ⇔ DISTRIBUTED / SIMPLE

An excitation system could involve a *single* entity capable of complex actions or multiple *distributed* components that are individually simpler but whose abilities are enhanced through joint action. The appropriate approach depends on the sonic object(s) and the performative motion(s) needed. A heterogeneous percussion ensemble is often better served by a distributed system, while plucking a string in various ways benefits from a more complex multi-DOF actuation system. The purpose of the project and the visibility requirements will also influence this choice.

SPECIFIC ⇔ GENERAL

Actuation systems may be designed to play a *specific* instrument (e.g., a pianola), or they may be able to be used more *generally*, such as playing several different instruments (e.g., modular percussion). An aspect of *generality* is *modularity*, where actuating systems can be combined and configured in different ways to play different instruments. An example is a modular percussion system that can be oriented and combined in different ways to sound myriad sonic objects (Figure 4.1.11).

FIGURE 4.1.11. **Modular percussion by WPI's MPR Lab and EMMI.**

FIXED ⇔ ADJUSTABLE

The components of a musical machine (including sonic objects and actuating systems) can be *fixed* in place or *adjustable*. *Parthenope*'s discs are fixed in place, while *PAM*'s tangents and the *modular percussion* actuators (Figure 4.1.11) are adjustable.

One may be tempted to conflate *specific ⇔ general* and *fixed ⇔ adjustable* dimensions, but they differ in applicable ways. An actuating system designed for a *specific* harp could have plucking mechanisms that are *fixed* or *adjustable*. A percussion striker designed to play many different instruments (*general*) may be *fixed* (requiring the sonic object to be moved), or it may be *adjustable* (moveable relative to the sonic object). The *fixed ⇔ adjustable* dimension is particularly germane to musical performance requirements. The difference between a fantastic and an unacceptable performance is a fraction of an inch when plucking a string or striking a chime. Being able to make adjustments in these scenarios is critical. Further, components that are nominally the "same" are not physically identical. When building a robotic glockenspiel, I realized significant differences between solenoids that were the same model. Accounting for these differences required *adjustability* in the instrument's structure.

Requirements

With a concept of the instrument defined, what must it be capable of? Prioritize musical and artistic aspects. What pitches and rhythms do you want to sound? What visual identity do you want to establish? Answering these questions will produce a set of requirements in musical performance, visual form and gesture, and modes of interaction. **Establish requirements independent of specific design ideas.**

> **Lesson**: establish musical, visual, and interaction requirements first before committing to a particular design(s).

Setting requirements first directs your focus on outcomes and leaves how to get there open to possibilities. Students often start with a design in mind and discover its musical capabilities after building the system. Often, the machine's abilities are limited, falling short of unarticulated expectations and causing disappointment given the time and effort expended. Establishing requirements at the beginning gives the project direction and helps ensure it is tractable, so time and effort are well spent and expectations are met. When setting requirements, describe how they will be evaluated to determine if they have been met.

MUSICAL PERFORMANCE

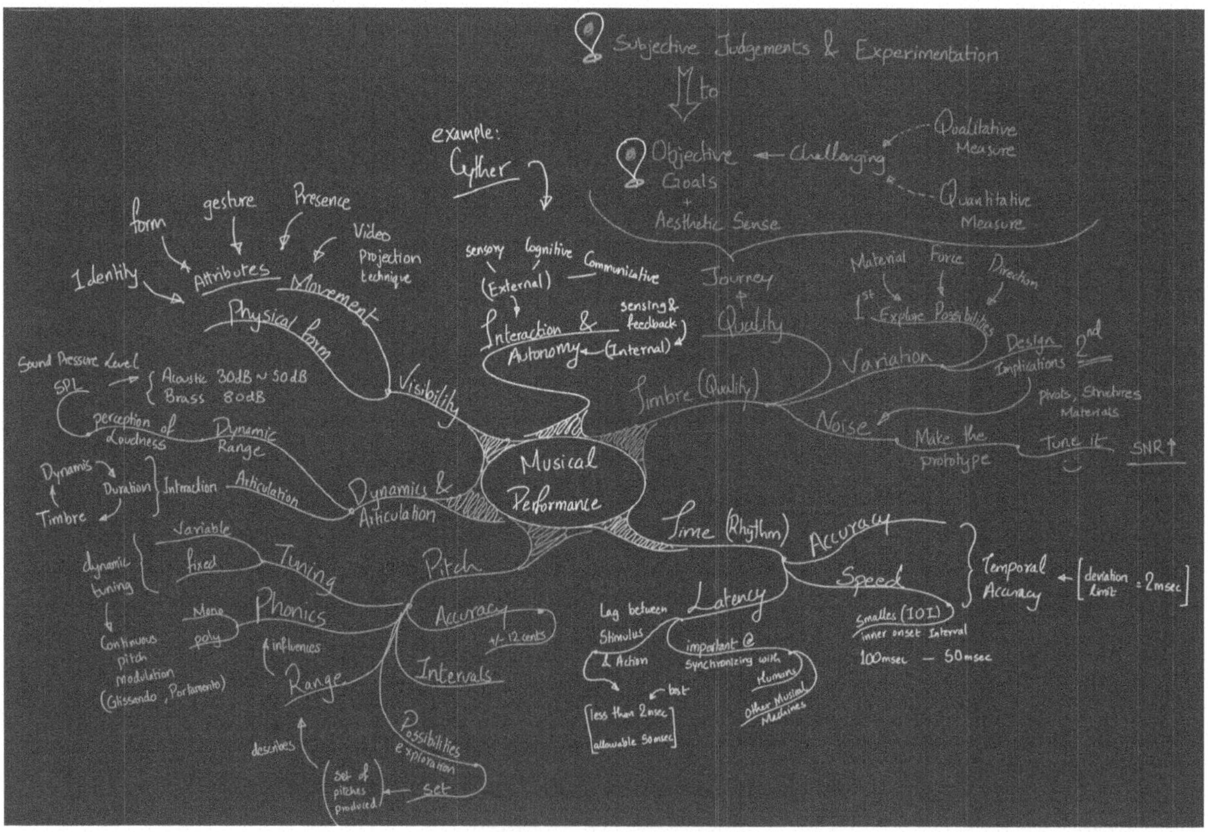

FIGURE 4.1.12. **Mindmap by Yash Garje.**

Among the myriad dimensions that characterize musical performance, timbre, time / rhythm, pitch, dynamics, and articulation are essential. Establishing goals for accuracy, range, and quality in these categories guides design and evaluation processes.

Timbre

What kinds or qualities of sounds will your instrument produce?

It is easy to get lost in the technical complexities of a musical robot and lose sight of what matters most: *the identity and success of a musical machine depend on its sound quality.* We are drawn to instruments by their timbres, a fact that has only intensified with the electronic revolutions in both popular and art music that began in the twentieth century. A beautiful-sounding instrument activated by simple computational and electromechanical means is usually more compelling than a technically complex one that is sonically blasé. Developing **timbral quality** relies on subjective judgments and experimentation. Set up test rigs with different kinds of strings, hit drums with assorted strikers, and move a piezoelectric transducer around an instrument's surface. Let experiments guide how you adapt the physical characteristics of your instrument.

Subjective experimentation also has issues. We lose our bearings after elongated periods of exploring sound, becoming less sure about our aesthetic judgments. We may not be sure what we like in the first place, or we may like multiple instances equally well. The first version of a string instrument might produce sounds that are biting and twangy, while the second speaks in a warm and mellow way. If you like both, which one are you going to choose? You might have esoteric tastes that do not align with the general public's. Good for you for being unique, but that might not help you in your quest to get your musical robot in the hands of other musicians.

Introducing objectivity into the process can mitigate these problems. While objective measures of timbre are challenging both quantitatively and qualitatively (Krumhansl, 1989), sonic references, including physical instruments, recordings, synthesizers, and signal processors, can provide timbral goals to which to aspire. These can involve quantitative measures of timbre used in spectral analyzers (see Figure 4.1.13), sonic models (e.g., a specific drum sound), or measures such as (in)harmonicity. For example, you might aspire to a spectrum characterized by a strong fundamental, a clear second harmonic, and a lower relative level of surrounding inharmonic partials that function as accompanying noise.

Qualitatively, sounds can be described using linguistic metaphors such as "bright" or "full" that provide timbral objectives to design toward. Quantitative and qualitative references root the sound design process in the musical context within which you work, facilitating refinements and ensuring you do not lose your aesthetic bearings.

Timbral variation makes acoustic instruments more engaging. The sound of a snare drum's center may be appealing, but character, nuance, and variety are realized as one strikes the edge of the head, the shell, the rim, and all points in-between. Exploration and experimentation teach you about possible sounds and the materials, forces, and directions that activate them. These activities can help you form actuation and structural designs. Doing it the other way—building the electromechanical system first and exploring timbral variation later—is risky. What if the instrument does not sound very good after all the time and effort spent building and designing, and the only fix is starting over? Sonic exploration can help you avoid such costly and frustrating mistakes. Find your sound early on!

Noise is an inherent part of sound, from a flutist's breath to the buzz of a guitarist's tube amplifier. Musical robots generate noise via the electronic circuits and mechanisms that propel their movements. When sonic byproducts are sufficiently distracting, they become *noise*. It is impossible to eliminate mechanical noise entirely, and, in most cases, such

FIGURE 4.1.13. A spectrogram (*y*-axis is frequency, *x*-axis is time, and color is intensity).

a scenario is not desirable: a bit of mechanical noise provides character and definition to a timbral envelope reflective of the instrument's materiality. At the same time, noise must be reduced to a point where it is the icing and not the cake. Choosing suitable materials, pivots, and structures is fundamental to getting an acceptable signal-to-noise ratio. One school of thought followed in the Music, Perception, and Robotics Lab at WPI is to get the prototype working and then iteratively refine the design to reduce noise. It is better to have a noisy machine that works than one that is silent because it only exists on paper. Future efforts can produce quieter outcomes. With that said, noise-reducing elements such as bearings and brushless motors can and should be specified at the beginning, saving time and cost of future iterations. If sound-attenuating features do not compromise basic functionality, incorporate them as soon as possible. Whatever the approach, make a requirement that establishes the kinds of noise and their acceptable quantities (usually expressed in dB SPL).

Time

How fast can the machine play? How temporally accurate does it need to be? What latencies are acceptable?

Music is a temporal art; the **speed** and **accuracy** of note onsets define the musical potential of a machine. **Temporal accuracy** is measured by determining the deviation of a note onset from the position specified in the instructions (e.g., traditional score, DAW sequence, code). Human musicians produce nominally synchronous notes within a 30–50 ms range (Rasch, 1988), providing a boundary for allowable temporal deviation. Ideally, temporal deviation should be 2 ms or less, given that is the duration at which humans can distinguish between two tone onsets (Hirsh, 1959). Speed is defined by the smallest inter-onset interval (IOI) that the system can produce. While faster is generally better, two temporal thresholds are particularly relevant in music. The first is 100 ms, established in the psychological literature as the shortest possible duration within a metric cycle (London, 2004, pp. 28–29). Given this duration's musical and psychological importance, musical machines should ideally be able to play at this rate. The second threshold is 50 ms, or 20 Hz, which is the frequency at which pitch sensations emerge in humans. As actuators approach and surpass this threshold, the domain of musical expressivity transforms from rhythm to timbre and pitch. IOIs may vary depending on pitch interval, mechanisms, and instrument structure; these aspects should be factored into the design and evaluation processes.

Latency is the temporal delay between system events, such as the press of a key and the appearance of a character on a screen. The temporal interval between an instruction (e.g., command, signal) and an action (e.g., solenoid hitting a drum) introduced by software and hardware components has significant musical consequences. Latency is an aspect of temporal accuracy. A machine may produce accurate temporal intervals (e.g., each is 500 ms), but they are offset from "absolute" positions (e.g., as provided by a system clock) because of latency. This latency is critical when synchronizing with other players, human or machine. Quantifying allowable latency follows the guidelines given for temporal deviation: latency should be as small as possible and no more than 50 ms. Realistically, some latency will always result from communications, data processing, and physical movements. The more you can reduce latency, the easier real-time interactions will be.

Pitch

> What pitches and pitch modulations are possible? How can the instrument be tuned? Will it be monophonic or polyphonic? What measure of pitch accuracy is required?

When making an instrument that produces pitch, consider pitch set, range, tuning, glides (i.e., glissando and portamento), intervallic leaps, polyphony, and accuracy. Determine which pitches the instrument can produce in terms of **set** and **range**. A **pitch set** is a collection of pitches or pitch classes, often articulated according to musical **scales**. The chromatic scale is common to Western instruments, allowing myriad pitch groupings. For example, it allows the diatonic scale to be formed at each pitch, facilitating access to all possible key signatures. Alternatively, a pitch set can be determined mathematically (e.g., by dividing a pitch range—which can but does not have to be an octave—into *x* intervals or by using interval ratios, as in just intonation). The pitches that an instrument can produce need not be a complete or known scale: subsets, variations, and divergences are enticing alternatives. Will the **tuning** of the instrument be fixed or variable? If variable, will tuning occur before or during a performance? Knowing the pitch set allows the **pitch range** to be established, dictating how many times that set can be repeated. In general, the minimum range should be an octave, given the psychological importance of the interval. Two octaves is a more reasonable goal, which opens territory for melodic writing and harmonic voicing. An instrument's pitch range influences its physical characteristics (e.g., the length of the neck of a string instrument).

When considering pitch range, decide whether the instrument will be **monophonic** (one note at a time) or **polyphonic** (multiple notes at a time). For string instruments, increased range usually means more strings, which enables polyphonic capabilities. Imagining the role of the instrument can help determine this choice. Will it weave melodic passages, provide harmonic accompaniment, or both?

What kinds of **modulations (glides)** are going to be possible? These might be continuous, as in a portamento, or discrete. **Intervallic leaps** are afforded by characteristics of both the player and the instrument. On a human-played monochord, intervallic leaps are determined by the size of the player's hand and the speed at which they can move it laterally. Multiple adjacent strings tuned to different pitches allow larger intervallic leaps to be achieved by traversing less distance. Changing pitch by modulating string tension requires more time to produce larger pitch intervals. It is thus imperative to establish what kinds of pitch intervals will be possible and how quickly they need to (or can) be realized.

To determine acceptable **pitch accuracy**, we again turn to the psychological literature. Listeners with relative pitch can distinguish between two different pitch intervals when larger than 3–25 cents (a cent is 1/100 of a semitone), depending on the method employed and the interval quality (Thompson, 2013). From this research, a reasonable goal for a machine should be to produce a pitch within ±12 cents of the ideal frequency.

Dynamics and Articulation

> What is the instrument's dynamic range?

The **dynamic range** of an instrument plays a fundamental role in musical expression, emotional communication, tension, resolution, contour, nuance, variation, style, topography (the "size", arrangement, and "closeness" of musical voices), and timbre. Generally, more range is better. Dynamics correspond to the perception of loudness, which depends on parameters including sound intensity, frequency, time of exposure, and so on. Of these parameters, sound intensity, expressed as sound pressure level (SPL), is key and is proportional to perceived loudness. The dynamic ranges of acoustic instruments provide guidelines for designs and evaluations. The dynamic ranges of most pitched acoustic instruments span from about 30 dB (violin, viola, trumpet, oboe) to 50 dB (clarinet), while percussion instruments such as the bass drum can extend to 80 dB (*Acoustical Characteristics of Musical Instruments*, n.d.). Human physical capabilities are considerable; matching their results can be challenging. Doing so requires actuators that can be controlled precisely and handle large amounts of power, usually accompanied by lofty price tags. Narrowing dynamic range goals may be necessary given your circumstances, but such allowances should be minimized as much as possible.

> What kinds of articulations can the machine produce?

Musical **articulations** describe how notes are shaped through the interaction of dynamics, duration, and timbre, emphasizing the attack of and transition between notes. In European

Art Music, articulation conventions include *legato* (smooth and connected), *staccato* (sharp and punctuated), *tenuto* (full duration), *marcato* (marked, emphasized), and *accent* (emphasis on a pitch or chord). Articulations are vital for musical machines as they disrupt the homogeneity that characterizes mechanical production. Possible articulations depend on the sonic object (e.g., playing legato is far easier on a string instrument than a woodblock) and the player. Human wind players manipulate embouchure; string players bow, pick, pluck, and damp strings; percussionists roll, drag, and flam. The extent to which these methods can be produced should be considered, as should new methods enabled by machinic abilities.

Summary of Musical Performance Requirements

- timbre: quality, variation, noise
- time: accuracy, speed, latency
- pitch: accuracy, set (i.e., scale), intervals, range, mono / polyphonic, tuning, modulation
- dynamic range
- articulations

VISIBILITY

FIGURE 4.1.14. **Felix's Machines at the Brighton Festival (2013). Image from https://www.facebook.com/felixsmachines/photos.**

What is the visual experience of the machine going to be like?

A musical robot's physical form and movements affect the engineering of its subsystems; they are also a fundamental part of its identity, influencing how it is used and perceived. One of the attractions of musical robots is that they bring physical presence, form, and gesture to the stage in a way that loudspeakers and laptops cannot. What is the machine going to look like and how is it going to move? It could be imposing or diminutive, complex or simple, fluid or segmented. What materials is it going to be made of? It could feature sleek surfaces of polished aluminum, ornate wood carvings, or a nest of exposed wires. What colors will it be? Will the technologies that power it be visible or hidden?

Envision the physical contexts that your machine will play in. It may take center stage, engaging the expectations of concert-goers and video voyeurs. It may inhabit the corner of a studio, used for composition and recording. It may interact with others in an art gallery installation, with visual aspects affecting communication, sound production, and associations with musical instruments, non-musical technologies, and culture. The ramifications of these choices are enormous in terms of size, cost, complexity, transportation, and development time.

One way to make movements more visible is by using video projection. Cameras can be placed around an instrument to magnify parts and movements, making the puny suddenly grand. I have used this technique in live performances of *Tempo Mecho* (2019) for PAM, modular percussion, and percussive aerophone (Media Example 4.1.1 ▶). An advantage of this approach is that you can use small actuators positioned close to a sonic object, which may be necessary to satisfy musical requirements. The disadvantages of this approach are that more equipment is necessary (cameras, stands, cables, video mixers, screens), which costs money and time to transport, set up, configure, and take down. This may be a contradiction for some, who see the purpose of a musical machine as a way to make sound production physical, tangible, and visible. Mediating technology such as cameras and video switchers may detract from this experience. There is no right or wrong here; it is one of the many cases where compromises must be made. Consider these matters from the beginning by establishing visibility requirements: it is hard to make small, closely positioned solenoids more visually expansive after the instrument is finished.

Your ideas for the machine's visual identity can be abstract at this point; the details will be worked out later when choosing actuators and designing mechanisms. What is important now is to formulate the qualities that will characterize the project. Collect images and videos of aspects you want to incorporate or inspire you. Keep these references handy as you move through the design process.

INTERACTION AND AUTONOMY

To what extent can the system function by itself? How is it going to sense and respond to others?

In musical machines, interactivity and autonomy exist in *external* and *internal* domains. *External* questions involve *who* or *what* affects the machine's musical actions. Instructions can be predetermined (e.g., in a score, MIDI sequence, or computer program), result from human actions (e.g., pressing a button on a MIDI controller), or algorithmically generated in real time. Who will the system act with? Actions could proceed with or without regard to the surrounding musicians and audience members. The machine could perform by itself on stage or join a group populated by humans and/or other machines. If it is going to be part of a human–machine ensemble, how will it interpret and respond to the expressions of others? It could be subjugated to the desires and preferences of a human(s). It could be allowed to travel its own paths. It could be some combination of the two.

External interactions require the ability to sense, cognize, and communicate. Sensory possibilities could be musical (e.g., rhythmic groupings, meter, beat, key), contextual (e.g., what other performers and instruments are participating), and psychological (e.g., interpreting facial expressions). What knowledge will the machine have access to prior to experience (e.g., harmonic function in major/minor tonality, different kinds of clave rhythms, and the meaning of visual cues) to facilitate these interpretations? Will it form memories as it performs, and if it does, to what extent will these memories affect future decisions? Will its behavior change because of interactions? Will it learn? Interaction requires that interpretations are communicated to others. The scope of these *external* musical communications involves conveying intentions, desires, interpretations, timing, and cues to other collaborators. They signal who is leading, following, soloing, setting the tempo, or establishing the music's key. How such information is conveyed may occur within the sonic output of the system, or it might involve physical gestures, text, images presented on a screen, or a matrix of lights.

Interactions also happen *internally* within the components and subsystems of a musical machine. A percussion machine attempting to produce a specified dynamic level may require sensing the position and velocity of the drumstick relative to the drumhead, which is communicated to a microcontroller controlling the drumstick motor. A string instrument that tunes by adjusting string tension may require sensing the fundamental frequency of the vibrating string, which it uses to control the position of a motor coupled to a tuning machine. As in the *external* case, sensing and feedback are required for these interactions to occur and will determine the extent to which the instrument can function autonomously.

In summary, interaction and autonomy on the *external* level involve how the machine functions relative to the actions of others, which involves musical preferences, ideas, and goals. Interaction and autonomy on the *internal* level involve physical reactions to sensed states: a sensor determines the position of *object A* (drumstick, moving bridge, tuning mechanism), which results in the movement of *object B* (motor), which modifies the position of *object A*.

In designing interactive capabilities of a musical machine, the "mind" can be separated from the "body." One team can build a mechatronic percussion system that responds to MIDI commands, while another builds an AI improviser that senses the utterances of others and responds in kind. These systems can be developed independently and integrated in the future, even by different groups of people. At the same time, there are advantages to thinking about interactivity during a machine's mechanical, electrical, computational, and structural design. The motivating idea of The MPR Lab's *Cyther* was a human-playable robotic zither. The interactions imagined between a human and the machine influenced design aspects such as the spacing between strings and the location of excitation and damping solenoids. Thinking through modes of interaction is useful for non-human-playable examples as well. The MPR Lab's *Parthenope*'s disks were configured with a composer's potential harmonic desires in mind. Providing physical cues to musicians was one of the motivating ideas behind the robotic marimba player *Shimon* (Hoffman & Weinberg, 2011). As interactivity is an integral part of music, it should be considered early in the design process. Such ruminations could meaningfully affect structure, form, movement, components, and communication choices.

References

Acoustical characteristics of musical instruments. (n.d.). DPA. Retrieved May 19, 2020, from https://www.dpamicrophones.com/mic-university/acoustical-characteristics-of-musical-instruments

Barton, S., Prihar, E., & Carvalho, P. (2017). Cyther: A human-playable, self-tuning robotic zither. *Proceedings of the International Conference on New Interfaces for Musical Expression*, 319–324. https://doi.org/10.5281/zenodo.1176266

Bianchi, I., & Savardi, U. (2008). *The perception of contraries* (1st ed.). Aracne.

Bianchi, I., Savardi, U., & Kubovy, M. (2011). Dimensions and their poles: A metric and topological approach to opposites. *Language and Cognitive Processes*, 1–34. https://doi.org/10.1080/01690965.2010.520943

Cog Project Overview. (n.d.). Retrieved June 24, 2020, from http://www.ai.mit.edu/projects/humanoid-robotics-group/cog/overview.html

Hirsh, I. J. (1959). Auditory perception of temporal order. *The Journal of the Acoustical Society of America*, *31*(6), 759–767.

Hoffman, G., & Weinberg, G. (2011). Interactive improvisation with a robotic marimba player. *Autonomous Robots*, *31*(2–3), 133–153.

Krumhansl, C. L. (1989). Why is musical timbre so hard to understand? *Structure and Perception of Electroacoustic Sound and Music*, 43–53.

London, J. (2004). *Hearing in time: Psychological aspects of musical meter.* Oxford University Press.

Rasch, R. A. (1988). Timing and synchronization in ensemble performance. In J. A. Sloboda (Ed.), *Generative processes in music: The psychology of performance, improvisation, and composition* (pp. 70–90). Oxford University Press, Clarendon Press.

Solis, J., & Takanishi, A. (2007). An overview of the research approaches on musical performance robots. *International Computer Music Conference Proceedings, 2007.* http://hdl.handle.net/2027/spo.bbp2372.2007.179

Solis, J., & Takanishi, A. (2011). Wind instrument playing humanoid robots. In J. Solis & K. Ng (Eds.), *Musical robots and interactive multimodal systems* (Vol. 74, pp. 195–213). Springer. https://doi.org/10.1007/978-3-642-22291-7_12

Thompson, W. F. (2013). Intervals and scales. *The Psychology of Music*, *3*, 107–140.

CHAPTER 4

THE PROCESS OF MAKING A MUSICAL MACHINE

Part 2: Design and Making

Design

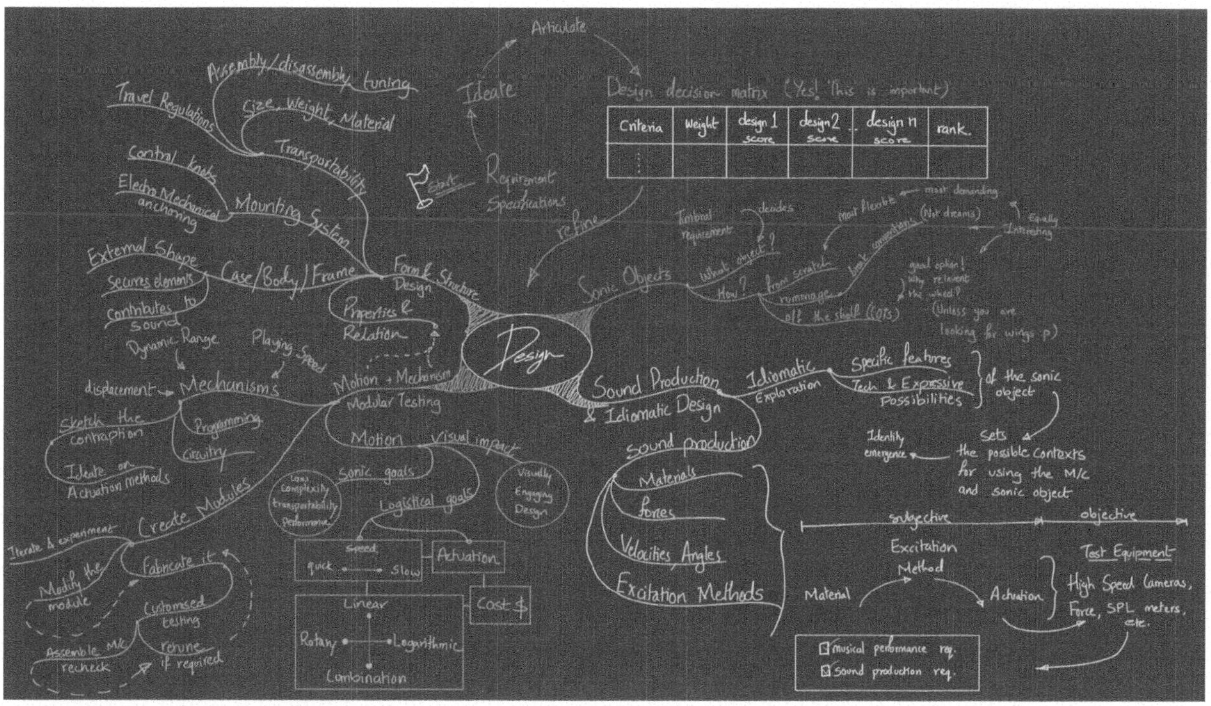

FIGURE 4.2.1. **Mindmap by Yash Garje.**

> Develop multiple designs that realize the stated requirements. Compare options and choose the one that best fits the established criteria.

Addressing the questions and directives of the preceding sections will help you produce a set of requirements. These clarify what the machine will look like, how it will act, and the sounds it can make. The next step is to explore and articulate how these requirements can be achieved.

There is more than one way to realize your idea, and, likely, the first version you come up with will not be the best. Formulate different designs that meet your requirements and establish criteria to choose between the possibilities, articulated in a **design decision matrix**. The matrix rows represent the criteria, and the columns represent the design options. Prioritize criteria using a *weight* column. Score a design by multiplying each criterion's rating and weight and sum the results. For example, design approaches for a percussion machine are compared in Table 4.2.1:

TABLE 4.2.1. **A design decision matrix for a percussion machine.**

Criteria	Weight (%)	Design 1		Design 2		Design 3	
		Rating	Weighted score	Rating	Weighted score	Rating	Weighted score
Playing rate	40	5	2	5	2	2	0.8
Dynamic range	30	1	0.3	4	1.2	5	1.5
Visibility of movement	25	1	0.25	4	1	3	0.75
Configurability	5	4	0.2	4	0.2	3	0.15
Total			2.75		4.4		3.2

This fundamental process is often ignored by enthusiastic builders who settle on the first idea encountered. This is an opportunity to stretch your creativity. Approaching design from different perspectives can help you avoid problems and reveal innovations.

The following sections are presented in a particular order, but it is not the only order possible. An alternative permutation may align better with your idea. For example, if the instrument's structure is the definitive aspect, start there and address sonic objects and actuation methods later. Consider the following sections as modules that can be moved around to suit your design ideas.

Sonic Objects

What object produces sound? Will it be made *from scratch*, acquired through *rummage*, or purchased *off-the-shelf*?

Complex mechanisms and fancy algorithms will not make a feeble drum sound vibrant. Choosing a sonic object is a critical part of the design process. Review your timbral requirements and identify your options. Navigate the possibilities by determining whether to acquire the object *off-the-shelf*, via *rummage*, or make it *from scratch*.

OFF-THE-SHELF

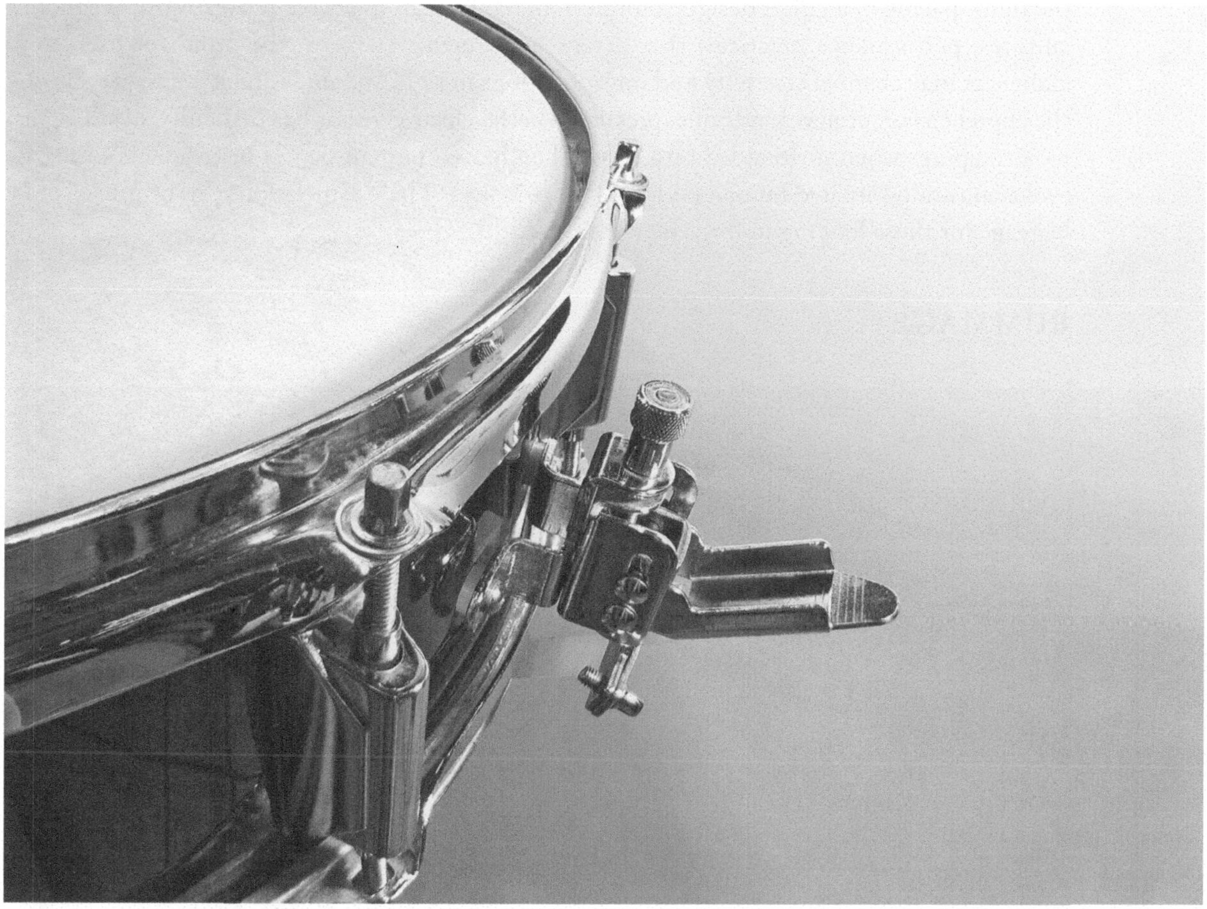

FIGURE 4.2.2. **Off-the-shelf drum hardware.**

Purchasing commercially produced instruments *off-the-shelf* (aka Commercial Off-The-Shelf, or COTS) has advantages. The craft involved in making acoustic instruments involves choosing, procuring, honing, and shaping materials in particular ways, which take time and effort to learn. These methods have been refined over hundreds of years in some cases, benefiting from the collective expertise and innovation of musicians, luthiers, and fabricators worldwide. You are awarded instant access to the fruits of this labor by acquiring *off-the-shelf*. As reliable, high-quality, and finished products, they allow designers and builders to devote time that would have been spent on the instrument to other parts of the system. They often have cost benefits due to economies of scale in manufacturing and distribution. Their *sonic properties* can instantly be evaluated. They provide direct access to musical canons specifically written for them. Their cultural meaning and familiarity allow users to draw connections between, make comments about, and evoke associations relative to specific works, artists, and characteristics. If what you need is a wheel, why reinvent it?

The *off-the-shelf* strategy has disadvantages. The fact that these instruments were designed to be played by humans means they are constrained by human physiology. Choosing a guitar as the sonic object of a robotic string instrument constrains the actuating mechanism to the spacing and number of frets and strings, the shape and weight of the object, and the method of transduction. The guitar was made for humans who are limited in the number, size, and shape of their hands. Machines have no such restrictions; why should the instruments they play be constrained? Starting with a pre-existing instrument is not always easier. Making numerous independent monochords may be less onerous than trying to cram a multi-actuator system into the tight spacing of a guitar designed for human hands. The *associations* (e.g., musical canons, cultures, performance practices) that extant instruments evoke in the minds of users and audiences may channel creativity and interpretations in rigid and undesirable ways, preventing the apprehension of new kinds of expression. As the electric guitar has difficulty shedding its rock trappings when surrounded by a classical orchestra, human-played instruments similarly evoke culturally-situated human performance practices. These associations can be unwelcome baggage for those looking to forge new artistic paths.

RUMMAGE

FIGURE 4.2.3. **Rummage: EMMI's CADI playing a found object.**

A builder who *rummages* makes instruments out of objects designed for non-musical purposes, circumnavigating the conventions of traditional instruments and engaging with cultural and economic associations of found objects and the conceptual implications of their appropriation. Objects acquired through rummaging do not need to be made (though they likely require modification), presenting some time-saving and expertise-avoiding benefits. The treasure troves of the rummager include flea markets, recycling facilities, tile showrooms, home goods retailers, and, of course, rummage stores. Odd looks may come your way when exploring the sonic potential of glass vases in aisle seven, but the efforts can be worth it.

Rummaging has disadvantages. It takes time and patience. It involves travel and evaluating many possibilities to yield a few suitable candidates. It requires knowledge of the right places to go, and even those locations have inconsistent inventory. Found objects carry associations with their colloquial usage, which may or may not agree with your system's aesthetic intent or structural design. As extant objects, objects acquired through rummage require actuation and structural systems to be designed *around* their features.

FROM SCRATCH

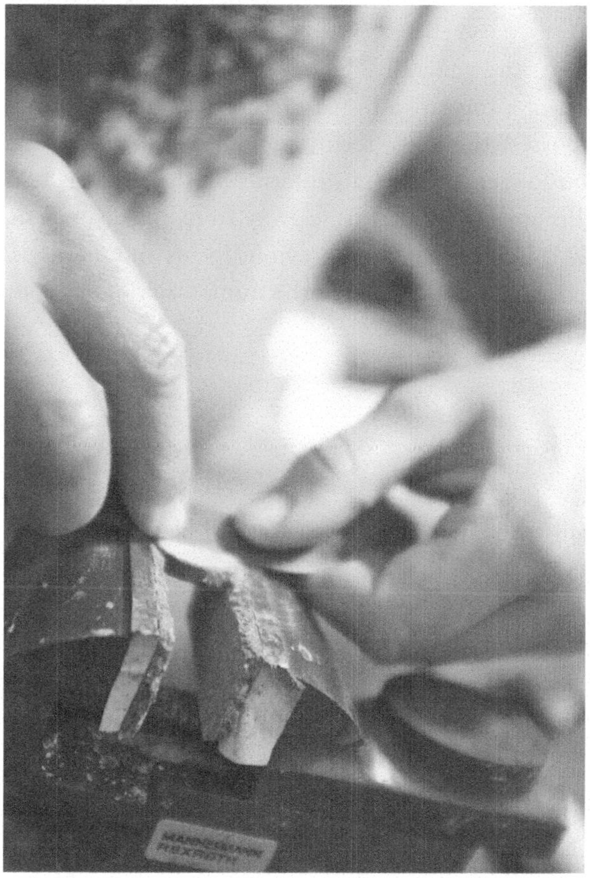

FIGURE 4.2.4. **Making from scratch (image By Benjamin Fauvel—Own work, CC BY-SA 4.0, https://commons.wikimedia.org/w/index.php?curid=100190606).**

Some sonic adventurers are best positioned at the start, when identities are nascent and malleable. The promise of new instruments is driven by the promise of new music. What the instruments and music will be is not yet codified. Venturing into uncharted territory is exhilarating, revealing products of imagination that have come to life. At the same time, these expeditions can be disappointing; some efforts fail, so there is risk. The intrepid creatures that seek such adventures are drawn to building *from scratch.*

Building an instrument *from scratch* has great potential but is also the most challenging and laborious option. Unless you are experienced, the product will be unknown until the instrument is built and is unlikely to yield what is achievable by professionals. With that said, building *from scratch* offers rewards not found elsewhere. The process engages your creativity as established norms are reshaped or removed. It can reveal new *sounds*, which are attractive when building a novel *instrument* and not just an actuation system. Making an object or system *from scratch* will help you learn about materials, acoustics, and fabrication, and will give you an opportunity to improve current designs.

Designing an instrument *from scratch* may reveal possibilities bypassed by the developers of familiar instruments. Standard instruments are restrained by cultural conventions in tuning, harmony, timbre, dynamics, and articulation. They are also products of the technologies of their time, often foregoing electronic amplification, signal processing, and spatialization. The *from scratch* builder can shed or embrace preconceived notions of instrumental form and practice to taste. Such experiences could lead to the development of new kinds of art (including music, visual form, and performance), which provides the most compelling argument to start *from scratch*.

The disadvantages of the *from scratch* case are generally the opposite of the advantages of the *off-the-shelf* case. Bypassing the fruits of cumulative human expertise returns you to square one on most points. If you do not possess the tools, techniques, and scale of professional manufacturers, it will likely take you longer to make an inferior product that is less robust and more expensive. Engaging in this approach takes time away from endeavors you may care more about, such as writing music. The freedom from previous forms and methods of a *from scratch* approach is liberating but also challenging, given the numerosity of possible choices.

IDIOMATIC DESIGN

What is idiomatic to the sonic object or instrument that the machine will play?

With a sonic object in hand, consider how its unique features afford different ways that it can be played. Andrew Koenig describes an idiomatic design as one "that takes context into account" (1996, p. 96). A design should not be made in a vacuum, independent of the ramifications of its implementation. In music, *idiomatic* refers to "the specific features, the technical or expressive possibilities and unique qualities that characterize each individual instrument or type of voice" (*Idiomatic Practice | The Idiomatic Orchestra*, n.d.). It is not just the qualities of an instrument that define it; the contexts in which it is used also play a role in establishing identity. This idea can be extended to the *way* a particular instrument is played. An *idiomatic* passage is natural for a musician to produce; it suits the way their fingers move and the shapes that their hands can make. This depends on the characteristics of the player and the sonic object. That which is idiomatic for humans emerges as the characteristics and capabilities of players interact with the characteristics and capabilities of sonic objects via the development of *techniques* (e.g., flutter-tonguing, *sul ponticello*, multiphonics) and their sonic results. As idiomatic qualities coalesce, identities of *instruments* emerge, embodying players, techniques, and sounds in physical design. Idiomatic characteristics can provide a model for builders seeking to emulate human performance but should be recognized as one option in a set of possibilities. Thinking beyond what humans have done with a particular instrument is one of the opportunities (some may argue obligations) of musical machine designers.

What kinds of techniques are possible on an instrument whose design is freed from the constraints of human physiology? One of the most exciting possibilities of musical machines is their ability to do what humans cannot by assuming new sizes, shapes, components, materials, and configurations. Thinking beyond the boundaries of what is idiomatic for human players is a catalyst for innovation and the creation of unique identities. The question becomes, what will be idiomatic for these new kinds of player-instruments?

Excitation

How does the instrument produce sound? What materials, forces, velocities, angles, and positions are necessary to excite the sounding body?

Sonic objects must be **excited** into vibration to produce sound. What implements (e.g., drumstick, pick) and methods (e.g., plucking, blowing) will achieve this state? How will vibrations be damped to control note durations? How will different pitches be produced? Answering these questions involves specifying physical quantities such as velocity, force, torque, and spatial position of the sonic object, exciter, damper, and pitch-producing components. Requisite quantities are ideally obtained by taking measurements with force sensors, high-speed cameras, accelerometers, gyroscopes, and SPL meters. When such equipment is unavailable, estimation and experimentation can get you in the ballpark.

By obtaining the values needed to produce sound, you are establishing a set of *sound production requirements* that follow your *musical performance requirements* (e.g., speed, dynamic range) but are specific to the sonic object you have chosen. If an alternative design features a different sonic object, the musical performance requirements persist, but sound production requirements might change.

Summary of Sound Production Requirements

- excitation
- damping
- pitch production
- specifying physical quantities (velocity, force, torque, position)

Motion

How is the machine going to move to achieve your sonic, visual, and logistical goals?

What kinds of motion can produce the sounds you require? How will the machine's movements balance your sonic and visual goals? What kinds of movement do you find compelling?[1] Movements can follow linear, logarithmic, or rotational paths. They can be small and disjunct or fluid and grand. They can unfold simply or according to complicated functions concatenated in time. They can move quickly, slowly, and anywhere in-between. The choices involve trade-offs. Larger movements take longer to complete, resulting in slower musical performances and a narrower range of musical articulations. Varied shapes require greater complexity in mechanical design (e.g., more DOF) and/or software-based motion planning, both of which require expertise. These machines cost more money to build, require more space to house, and are more difficult to transport. Smaller movements can improve musical abilities but might be less interesting to view in performance. Positioning actuators closer to the sonic object enables faster notes but also makes their movement less visible. I have both made and seen musical machines whose movements were so minute that they were difficult to be perceived, much less appreciated, by an audience. These trade-offs are represented in the following dimension:

FIGURE 4.2.4A **Musical Performance–Visibility Dimension.**

Note that the attributes that characterize these poles are not always contraries. It is possible for a machine to perform at a high musical level and also produce visually compelling movements. It is also possible to have a complex or expensive structure that is not particularly visually engaging.

Instead of compromising between the poles of functional music performance and visually engaging movements, it is also possible to separate them. For example, in making a flute-playing robot, one system could be dedicated to opening and closing tone holes, while another moves an anthropomorphic head and body for the purposes of expression (Li et al. 2019). This is common in human musical performance (i.e., a guitarist's fingers move a relatively small amount while a great deal of expression comes from body gesticulations and facial gestures). The extent to which one can or would want to separate this continuum depends on the nature of how sound is produced by the instrument (i.e., the possibilities for percussion are different from those for flute).

The balance between visually engaging movements and musical performance is a primary design decision. One pole is not better than the other, but the proportions between them should be aligned with the instrument's purpose and defining *properties and relations*. If the machine is meant to jaunt about the stage of a rock arena, then its movements should be visible. If the instrument lives as a sonic tool in a composer's studio, its lack of visual presence may be forgiven. A system that is *single / complex* will be capable of different kinds of motion than one that is *simple / distributed*. Prioritize requirements and see where compromises can be made.

MECHANISMS, ACTUATORS, AND CIRCUITS

What mechanisms, actuators, and circuits can produce the motions required?

With a sense of how the machine will move, the next step is to choose the mechanisms and actuators that will effect that motion. For those requiring a primer (or would benefit from a refresher), the Companion Website ▶ contains sections devoted to *The Mechanical*, covering physical principles, mechanisms, conversions, connections, and methods. These pages provide a reference you can keep side-by-side as you work through the design process.

Start from the requirements you established. The instrument's dynamic range and playing speed have been specified, which dictate mechanical considerations such as the range of displacement, velocity, and acceleration of the excitation object. Simultaneously, consider your visual requirements, which may or may not call for a different mechanism. Think about what linkages, gear trains, and cams are capable of, but also consider what they look like when moving.

What kinds of actuators will the mechanisms require? These could be electric solenoids, pneumatic cylinders, servos, DC motors, voice coil actuators (VCAs), or variable stiffness actuators (VSAs). The *Actuators, Circuits, and Code* sections of the Companion Website ▶ provide technical explanations and projects to get your machine moving. The pros and cons of each actuator should be deliberated relative to the requirements and design possibilities established in the previous steps. Electric solenoids are fast, but they typically have small strokes (the distance which the armature can travel), which can limit visual aspects of the system. Pneumatic actuators can be powerful, but the ways in which they can be modulated to increase dynamic range and create musical articulations can be challenging to work with compared to electric solenoids. Electric solenoids run on compact, quiet power supplies, while pneumatic systems require air compressors, which are more at home on construction sites than in concert halls. The criteria for actuator selection should come from the requirements and designs rather than from your personal convenience (i.e., what happens to be lying around your workshop) or comfort (i.e., only using actuators that you are familiar with).

A similar approach can be used to evaluate choices in electrical circuits, programming, and communications required to get your motors running. You can control a stepper motor with an H-Bridge circuit or a motor controller. You can program a musical sequence by using delay functions or hardware timers. The *Actuators, Circuits, and Code* sections of the Companion Website ▶ provide lessons in these areas. With that knowledge, you can choose the components and configurations that best fit the project's requirements.

Make One Module

> Make a core module -> test it -> refine the design.

Musical machines (particularly non-anthropomorphic ones) often have a core module that is replicated many times. For example, a module's purpose may be to pluck a string. That module is then replicated and configured to play other strings on an instrument. This idea is evident in acoustic instruments, such as a piano's action. The module is one of the first points where the technical aspects of the project converge, consisting of a mechanism(s), an actuator(s), electronic circuitry, and a program that work together to produce a desired effect.

After designing a module, make a functioning prototype, test it, and refine the design. Repeat. Too often, students build an entire machine without following this principle. As first attempts are rarely (if ever) flawless, the result is a "complete" product devoid of basic functionality. As a result, builders are forced to scrap many modules instead of just one. This wastes time, effort, materials, and money and is an expeditious way to add frustration and stress to your life. Learn from these mistakes without practicing them yourself:

> Get the initial module right before replicating it.

Refine the module until it performs according to your requirements. Ensure it produces sound on the instrument. This is another common mistake. An actuator may seem to be working correctly on the bench, but when it is applied to an actual instrument, the sonic result is underwhelming. The production of musical sound is a delicate and nuanced affair. Human musicians spend hundreds and sometimes thousands of hours learning these nuances. Guitarists balance the proportions of pick, thumb, and finger to activate a string along specific vectors to get the sound they want. Violinists vary the pressure and position of stopping fingers to create vibrato, portamento, and harmonics. Percussionists control striking force and stick rebound to modulate levels of damping. Woefully unaware of these intricacies, mechanical actuators often miss the sonic boat on the first attempt. Sometimes, all that is required is a slight change in position or force to yield massive improvements in sound. Sometimes, the actuator module needs to be redesigned significantly. The only way to know is to experiment with a prototype on the actual instrument. Address issues raised in testing in the next design iteration and test again. Get one module to perform predictably well and then make a bunch of them.

Structure

> What structure will allow you to realize your sonic, visual, and logistical goals? What is the relationship between the actuation system and the sonic body?

Once the actuation system has been finalized, the structure that connects or orients it to the sonic object must be defined. This is where subsystems coalesce, bringing the necessities of actuation and motion together with grander notions of purpose, visibility, emulation, and novelty.

The structure's characteristics are often dictated by the *properties* and *relations* described in Chapter 4, Part 1. A machine that purports to *emulate* a human guitar *player* will need to adhere to the *form* and movements of the latter. The structure of a project that seeks to *integrate* sonic objects and actuating mechanisms as *inconspicuously* as possible will be constrained according to those design objectives. If you did not spend much time thinking through the *properties* and *relations* mentioned earlier in the chapter, now is the time to do so. The structure of an instrument meaningfully affects its visual identity and influences how the music it makes is perceived.

Many structural choices are specific to a particular design, but some elements apply in most cases. There needs to be some physical system that holds everything together. This may be a **case** in which components are housed, thereby defining the instrument's external shape (Figure 4.2.5).

FIGURE 4.2.5. **Automated guitar from Ragtime WEST (ragtimewest.com).**

Sometimes, a structure is a **body** that not only secures components in place but also contributes to the sound produced and/or influences how players interact with the instrument, as with the string instruments *PAM* and *Cyther* (Figure 4.2.6).

FIGURE 4.2.6. **MPR Lab's Cyther.**

Sometimes, a frame is required to configure the instruments in a specific way, as with the MPR Lab's *Percussive Aerophone* (Figure 4.2.7).

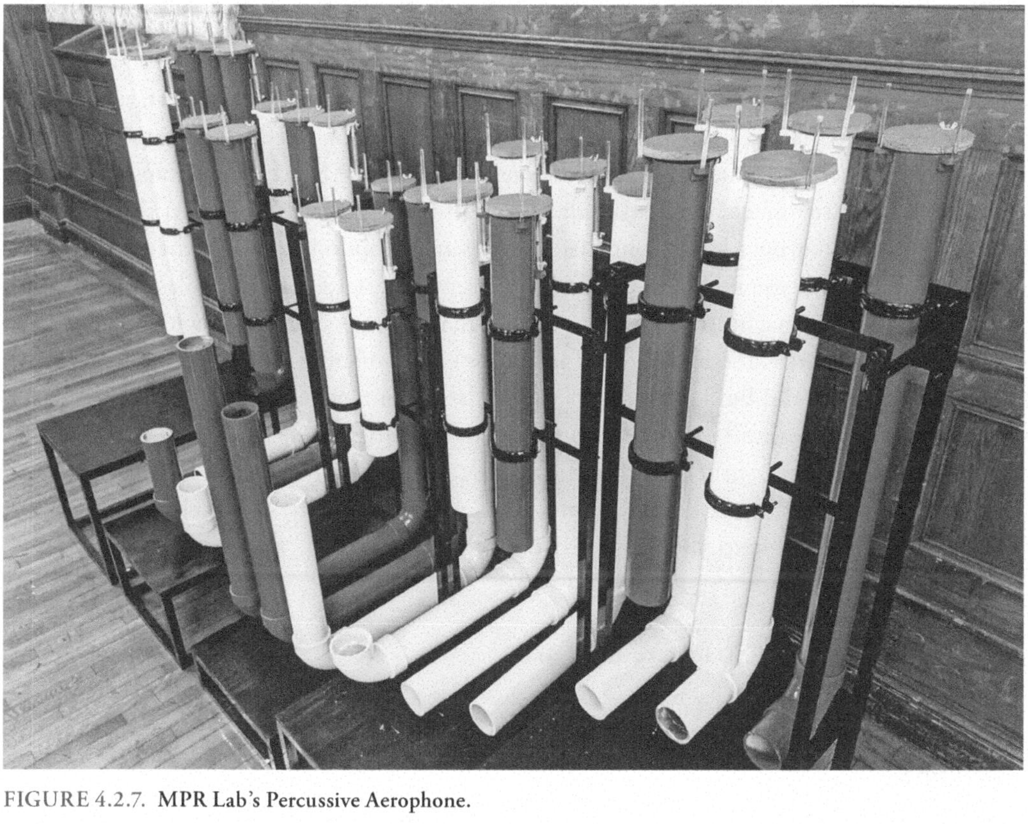

FIGURE 4.2.7. **MPR Lab's Percussive Aerophone.**

Sometimes, a **mounting system** must be designed so that the mechatronic actuating system can be variably coupled to different instruments, as with the MPR Lab's *modular percussion instruments* (Figure 4.2.8).

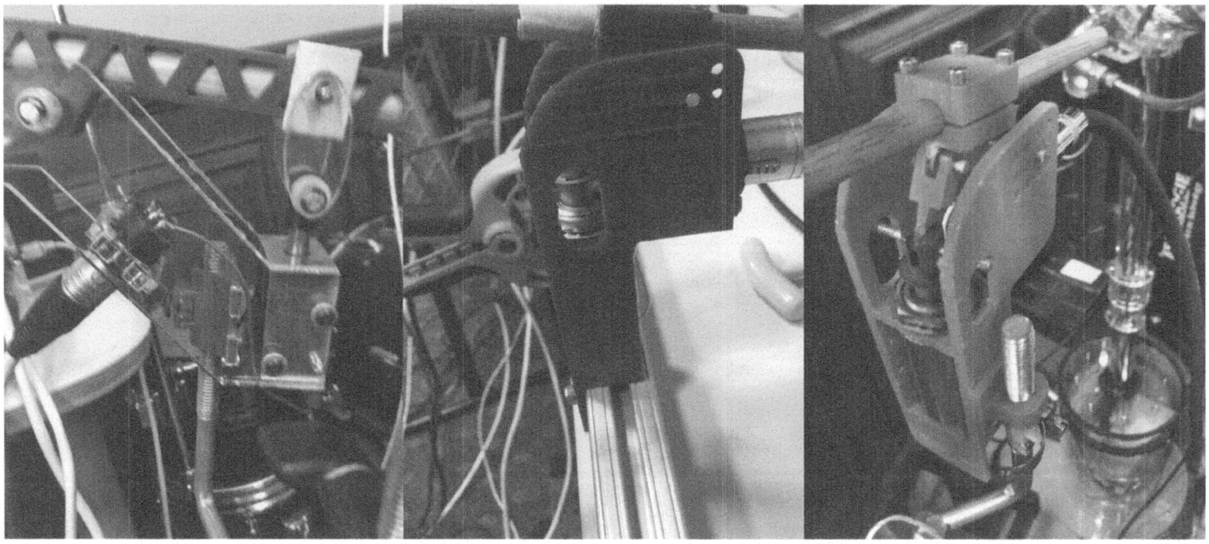

FIGURE 4.2.8. **Mounting options for the MPR Lab's modular percussion.**

Structural design must also account for electronic circuity (e.g., microcontrollers, printed circuit boards [PCBs], power supplies) that controls the actuators (often, these components are enclosed for safety reasons). Sometimes, these elements are located on the instrument's body; sometimes, they are housed in a separate enclosure and connected via cables. Jacks, usually data and power at minimum, also require space, as do the wires that connect all components (do not underestimate the space wires require). It is also possible that the instrument is self-contained using batteries, programs stored on a microcontroller, or through wireless communications.

Consider how and whether the machine will be **transported**. Will the instrument live in a lab or a performance hall, or will it travel? If it is going to travel, can it fit in your car, or do you need to rent a truck to move it? Can you fly with the instrument? Good transportation design involves protection so that the bumps of the road do not make unplanned modifications (a loose cymbal in my car once did some unscheduled reconfiguring of PAM). It is preferable (and usually cheaper) if the instrument fits in a standard case (this was a consideration in the design of *Cyther*, which fits in a keyboard case), but custom cases are an option too. How many people are required to move it? Will you need equipment from a venue to present it (i.e., tables or stands)? If you are going to make a machine suited for the stage, make sure you know how it will get there!

In building a structure, the task is to bring abstract and concrete ideas together in a design that can be articulated and built. Even with specific requirements and thoughtful consideration of *properties*, *relations*, and pragmatic concerns, the possible forms and structures abound. Collaborations with those in other relevant disciplines (e.g., sculpture, architecture) are worthwhile. You will likely have to make some compromises along the way. What characteristics of the motivating ideas are essential, and what can be changed?

Summary of Form and Structure Design

- properties and relations
- case / body / frame
- mounting
- transportability

Realization

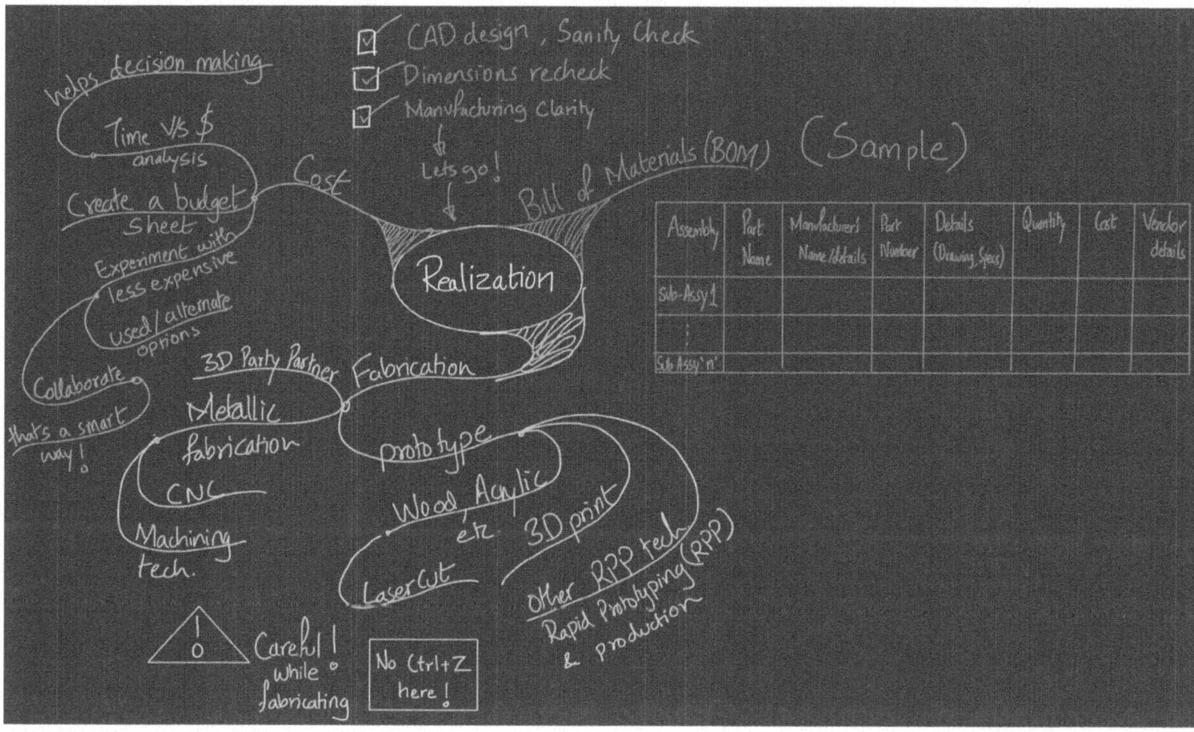

FIGURE 4.2.9. **Mindmap by Yash Garje.**

After much research, thinking, and design, it is time to realize your idea! Start by identifying the components or features that need to be included. Sketch some ideas with pencil and paper, beginning with those aspects. The dimensions do not need to be perfect at this point; you are just trying to get a sense of how the pieces of this puzzle will fit together. Once the design has solidified, calculate the exact dimensions needed. One option is to use computer-aided design (CAD) software to model the system, which produces files used by fabrication machines such as 3D printers, laser cutters, and computer numerical control (CNC) machines. It is also possible to dimension a design manually through fastidious measurement. Start with the dimensions of the components that must be included (e.g., actuators) and work from there. Use those figures to establish dimensions for the other structural components. Determine how the parts and subsystems will be made and how they will connect to each other.

BOM

Once you have identified parts and raw materials, assemble them into a *bill of materials* (BOM) that can be used for collaboration, manufacturing, and documentation. This list should detail the components that are needed to realize the product. While there are many different formats for a BOM, essential categories include the assembly, part name, manufacturer, manufacturer part number, specifications relevant to the project, quantity, cost, and vendor. List this information on a spreadsheet (Table 4.2.2). If someone wanted to make your machine but was unable to ask you questions, what information would they need? (The same questions apply when composing a musical score.) This is what should go in your BOM.

TABLE 4.2.2. **An example BOM.**

Assembly	Part	Manufacturer	Part #	Specs	Quantity	Cost	Vendor
Picking	Push solenoid	Guardian Electric	TP12X13-C-2-4D	24 V, cont. duty, 41 N holding force	1	$55	newark.com
Control	Microcontroller	Arduino	Uno	5 V, 14 digital pins	1	$21	adafruit.com

FABRICATION

Lesson: Make a mockup of your design in cardboard or wood first.

It is often a good idea to start by making a **mockup** of the mechanism out of materials that are cheap and easy to work with, such as cardboard. Staples, tape, and clips (binder clips are useful) can hold things together; push pins and toothpicks can act as joints. These models can illuminate the range and types of motion that will be produced and help to specify the required dimensions. They show problems such as parts unintentionally running into each other, strikers not reaching surfaces, actuators spaced too closely together, and frames impeding motion. They also can give a sense of the visual engagement of the design as it appears at scale and in motion. You do not want to learn about such aspects after you have ordered your custom titanium parts from a fabricator. Making a mockup can seem like an unnecessary use of time and effort, but these models can save time, effort, money, and frustration in the long run.

After the mockup is finished and you have adjusted your design accordingly, the next step is to make a **prototype**. The prototype stage typically involves multiple revisions, so stick with something easy to work with that can be finished quickly. Wood has played a role in musical instruments for thousands of years; it is readily available and modifiable with standard tools such as saws and files. Its combination of rigidity and flexibility makes it an excellent resonator. Structural parts of an instrument are often made from wood (e.g., the "neck" of a string instrument). It is an expeditious prototyping material that circumvents the need to make a model in software or book time in a machine shop or rapid prototyping lab. An amateur craftsperson can achieve "ballpark" results out of wood, but achieving higher levels of precision and more sophisticated aesthetic design requires experience, knowledge, and techniques (as possessed by professional luthiers). Machine fabrication is an attractive option if the latter is not your calling.

Manufacturing and prototyping technologies have had a game-changing influence on the development of musical machines. They have allowed those with little fabrication experience to realize physical structures and mechanisms that require precise dimensions, numerous parts, and complex shapes. Of these technologies, 3D printers and laser cutters are particularly accessible and useful. 3D printers are best suited for smaller three-dimensional parts. They take up little space and are reasonably affordable, meaning you will find them in bedrooms as much as in rapid prototyping centers. Laser cutters are useful for making larger two-dimensional planes that are often used as external surfaces of a structure. Their beds can be large, they are relatively fast, and they can cut many different types of material (depending on the model; check the specs of your machine), including acrylic, wood, and even metal. Translating your CAD model into instructions the machines can process is typically straightforward. Many manufacturers (and users) offer online tutorials to help you get started (and hone your skills).

These technologies also have downsides. While 3D printers can use a variety of materials, most of them print in plastic. This results in parts with limited durability, depending on the design and use of the part. Avoid using 3D-printed plastic for components that will endure significant stress. The beds of 3D printers are usually small, limiting the size of the parts they can produce. Prints can take a long time and some are unsuccessful, lengthening the process. Laser cutters are larger, but they are more expensive, require ventilation, and are usually confined to machine shops. The restriction to two dimensions will limit the parts best suited for a laser cutter.

If the question is how to make more durable components, the answer is usually metal. Some metals, such as aluminum or brass, can be cut by hand with tools such as a hacksaw, but such efforts can be laborious and imprecise. A better option is to use machining tools such as band saws, grinders, lathes, and CNC machines. Generally, working with metal is a subtractive process where you start with a block of material and systematically remove portions of it to achieve the desired design. This process can be automated using a CNC machine, which requires a CAD model to be translated into G-code (a programming language that produces instructions for the CNC machine) via CAM (computer-aided manufacturing) software. The technical knowledge and expertise required to do this will be an obstacle and, thus, a disadvantage for some. These tools are powerful, dangerous, and sometimes complex, requiring proper training, supervision, and maintenance. Like woodworking, machining is both an art and a science that requires significant time to gain proficiency. It is a worthwhile option if you have the interest, time, and access to do so. If not, it is best to collaborate with someone with experience or send your design to a third-party manufacturer.

COST

> What is required in terms of time and money to realize your idea?

Developing a budget for any project is necessary and will influence the design choices mentioned throughout this section. A $500 budget will exclude the 41-DOF anthropomorphic robot guitar player made of brushed aluminum that rocks the stage of your dreams. With that said, a lack of funds need not discourage ambitious designs, though they will require thriftier approaches in procurement and fabrication. Instead of buying brand-new solenoids whose specifications perfectly match the stated design requirements (the right way to do it), you might have to acquire components from surplus warehouses. Instead of ordering new stock, visit machine shop waste bins and salvage yards. Instead of having metal components custom laser cut by a third-party fabricator, you may have to learn how to use a CNC machine (or find someone who does). The good news is that given the possibilities of less-expensive options, building a musical machine does not have to be an exorbitantly expensive affair.

Time is also a cost, and there is a relationship between the cost of time and the cost of money. Is an extra five hours of your time worth the $50 you will save by having the fabrication done by a third party? How much time will it take to learn how to design a PCB, and is that time worth it when you could have someone else with more expertise do it for a reasonable sum? The answers to these questions depend on monetary resources and how much time is available to devote to the project, what you want to learn, what you enjoy, and what you want to be. If you love designing mechanisms and want to play that role on a team, then it makes sense to devote time to learning CAD software. Time learning *all* of the disciplines involved in a musical robotics project—mechanical engineering, electrical engineering, programming, and music—is well spent. This does not mean becoming an expert in all domains. Know how the circuit works even if you do not design the PCB. Understand the mechanical systems even if you are not operating the CNC machine. Knowledge in areas that are not your strongest suits will allow you to communicate with others who are making decisions that affect the potential of your project. Draw boundaries for yourself and work with others who can handle what you do not (or cannot). Collaborating with others is one of the best ways to learn. Trying to do it all yourself can result in wasted time and inferior products. It can steer you away from your musical goals, which should remain front of mind.

Control and Feedback

> How is the system going to be controlled? What system generates instructions? What kinds of interaction are possible?

For those requiring a primer (or would benefit from a refresher), the Companion Website ▶ contains sections devoted to *Sensing, Communications, and Control*, offering technical descriptions and projects to help you collect and interpret information and connect and control system components. These pages provide a reference you can keep side-by-side as you work through the design process.

With the physical machine realized, determine how the instrument will be controlled, referencing the *Interaction and Autonomy* requirements established earlier. One possibility is that predefined sequences or algorithms are stored on a microcontroller, allowing for autonomous operation with a flip of the *on* switch. Alternatively, musical instructions may be generated by a host computer. This option is attractive given personal computers' extensive processing power and multitudinous software programs, enabling real-time interaction and access to popular music-making environments, thereby increasing accessibility for a broader group of musicians. To deliver instructions, the computer will need to be connected to the musical machines using a communications protocol (e.g., Open Sound Control).

Integrating a host computer broaches the "mind/body" distinction. Some builders seek separation, making electromechanical "bodies" (e.g., a mechatronic percussion striker) that can interface with a variety of computer "minds" (e.g., a DAW or AI). Conversely, the higher-level algorithmic control system (i.e., the one making choices about which notes go where) may be integrated with the lower-level one(s) (i.e., that determines which solenoid to activate) in deep and non-dualistic ways. A string instrument that can monitor its output may detect rhythmic inaccuracies when attempting to perform a passage with large intervallic leaps. This low-level information could be fed to the higher-level system, resulting in a modification of the passage by dropping notes, changing the rhythm, or altering sequential pitch intervals. If equipped with memory, the higher-level system could make note of this limitation, influencing sequence-generating algorithms to avoid such phrases in the future. Now that the actuators, sound-production mechanisms, and structure are more concrete, determine if and how these lower-level systems are connected to the higher-level ones that issue musical commands.

The ability to sense bridges low- and high-level worlds. What kinds of information does your machine need to gather? Information about an actuator's position is often helpful, determined by devices such as potentiometers and encoders. The physical states of components, such as string tension, are ascertainable with strain gauges. Current draw can be measured and used in a motor control algorithm. An instrument's sonic output can be represented by a transducer (e.g., electromagnetic pickup, piezoelectric disc, optical sensor), analyzed, and fed back into the control system. Co-performers and audiences can be sensed through microphones, cameras, and haptic devices.

As sensory abilities bridge the gap between "mind" and "body," they open new possibilities for autonomy and interaction with others. A system that can only play back predetermined sequences functions "on its own," but in an acontextual way. It is incapable of interaction (one can react to it, but it cannot react back). Enhanced sensory capabilities facilitate musical connections: giving and receiving cues, interpreting and performing written notations (Xiao et al., 2019), adapting and improvising, and transforming and contributing new ideas. These are essential aspects of musical discourse, and their integration distinguishes the mechatronic from the robotic. Machinic capability in this domain is unnerving in some contexts, but I see good in it too. Increasing machinic autonomy augments the possibility of novel musical statements that interest and inspire us as audiences and creators. Interpreting, integrating, and responding to these statements may reveal a path toward new kinds of musical expression. To what extent will you explore this path in your design?

Evaluation

Objectively evaluating a system and its components is essential to the design and realization process. Return to your requirements and develop methods for testing each one. The specifics of the methods employed will depend on the nature of what is being tested. Calibrated SPL meters can define dynamic range; hardware and software tuners can be used to evaluate intonation; IMUs and video recorders can help determine velocity and acceleration; microphones and spectral analyzers can give insight into dynamic and timbral envelopes produced. Incorporate the following into your experimental method:

- **Measure and document** (e.g., text, photos) the name/model #, position, and equipment configuration where applicable. For example, "an Earthworks M30 omnidirectional microphone was placed one inch above the center of the drum perpendicular to the drumhead. The microphone was connected to a Focusrite Scarlett 2i2 preamp and interface that fed into a Macbook Pro running Audacity 3.0.2, etc."
- **Conduct multiple trials** to illustrate how much the data varies. If you built a tuning mechanism for a string instrument, do not just test the accuracy of a particular pitch interval once: run multiple intervals multiple times.
- **Randomize conditions** to avoid sequence effect, which can limit the validity of results. For example, instead of progressing through different velocity levels linearly each time, randomize the velocity levels played.
- When choosing what to test, **be as comprehensive as possible while simultaneously establishing boundaries that make the effort tractable.** Returning to the example of the string-tuning mechanism, you could test each ascending pitch interval in the chromatic scale from a particular pitch (e.g., C4 > C#4, C4 > D4, . . .). You could change the root note, starting at C#4, then D4, and repeating the sequence. You could test microtonal intervals. The possible permutations are infinite; set boundaries to make the process tractable while illuminating the system's capabilities relative to the requirement at hand. Identify

performance categories that are musically meaningful. For example, a percussion instrument may be capable of producing only three dynamic levels: "soft," "medium," and "loud." Make sure to test each of these levels rather than getting lost in the minutiae of just one.

- **Record all relevant information in each data entry** (e.g., in a spreadsheet row) needed in the analysis (e.g., block #, trial #, start time, measurement, etc.).

Be thorough, precise, and clear in your documentation. Write it so that anyone can understand what you discovered and how you discovered it, and replicate the process if desired. The absence of such rigor causes confusion for your teammates, yourself, and those in the future who attempt to continue your work. Do not assume you will remember how the experiment was conducted or what a row of numbers in Excel represents; you likely will not. Good documentation saves time, facilitates progress, and inspires new ideas. Be diligent about it!

Evaluation of a musical machine should not be restricted to only technical matters of latency and energy usage; it should also consider understandings of creativity and "success." In *Musical Robots: Overview and Methods for Evaluation*, Frid (2023) emphasizes the importance of the social, emotional, and interactive aspects of music. As Frid describes, incorporating evaluation methodologies from neighboring fields such as human–robot interaction (HRI), HCI, and NIME can be useful.

In all cases, evaluative methods and results should align with the machine's purpose. If the machine intends to be a compositional aid, its results are validated not only by its physical performance but also by the extent to which the composer's practice was affected. Worthwhile evaluations illuminate possibilities for future improvement. When the results of evaluations become new (or revised) requirements, a feedback loop is formed.

Summary

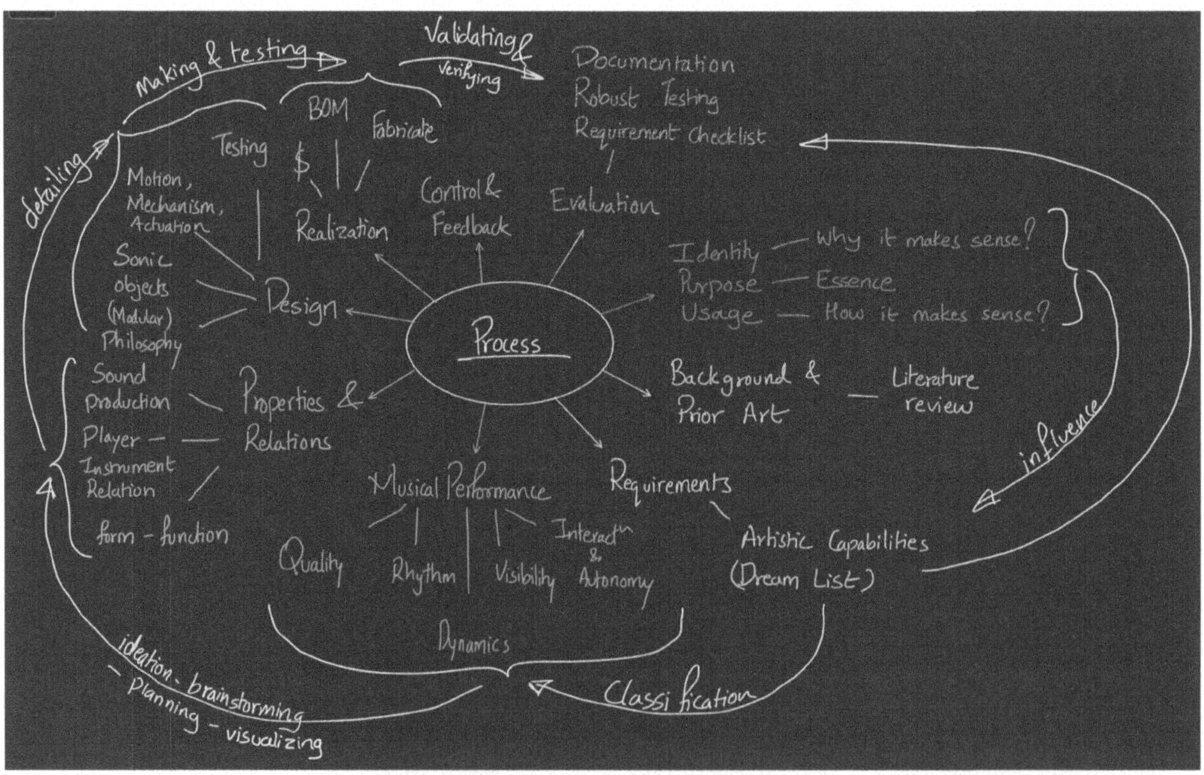

FIGURE 4.2.10. **Mindmap by Yash Garje.**

The design and realization of a musical machine require inspiration *and* process. While processes will differ for individual situations, a general framework ensures a balance between the artistic and the technical, the sonic and the visual, and the imaginative and the pragmatic. Start with collaboration: Who will you work with to realize your idea? Connect to others wherever and whenever possible, and the earlier in the process, the better. With your team assembled, articulate the essence of your idea, its purpose, and who will use/experience it. Conduct background research about theoretical and technical topics as well as related prior art, which can inspire your imagination about possibilities not yet realized. Extant examples can be situated in a set of relations including *sound production, players ⇔ instruments, form ⇔ function, emulative ⇔ novel, human ⇔ machine, conspicuous ⇔ invisible, single / complex ⇔ distributed / simple, specific ⇔ general, fixed ⇔ adjustable*, providing a context for comparisons, revealing key features, and inspiring design choices. Specify requirements in musical performance categories (timbre, time, pitch, dynamics, and articulation), visibility, and interaction and autonomy. Come up with multiple designs that could satisfy those requirements and compare their pros and cons in a design decision matrix. As you work on each design, consider sonic objects (which can be made *from scratch*, acquired through *rummage*, or purchased *off-the-shelf*) and how those objects will be excited (including the excitation object(s), mechanisms, actuators, and circuits). When you have settled on a design, make one module to see how the ideas on paper function in the real world. Go back and modify the design as needed. With the excitation modules finished, incorporate them into a larger structure (which might include a network of smaller elements) that realizes pragmatic and artistic objectives. With the design finished, draft a detailed BOM and determine fabrication resources. Build the physical system and establish how it is going to be controlled. Define or incorporate a communications protocol (e.g., MIDI). With a functional system, evaluate it using methods congruent with your requirements and objectives. Identify what can be revised in the current iteration and repeat previous processes as needed. Make choices about what should be left for future work.

References

ANIMUSIC. (n.d.). ANIMUSIC. Retrieved April 15, 2020, from https://www.animusic.com

Frid, E. (2023). Musical robots: Overview and methods for evaluation. In R. Savery (Ed.), *Sound and Robotics* (1st ed., pp. 243–284). Chapman and Hall/CRC.

Idiomatic Practice | The Idiomatic Orchestra. (n.d.). Retrieved July 22, 2021, from http://theidiomaticorchestra.net/idiomatic-practice/

Koenig, A. (1996). Idiomatic design. *Communications of the ACM, 39*(11), 96–99. https://doi.org/10.1145/240455.240481

Li, J., Hu, T., Zhang, S., & Mi, H. (2019). Designing a musical robot for Chinese bamboo flute performance. *Proceedings of the Seventh International Symposium of Chinese CHI*, 117–120. https://dl.acm.org/doi/abs/10.1145/3332169.3332264

Xiao, Z., Chen, X., & Zhou, L. (2019). Real-time optical music recognition system for dulcimer musical robot. *Journal of Advanced Computational Intelligence and Intelligent Informatics, 23*(4), 782–790. https://doi.org/10.20965/jaciii.2019.p0782

CHAPTER 5

PERCUSSION

Part 1: Context and Ideation

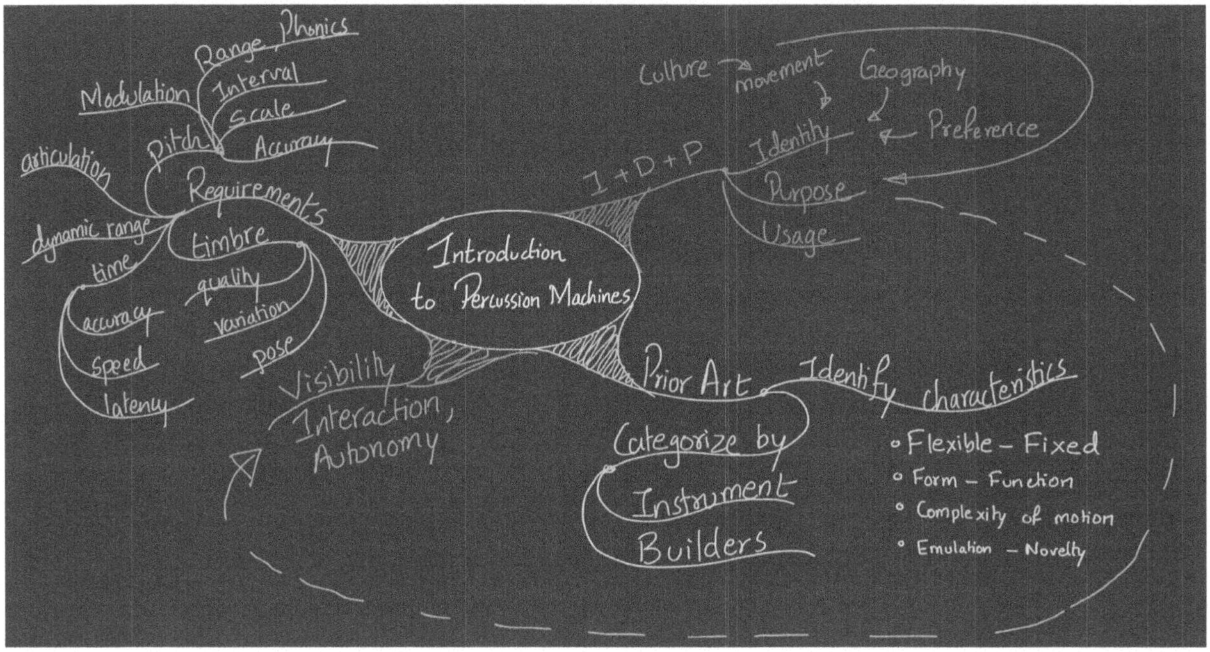

FIGURE 5.1.1. **Mindmap by Yash Garje.**

Percussion instruments are simple in some ways: find a sonic object, a striker, and hit one with the other. While the design of these instruments and the methods of playing them can become significantly more complex, the approachability of percussion instruments makes them a logical starting point for exploring the world of musical robots.

Identity, Purpose, and Usage

What do you think of when you hear the word *percussion*? How about a *drum*? Specific objects, such as a snare drum, likely come to mind, but so do associations with styles of music, playing techniques, and musical roles. The particularity of these associations depends on culture, geography, and individual preferences. For those connected to rock, drums connote a *drum set*, a heterogeneous amalgam consisting of a bass drum, snare, toms, hi-hat, and cymbals surrounding a single player. The purpose of the drums (and the drummer) in "popular" music is typically to establish the beat, tempo, meter, and groove, as well as provide anticipatory gestures at sectional transitions (e.g., fills, risers). They articulate a structure relative to which other musical elements are oriented. The drums in rock are a source of power and energy,

Musical Machines. Scott Barton, Oxford University Press. © Oxford University Press 2026. DOI: 10.1093/oso/9780197599891.003.0005

which necessitate athletic gesticulations from the drummer. These movements become an integral part of the instrument's identity.

This same basic collection of sonic objects—the *drum set*—is found in jazz, sometimes with minor differences, such as fewer toms or a smaller bass drum. The difference between drums in jazz and rock is less in the objects themselves and more in *how* they are played. The kick and the snare are fundamental in rock, while in jazz, the foundation is often the ride pattern. Rock drummers play "heel-up" to produce loud, staccato notes, while jazz players play "heel-down" to create softer notes with more sustain (Schlueter, 2018). Rock players often exert significant force on snare hits to be heard over blaring guitar amps, limiting the rate at which such hits can sound. In jazz, the snare is a location of nuance and contour, requiring alternative articulations. Distinctions in playing technique are accompanied by divergences in physical movements and appearance. Listening to and watching a rock drummer is meaningfully different from listening to and watching a jazz drummer, even though both interact with roughly the same physical objects.

In Western Art Music (i.e., "classical" music), drums are a subset of a more extensive and diverse set of percussive instruments. They are not typically configured as a *set* (other than in pieces that synthesize musical styles), which alters how they are played and perceived. How they are used and arranged depends on the requirements of a particular work. In Western Art Music that eschews explicit statements of pulse and meter, percussion is often used for timbral or textural effects, accent, or to illuminate temporal relationships. These roles starkly contrast with those assumed in popular forms such as rock and jazz. Journeys around the world (e.g., Africa, India) show further distinctions between the qualities and configurations of percussive musical objects and the techniques in which they are played.

Instrumentality extends beyond the boundaries of a physical object. An instrument is defined not only by its material disposition but also by the ways it is used, the people who use it, and the music it makes. Instrument designers are well-versed in musical conventions, which they take into account when making their choices. At the same time, some aspects are outside the control of instrument designers. Those who shaped the guitar didn't necessarily have Hendrix or rock n' roll in mind as they developed the instrument. This is often a good thing: new art is formed from the dialogue between instrument design and creative exploration.

As we endeavor to design a robotic percussionist, we begin with a question: *What is the machine's musical or artistic purpose, and for whom is it designed?* It could be an autonomous performer on stage or screen. It could provide a percussion collaborator for human instrumentalists to learn, practice, and play with. It could function as a cooperative prosthesis for humans with physical limitations (Bretan et al., 2016). It could allow composers to hear rhythmic ideas that are otherwise evasive because of the complexity of the parts or the availability of musicians. Consider the technical skills users possess. Some will be fluent programmers; some will be familiar with DAWs; some will be technologically adverse.

What role will the machine play in the music? It could provide a rhythmic anchor, textural or timbral diversity, or virtuosic solos. It may be capable of complex polyrhythms, extended techniques, or superhuman speeds. Its dynamic range will determine if it is better suited to play with small acoustic ensembles or with a rock band. Is it optimized for a particular style or can it play multiple styles flexibly? Framing musical questions in terms of *properties and relations* can help guide technical requirements and the project's design. For those seeking to play an extended rock drum set or diverse orchestral percussion battery, a *distributed* approach of smaller, simpler mechanisms may be best. For those desirous of writing exploratory, nuanced works for solo timpani, a *single* actuating system capable of *complex* movement might be a suitable option.

A machine's purpose must be considered in terms of both its visual identity and its performative capabilities. Is it important that its movements are visible to coordinate other musicians or to interest an audience? This will affect the size of the machine and its movements, which in turn will impact how it can or should be positioned on stage. Will the machine's form and movements evoke notions of human percussionists, or will they embrace machinic qualities? This question relates to who will use and interact with the machine. If it is to be a member of an otherwise human ensemble, it might look different than if it was surrounded by other machines or if it was by itself. If it is to be used exclusively in the studio, or if it will be exhibited as part of a gallery exhibition, visual identities may change.

Defining purpose also requires entertaining pragmatic questions. Aspects such as transportability, setup, breakdown, and cost will affect who uses the machine and how it is used. Asking these kinds of questions at the start will help inspire initial design ideas and hone in on a direction.

Prior Art

Percussion's low barriers to entry and central position in the musics of the world have attracted the attention of machine builders. Percussion instruments can be categorized by properties and relations described in Chapter 4: The Process of Making a Musical Machine, including *flexible ⇔ fixed*, *form ⇔ function*, *complexity of motion*, and *emulation ⇔ novelty*.

FLEXIBLE ⇔ FIXED

Flexible percussion machines are electromechanical devices that can play a variety of percussion instruments and, thus, are *general*. This requires that they are *distributed* and *adjustable* so that they can be configured in multiple ways. The designs of these systems tend toward simplicity, allowing them to be easily replicated in a percussion ensemble. Kapur et al. (2016) define two system topologies for modular musical robots: (1) *MIDI Bus*, where a single power and data line from the host feeds independent modules comprising electronics, drivers, and actuators, and (2) *Star*, where a single electronics-driver-power hub is connected to multiple independent actuators (e.g., MPR Lab's Modular Percussion, Polyend's *Perc*). The downside of the MIDI bus topology is the electronics complexity of each individual module, while the Star topology limits the number of actuators that can be connected. *Novalis* (Kapur et al., 2016) functions as an actuator hub comprising a power supply, microcontroller, and driver boards, which supports up to 48 actuators that can be flexibly configured. Modular percussion robots can strike (HammerBots, CADI, Perc, MalletOTon), apply friction (SpinnerBots), scrape (RecoBots), shake (SistrumBots, VibroBots, CADI), bow (BowBot), and pluck (PluckBots) sonic objects. They can be mounted to or positioned near instruments using specialized devices (such as percussion claws, like those made by LP), parts found in hardware stores (clamps and rods), or stands (mic, drum, or camera stands).

Flexibility is an advantage as it affords sonic exploration and discovery. Instead of designing and fabricating an entirely new machine, a module can be configured to the instrument of choice and mapped to software or hardware controllers. My compositions have featured flexible percussion robots on traditional drum sets, on a battery of found objects arranged on a garden trellis, and in places in-between. The percussive explorations using LEMUR's Modbots have been similarly diverse.

Regarding disadvantages, flexible machines' ability to configure around nearly any percussion instrument means they are not tailored to the sonic possibilities of any particular one. Unlocking the full potential of a percussion instrument such as a snare or cymbal requires different playing techniques. *Flexible* percussion instruments require setup and configuration each time they are moved (and sometimes each time they are used). They need to be positioned precisely relative to the surface of the percussion instrument to produce the timbre desired and to maximize articulation speed and dynamic range. They require the transport or procurement of stands or frames to attach to. It takes time to set their physical position, connect all of the wires, and reverse such procedures when the show is finished. These issues are partly due to the varied nature of the world of percussion itself and, thus, are difficult to avoid. The required efforts are not unreasonable, but they are time-consuming.

Another primary issue with many *flexible* percussion instruments is their limited dynamic range. Mechatronic percussionists, particularly those powered by electric solenoids, can typically achieve only part of the enormous range (up to 80 dB) that humans can produce. The limitation is fundamentally pragmatic, for machines can generate far greater forces and velocities than humans. The issue is that actuators with such abilities are often too large and/or expensive to be integrated into designs that require many strikers.

Examples of flexible percussion machines include LEMUR's Modbots (Singer et al., 2004), EMMI's CADI, The MPR Lab's modular percussion robots, and Polyend's Perc (*Perc Drumming Machine*, n.d.) (Figure 5.1.2).

FIGURE 5.1.2. **Top left: Polyend's Perc (Albaono, 2016); top right: MPR Lab's modular percussion; bottom left: LEMUR's Modbots (Singer et al., 2004); bottom right: EMMI's CADI.**

Fixed percussion machines are typically designed for *specific* instruments. They are often less *adjustable* and are *integrated* with a sonic object. *Fixed* designs address some of the limitations of *flexible* percussion machines. Instead of tediously configuring an actuator relative to a sonic object every time, the actuator position is measured and set in place. This reduces setup and takedown time, allowing the dynamic range to be calibrated precisely. It also enables one to find the "sweet spot" of a drum that produces the desired timbre, which can take longer than many pre-performance setups allow.

The disadvantages of *fixed* percussion machines are largely the opposite of *flexible* ones. Orchestrational modifications are difficult, if not impossible: if actuators are permanently positioned around a set of toms, you will not be able to move them to a set of ceramic bowls if so inspired. If a snare's "sweet spot" changes because of the venue and musical work performed, it is rendered inaccessible if the striking position is constrained. Transporting the instrument may be more difficult. Moving an instrument "as is" typically requires a custom case, which is expensive, or a lot of space and adequate restraints to prevent movement (or anything else from moving into it). Separating and packing the system into compact, stackable cases filled with cushioning material allows for smaller transport vehicles and alleviates concerns about items being damaged.

Examples of *fixed* percussion machines include EMMI's MADI (Multi-mallet Automatic Drumming Instrument), The Robotic Glockenspiel (Long et al., 2016), and many of Logos's percussion instruments (https://www.logosfoundation.org).

FORM ⇔ FUNCTION

The design of some percussion machines prioritizes musical *function*, namely, striking the instrument according to specified musical performance criteria with little extraneous movement or structure. The most straightforward example of this approach involves positioning actuators (usually linear ones such as solenoids) directly above or below the surface of the sonic object (Figure 5.1.3), allowing rapid and precise articulation. There are no extraneous mechanisms for producing physical gestures or visual effects.

FIGURE 5.1.3. **Actuators positioned directly relative to sonic objects. Logos's <Vibi>** (*Logos Foundation—Automotons and Robots*, **n.d.**).

<Vibi> (Figure 5.1.3) is an example of an actuation system that is inconspicuously built around an instrument designed to be played by a human. Focus is on the instrument, its associations, and the sounds it makes rather than on the technology it is equipped with. The machines that play Jazari's drums (Media Example 5.1.1 ⊙) are more conspicuous but are minimized to what is necessary to achieve musical functionality. Polyend's *Perc*, previously mentioned, could also be included in this category, given its direct actuator positioning, though its spherical structure shows attention to the *form* of the actuator.

Attention to *form* is evident in examples of percussion machines that utilize traditional drumsticks or mallets, positioned as they would be by a human percussionist. These designs typically involve a 1-DOF mechanism powered by a single actuator attached to a drumstick or mallet and positioned near the drumhead (Figure 5.1.4), including Thelxiepeia (Baginsky, n.d.) and Percussibot Drummer (photonflood, 2009).

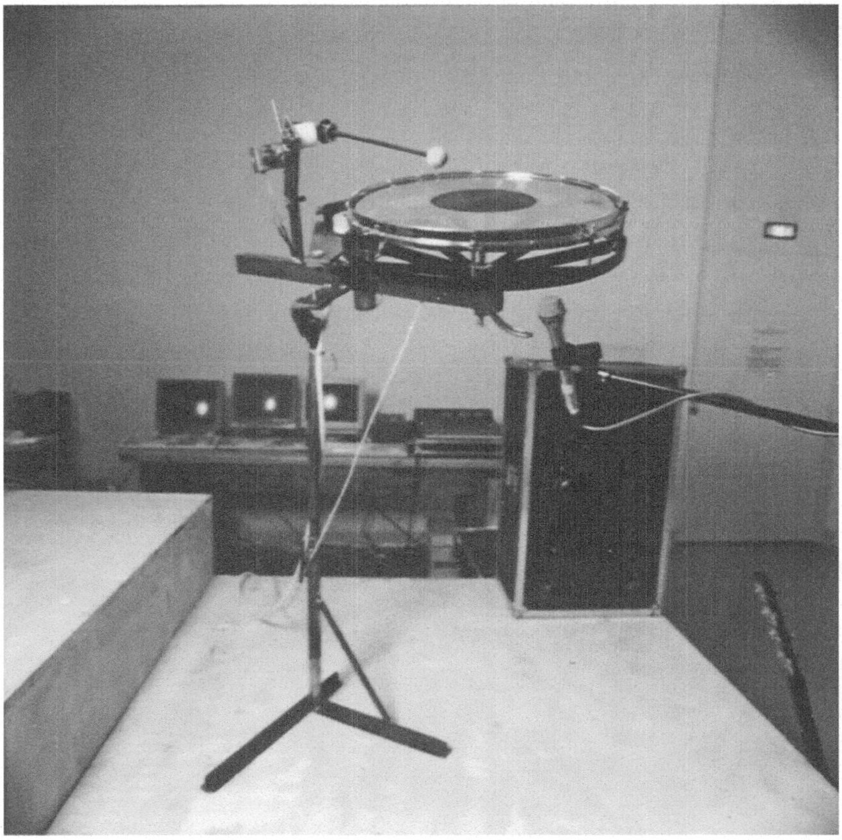

FIGURE 5.1.4. **Machinic percussion using a single drumstick: Thelxiepeia, drum robot of the Three Sirens Robot Band (photo by N. A. Baginsky).**

Some examples feature two separate drumsticks or mallets (e.g., Drums v.0.1, Figure 5.1.5), which increases the range and speed of musical gestures and strengthens human performance associations. *MechDrum* (Figure 5.1.14) is part of this group, which "favors function over form" and "prioritizes musicianship and sound quality above all else" (*MechDrum*, n.d.).

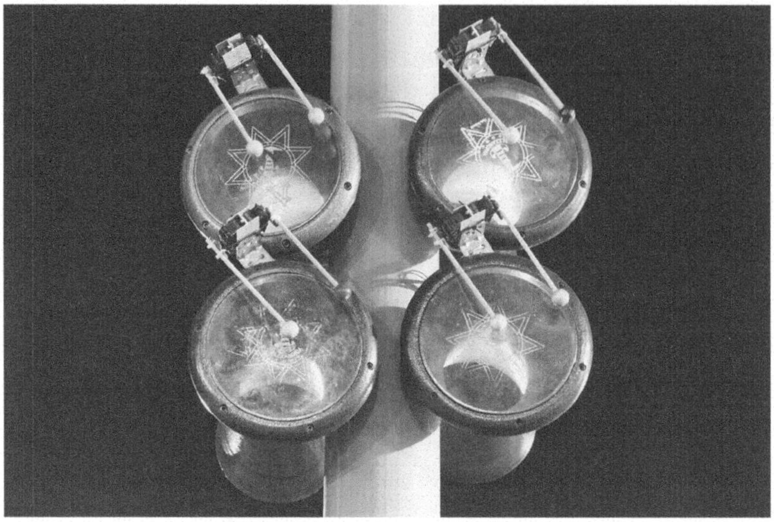

FIGURE 5.1.5. **Machinic percussion using two spaced drumsticks: Drums v.0.1 (image from Segal et al., 2010).**

The KarmetiK NotomotoN expands the number of stick-based beater assemblies to 18, which are positioned around the drum, as seen in Figure 5.1.6. The machine uses two actuators: the *TrimpTron*, which incorporates rotary solenoids, and the *KalTron*, which uses linear pull solenoids. The *KalTron* was capable of higher actuation rates, while the *TrimpTron* had a more consistent dynamic response, making the two useful for different musical purposes (Kapur et al., 2011).

FIGURE 5.1.6. **Karmetik NotomotoN (Kapur et al., 2011).**

The preceding approaches can be combined, featuring inconspicuous actuators close to the drum and stick-based strikers whose form and movements are more easily apprehended. Logos Foundation's <snar> (Figure 5.1.7) contains 13 solenoid actuators inside the drum and two drumsticks mounted on the instrument's exterior to produce rim shots. The number, physical position, and capabilities of these actuators expand timbral, rhythmic, and articulatory possibilities.

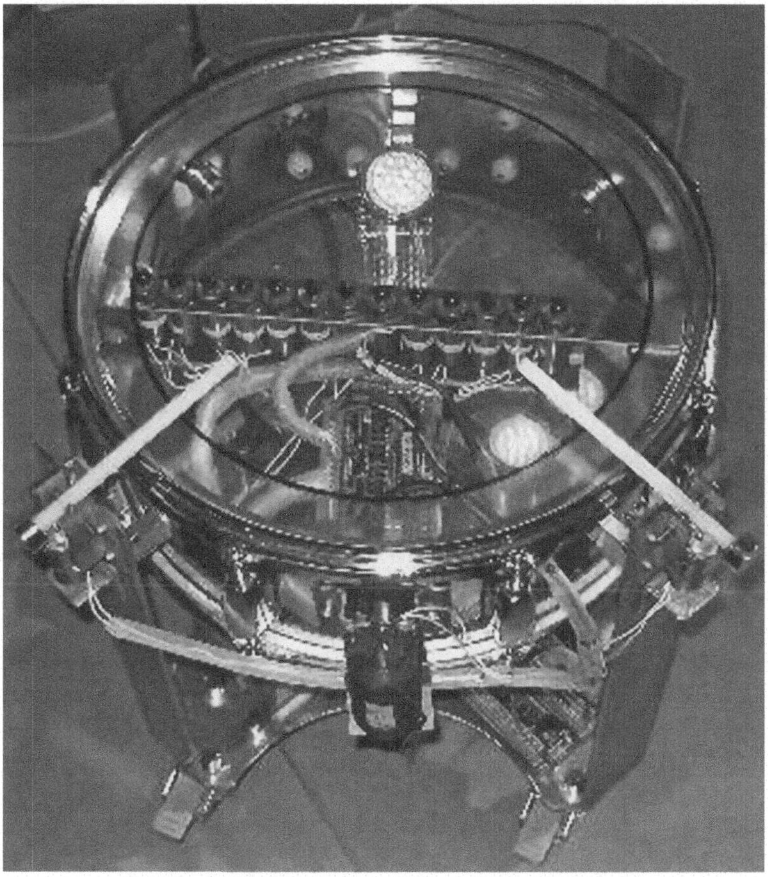

FIGURE 5.1.7. **Logos Foundations <snar>** (*Snar: An automated snare drum by Godfried-Willem Raes*, **n.d.**).

The preceding mallet-based percussion machines attend to *form*, but their designs are still largely *functional*. It is often easiest to position an actuator near the side of a drum, connected to a striker with a specific length, mass, rigidity, and material composition to produce the desired sound. The result looks like a traditional drumstick or mallet, which has the benefit of being widely available and reasonably priced. The distance between the tip of the drumstick or mallet and the drumhead is determined by the articulation and dynamic range desired, both of which are musical requirements.

Other percussion machines feature actuation systems that are less conspicuous but whose arrangement of components (both sonic objects and actuating technologies) form larger shapes, contours, and dimensions. The mechatronic and robotic percussion instruments designed and built by Godfried-Willem Raes and the Logos Foundation of Ghent, Belgium, over the last 50 years have defined this category. These include both unpitched (toms, snare, drum set, bells, siren, shakers, springs, "nature" noises) and pitched (vibraphone, carillon, saws, tubophone, xylophone) instruments. Geometrical shapes define the character of Logos's design style in the form of triangles (Figure 5.1.8) and other shapes, such as trapezoids (Figure 5.1.9).

FIGURE 5.1.8. Logos's fixed percussion instruments—triangular structures. Pictured above from left to right: <Llor>, <Per>, <belly>, and <temblo> (*Logos Foundation—Automotons and Robots*, n.d.).

FIGURE 5.1.9. Logos's fixed percussion instruments—other geometric structures. Pictured above from left to right: <Klung>, <Rotomoton>, and <Troms> (*Logos Foundation—Automotons and Robots*, n.d.).

Some instruments appear as mobile displays. They have large visible wheels (which are also functional) that support a tiered structure containing various instruments. While sonic objects are typically of the same type (e.g., Indian bells), their size and shape can vary, resulting in more and less rigid configurations (Figure 5.1.10).

FIGURE 5.1.10. Logos's fixed percussion instruments—mobile display. Pictured above are <*xy*> (top left), <Tinti> (top right), <vitello> (bottom left), and <Vacca> (bottom right) (*Logos Foundation—Automotons and Robots*, n.d.).

The design choices of these geometric examples are partly *functional* (e.g., a triangle is a compact way to arrange a group of sundry objects), but they also attend to the visual aesthetics of the instrument itself, the collection of instruments it belongs to, and the space in which it will reside.

Form prevails in percussion machines whose instruments are subsumed in elaborate structures that materialize as sculpture. The building blocks of these examples are similar to those discussed: the excitation systems consist either of an actuator directly above or below a sonic object or a 1-DOF actuator coupled to a mallet near the surface of the sonic object. The difference is that these automatic instruments are molded into more adventurous configurations. Sometimes, they are connected in expansive networks, as seen in the work of Felix Thorn (Figure 5.1.11).

FIGURE 5.1.11. **Felix's Machines installation at the Brighton Festival (2013) and Attenborough Centre for the Creative Arts, Brighton (2017), Photographer Julian Beresford (Felix Thorn, n.d.).**

In other instances, instruments and electromechanical actuation systems are sculpted into artificial agents, such as Ajay Kapur's *MahiDeviBot* (Figure 5.1.12). Kapur et al. (2007) describe *MahiDeviBot* as modeled after *Saraswati* of Hindu mythology. It features a diverse array of Indian percussion instruments, including frame drums, shakers, bells, and cymbals. The excitation mechanisms are based on LEMUR's *ModBots*, which were previously described.

FIGURE 5.1.12. **Ajay Kapur's MahaDeviBot at REDCAT Walt Disney Concert Hall, Los Angeles, 2010 (Kapur, 2011).**

MahiDeviBot is an example of how simpler mechanisms and instruments can be combined to form a more complex entity that engages with human culture. Kapur described his desire to make a human–robot performance system following the traditions of North Indian classical music, reminding us of the importance of step one in the design process: define the purpose of the machine.

The examples that prioritize *form* to the greatest degree captivate a viewer's gaze through unusual shapes, vivid colors, abundant replication, and imaginative configurations. Trimpin's work has defined this landscape, synthesizing kinetic sculpture, instrument design, musical composition, and performance. His exhibit *Klompen* features 120 Dutch wooden clogs suspended from the ceiling and sounded by embedded actuators (Figure 5.1.13).

FIGURE 5.1.13. **Trimpin's Klompen (image from the Nora Eccles Harrison Museum of Art, Utah State University).**

These works stimulate our imagination and show us the value of physical presence. One could record the sounds of these clogs and play them through speakers, but doing so would strip the work of its creativity and meaning. Encountering the clogs at the exhibit evokes physical connections, such as the sound and haptics of striking wood, as well as the feeling of wearing them on our feet. We become curious about the invisible forces that make them speak. The recorded sounds alone would inspire associations of a relatively small wooden object and little else. Absent would be the contours, cavities, and colors of the shoes; the visual and sonic effects of geometric configuration in space; the movement of the objects that result from being sounded; and the wonder of the appropriation and transformation of the everyday into the artistic. Works such as this engage us musically, artistically, and culturally, but also as beings who experience the materials of the physical world.

Attention to form does not require complexity or diversity. As in visual art, some works become expressive through simplicity in shape and uniformity of materials. *Rippler* (Zareei et al., 2015) is a mechatronic sound sculpture that features sheets of steel, oriented horizontally and vertically, lined with LED strips and sounded by push solenoids. The solenoids are activated, striking the sheets and causing vibrations that are perceived sonically and visually, the latter via the synchronized LEDs. The result embraces mechanical actuation noise and the inharmonic spectra of vibrating steel, translating glitch aesthetics from the digital to the physical domain. The reductionist design in sound and structure is a lesson that attention to form comes in many flavors.

COMPLEXITY OF MOTION

The excitation systems of the previous examples were *distributed* and *simple*. Actuator motions were typically limited to 1-DOF, so musical performance requirements were satisfied through multiplicity and distribution. New musical and visual possibilities emerge from more *complex* movements. Jim Murphy's *Nudge* features a rotary solenoid responsible for actuation, as well as a servo motor to adjust the height of the striker and a DC motor that rotates the system laterally (Jim Murphy, 2014; Murphy et al., 2014). The result is a machine capable of producing a range of percussive gestures and timbres. *MechDrum* features two 4-DOF strikers positioned above a drum's head (Figure 5.1.14). To produce nuanced performances with microvariations in timbre, dynamics, and timing, the motion of a human percussionist playing rudiments was recorded from multiple dimensions (Van Rooyen & Tzanetakis, 2015) and analyzed to develop a stochastic model applied to the machine's performances (Van Rooyen et al., 2016). This research highlights the importance of striker height, velocity, lateral position, and interhand differences in musical expressivity, which require multiple DOFs.

FIGURE 5.1.14. **Mechdrum (*MechDrum Video/Pictures*, 2017).**

Complex motions create visual interest. Robots designed for industrial purposes, such as manufacturing, often have multiple DOFs capable of executing complex motion paths. These machines have been utilized in artistic and musical contexts to leverage their capabilities for visual and performative interest. In "ABB Percussion Robot Performance" (Media Example 5.1.2 ▶), an industrial robot forms a trio with two humans who play cello and live electronics. The robotic arm is encircled by percussion instruments, including drums, cymbals, tuned PVC pipes, and a cowbell. The arm whips around to strike the assorted battery, extending to its full height in dramatic, sweeping gestures. The arm's extremities are two sticks in a Y-configuration, which, in combination with the machine's motion paths, limits the complexity of the musical statements made (complex movements do not always result in sonic complexity). Unlike *MechDrum*, the movements of the industrial robot are not intended to produce musical virtuosity. Instead, the visual contours traced by the machine in physical space excite our expectations, draw attention to coordinated musical statements, and amplify gestural intent. These are fundamental aspects of musical performance; the *complexity of motion* in a percussion machine is vital to consider.

EMULATIVE ⇔ NOVEL

Machine percussion instruments can be both *emulative* and *novel*. Some appear as percussion *players*, given the use of *off-the-shelf* instruments and striking objects (e.g., mallets), familiar striker and sonic object orientations, and likeness to human forms and performance. Other examples are distanced from such conventions, seeking new forms, configurations, paths of motion, and sounds enabled by machinic embodiment.

Emulative Percussionists

Emulation provides inspiration and a way to learn, so it is no surprise that emulative machinic percussionists abound. Musical output is the focus of some projects, which requires appeasing discerning listeners sensitive to the nuances of human performance. Form and physical gestures need not resemble human forms in this effort. This category of machines is effectively realized by recording the motions and sonic output of human performers, which are then analyzed and incorporated into the design of algorithms and electromechanical actuation systems. The aforementioned *MechDrum* is an example of this category.

Other machinic percussionists emulate human form in terms of shape, size, proportions, and movement. These examples are anthropomorphic and often find themselves in the company of other humanoid bandmates (e.g., Stickboy of Compressorhead and Ashura of Z-Machines, Figure 5.1.15).

FIGURE 5.1.15. **Anthropomorphic robotic percussionists. From left to right: Stickboy of Compressorhead (By Torsten Maue from Magdeburg, Deutschland—Musikmesse Frankfurt, CC BY-SA 2.0, https://commons.wikimedia.org/w/index.php?curid=32858470) and Ashura of Z-Machines (image by Hiroshi Ishii, https://www.flickr.com/photos/isinoid/10671548854/in/photostream/, CC by 2.0).**

General-purpose anthropomorphic robots have also been used to play percussion (e.g., MIT's Cog, see Media Example 5.1.3 ▶). Sometimes, human forms are referenced in more abstract ways (e.g., Haile and Shimon—GA Tech, Figure 5.1.16).

FIGURE 5.1.16. **Abstract Anthropomorphic robotic Percussionists. From left to right: Haile (image from Weinberg & Driscoll, 2007) and Shimon (image from Michael Irving, 2020).**

The features that connect these emulative examples include a form whose highest and most central point is a single head-like structure, arms, and the use of *off-the-shelf* mallets, sticks, and instruments. The movement of the machines is analogous to that of human drummers. There are DOFs at shoulder and elbow locations joined by arm-like cylindrical structures, which produce more or less human-like paths of motion when playing. Some can move their heads in synchronization with the music played (Stickboy, Ashura, Shimon), some have spiky mohawks (Stickboy), and some can even lift their arms and twirl drumsticks in-between their "fingers" in an homage to rock 'n' roll bravado (Ashura). Some are positioned behind a rock drum kit at the back of the stage, supporting collaborators who play guitar, bass, keyboards, or "sing" as they would in a typical band (Stickboy, Ashura).

The above machine percussionists engage with established norms. They provide an answer to the question, *what qualities and abilities are necessary for an audience to perceive a drummer?* Stickboy and Ashura (and Compressorhead and Z-Machines in general) ask the same question in the context of a musical genre: *What qualities and abilities are necessary for an audience to perceive a rock drummer*? By adhering to conventions (of human form and movement; of rock), these machines can evoke associations in the minds of their audiences. The excitement and drama of their physical presence are not only due to their size and range of motion but also because their form and gestures are well known. They gain access to a world of musical expression that they are otherwise excluded from. Authenticity is an integral part of certain rock sub-genres, which have conventions for physical appearance and mannerisms. These machines reinforce expected stereotypes (e.g., spiky mohawks, headbanging) to legitimize the music that they play. Watch Stickboy play "Ace of Spades" (Media Example 5.1.4 ▶) and then imagine the same drum part played by one of the modular percussion instruments mentioned earlier. Are the two equivalent as rock performances? The physicality of these machines makes performance, and not just sound production, primary. One then wonders, *to what extent does the performance affect the music heard?*

The above examples are emulative to varying degrees, but they are not strictly anthropomorphic (one would not confuse any of them with an actual person). They are unabashedly mechanical, with smooth metallic surfaces, exposed actuators and wires, and glimmering

LEDs. They move in jerky and linear ways compared to humans. They are *allusions* more than *replications* of human form and movement. These allusions enable the designers to reference aspects of human culture while simultaneously embracing machinic identity.

These anthropomorphic percussionists are novel in some ways. Haile's contours are space-aged (in contrast to its wooden structure); Shimon and Stickboy each have four arms. Ashura has arms but also simpler 1-DOF actuators coupled to drums, forming a hybrid. These examples illustrate how emulation can serve as a platform from which to make something new. The novel aspects of the design are palpable because they are delivered in the familiar package of a humanoid form.

Percussion machines can also be zoomorphic, taking inspiration from the non-human animal world. *RhumbLine: Plectrohyla Exquisita* (Schedel et al., 2021) comprises 19 "frogs" that use spindle motors to scrape wooden dowels, producing guiro-like effects. The frogs are distributed throughout a room (a "pond") and their sequence of calls is triggered by underfoot pressure sensors, MIDI signals, and connected motor controllers. The experience invites audience members to trace spatial maps defined by the configuration of the frogs, periodically changed by the curators. In addition to being *emulative*, the work is also a *simple-distributed* system where the proliferation of a fundamental building block enhances complexity and meaning.

Novel Machinic Percussionists

The other end of the *emulation* ⇔ *novelty* continuum is marked by examples that evade known forms, instead motivated by the expressive and performative potential of new kinds of shapes, configurations, motions, and sounds.

Some percussion machines forego sticks and mallets altogether, instead using liquid and gravity to effect sound. Two examples of such machines are *Liquid Percussion* by Trimpin (*Liquid Percussion with Trimpin*, 2008; Media Example 5.1.5 ▶) and *The Rain Drop Drum Machine* (Figure 5.1.17) by Andy Cavatorta, Marina Litvinskaya, and J. Viewz (Andy Cavatorta, 2018; Media Example 5.1.6 ▶).

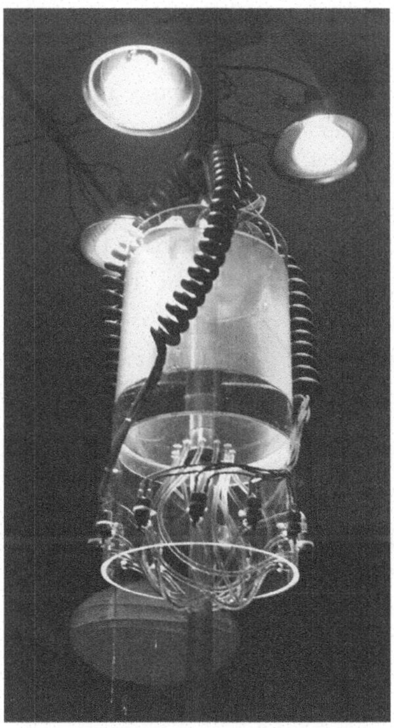

FIGURE 5.1.17. **The Rain Drop Drum Machine (Andy Cavatorta, 2018).**

Both instruments automatically discharge liquid from structures that are positioned *above* the sound-producing objects (and the viewers). Sound occurs when the liquid strikes either an acoustic drum (in the case of the Trimpin piece) or a piezoelectric sensor (Cavatorta et al.), triggering electronically produced sounds. Both instruments are *novel* in that they have little resemblance to the forms, materials, or motions of typical human percussion performance. While they can produce rhythms that are human-playable, they can also create rich, complex textures that approximate the flow and density of rainfall. Visually, the distance between a nozzle and the sonic object below it allows a viewer to see the liquid fall before it makes contact, shaping expectations of the sound to come. The works are as much visual art as they are musical instruments, and thus, they prioritize *form*.

These examples show how *novelty* and *emulation* depend on context and frame of reference. The liquid percussion instruments are *novel* in relation to conventional human form and practice, but in relation to the natural world, they are *emulative*. Rain falls from the sky and creates sonic textures as it impacts objects below. These instruments have much in common with this natural phenomenon in terms of function and form. There are also meaningful differences, including fewer dispensers (compared to clouds), the ability to control the flow precisely, and the capability to transform the quality of the sound produced. With that said, the fundamental idea has roots in the natural world.

Other percussion machines use strikers and instruments that are familiar but configurations that are not, making them *novel*. EMMI's *MADI* was inspired by the musical possibilities of varying striking positions and materials. The instrument contains fifteen strikers secured above a snare drum. The strikers are visibly apparent: their size, shape, materials, and number identify the instrument as much as the snare that they strike (Figure 5.1.18). They do not purport to mimic human percussion performance practice. Assorted materials form striker-ends, including wire brushes, felt, rubber, and acrylic. Striker positions span from the rim of the frame to the middle of the drumhead. The combinations of material and position make myriad timbres possible, affording musical statements that would otherwise be unattainable. For example, a drum roll typically performed by two sticks in the center of the drum can be voiced as a sequence made by acrylic on the edge > brush on the edge > rubber on the center. This sequence could be expanded, contracted, reshuffled, or randomized at any temporal interval. Such control is also possible using digital samplers, but the sonic products differ from those produced by acoustic instruments. The drum's head and body resonate in complex vibratory patterns after being struck, which are modulated as the drum is hit again. The tines of the brushes are displaced as they move, forming divergent aggregate shapes. The ontime of a striker affects the deformation of its foam tip, adjusting the damping of the drumhead. These aspects produce sonic nuance that is essential to musical performance, enabled by the marriage of computer control and physical design.

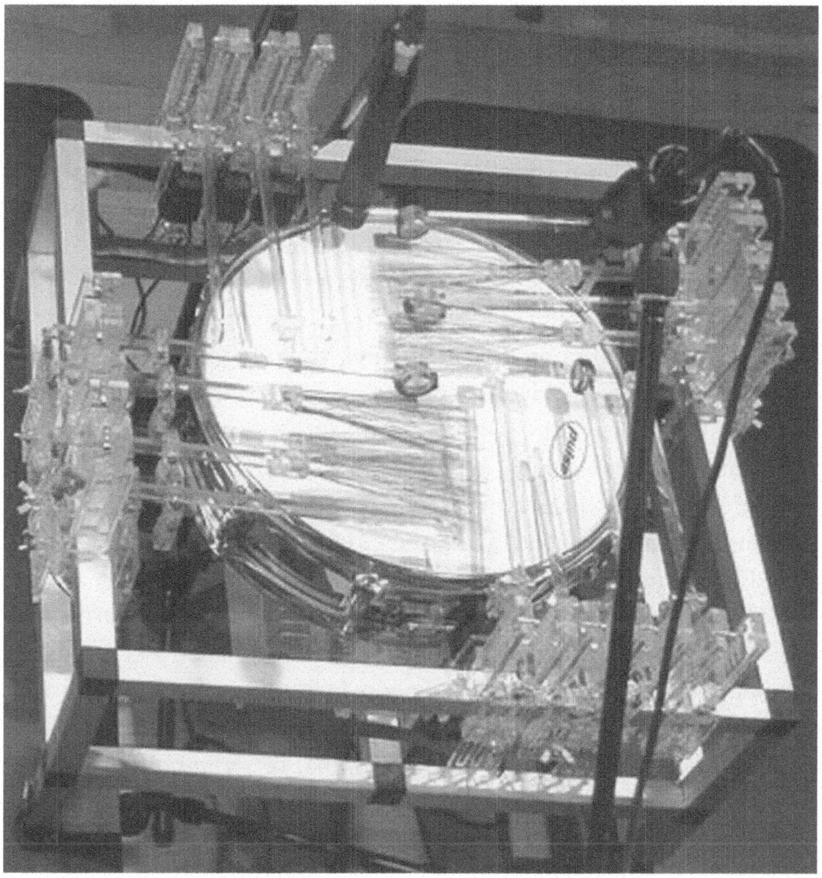

FIGURE 5.1.18. **EMMI's MADI.**

Requirements

With a more thorough understanding of prior machinic percussion instruments, the next step in formulating a design is to establish a set of requirements for musical performance, visibility, and interaction and autonomy.

MUSICAL PERFORMANCE

As established in Chapter 4: Process, musical performance requirements include:

- timbre: quality, variation, noise
- time: accuracy, speed, latency
- pitch: accuracy, set (i.e., scale), intervals, range, mono / polyphonic, tuning, modulation
- dynamic range
- articulations

Defining a specific timbre can be challenging in the world of percussion, given the numerous possibilities within the category. A crucial decision at this point is to determine whether the machine will produce a specific sound or a range of sounds. If the design involves a particular sound, what type of sound are you seeking? The sound may be linked to an *instrument*, facilitating connections with musical traditions. A rock guitarist seeking a machinic drummer is likely going to require the sounds provided by a kick drum, snare, toms, hi-hat, and cymbals. Thinking in terms of sonic *qualities* yields requirements that are more abstract (e.g., "sharp," "low," "long," "woody," "clear"). Qualities are important even when thinking of instruments (e.g., there is a panoply of snare drums, each with its own unique timbral identity). What type and degree of timbral variation do you require? Variation can be achieved by striking the object in different locations (e.g., at the center vs. near the rim of a drum), using various materials for the head of the beater (e.g., rubber, wood, cloth), damping the sounding object, and striking at different dynamic levels.

Regarding time, temporal accuracy is essential for percussionists, given their role in articulating rhythm and temporal structure. Following the guidelines in Chapter 4: Process, percussion machines should be accurate within 50 ms, with a stretch goal of less than 25 ms. Regarding rate, 100 ms IOIs are reasonable given the psychological research described in Chapter 4, with sub-50 ms rates as an attainable stretch goal.

Pitch considerations apply to some percussion instruments more than others. Within this category, pitch accuracy, range, intervallic leaps, (mono)(poly)phony, tuning, and modulation depend on whether the instrument is an *idiophone* (e.g., marimba) or *membranophone* (e.g., tympani). For idiophones, pitch accuracy and tuning depend on the material's characteristics and configurations. Membranophones can be tuned to specific pitches by adjusting the tension of the membrane. For both, specific pitches, intervals, range, and (mono)(poly)phony depend on how many sonic objects (e.g., bars) are present, how they are actuated, and how they are configured. Determining these requirements is musically consequential. Pitch modulation is possible for membranophones by dynamically varying the tension on the membrane, which can produce portamenti and facilitate tuning changes. Such is more difficult for idiophones, though possible in some cases (e.g., with tines).

The dynamic range of a percussion instrument is affected by the characteristics of the player, striker, and drum (see Dahl et al. 2011). As played by a human, a bass drum's typical maximum SPL at 3 meters is 115 dB with a dynamic range of 80 dB (an 80 cm Timpani has similar characteristics), a snare drum can produce 108 dB with a range of 55 dB, and a cymbal 105 dB with a range of 65 dB (*Acoustical Characteristics of Musical Instruments*, n.d.). These are values to aspire to but should be considered stretch goals, as the actuators accessible to most will likely fall short of these benchmarks. The dynamic range required by the instrument depends on its intended use. An instrument playing with small acoustic chamber ensembles will not need to play as loud as one playing in a rock band. At the same time, a chamber ensemble might require more range (i.e., softly, loudly, and everywhere in-between), while a rock drummer might require less.

Percussionists can employ a wide range of musical techniques. The Percussive Arts Society published a list of 40 drum rudiments that includes examples from drum corps, orchestral, European, and contemporary sources to build a shared vocabulary among percussionists (*Percussive Arts Society—Rudiments*, n.d.). These rudiments include rolls, flams, drags, paradiddles, and ratamacues (Media Example 5.1.7 ▶), which could be the basis for machinic percussion technique requirements. Alternatively, one may use percussion gestures voiced in a particular musical style or even by a particular player as a guide. This possibility highlights the value of musical analysis in establishing musical requirements. Such exemplars could also be used to evaluate the machine's performance after it is built.

VISIBILITY

When setting visibility requirements, keep the instrument's larger purpose in mind. Visibility is a higher priority if the machine is intended for the stage rather than a private recording studio. Is there a visual aesthetic, a "look," which the machine should align with or challenge? Use the characteristics identified in the prior art section as a guide. Will the machinic percussionists be presented "as-is" or can they be configured flexibly in accordance with the design of a stage, installation space, or private room? What is more important, *function* or *form*? Will the machines emulate human percussionists, or will they appear less familiar? Will their movements be constrained to only that which is required to produce the desired sound or also provide additional visual gestures and contours?

INTERACTION AND AUTONOMY

The diverse nature of percussion instruments affords different ways of interacting with them. Percussion machines can perform music independently, either through a predetermined score or algorithmic processes. They can also function as autonomous agents that human musicians or other machines interact with. For example, in my *Human-Robot Improvisations*, a guitarist improvises with a drum set that is played by modular percussion arms (Barton, 2013). Human and robotic drummers can even play the same instrument. I have explored this idea with percussionist Nate Tucker in improvisatory settings, where interactions are restricted to a single snare drum played by two modular percussion arms and two human arms. Having a human and a machine produce sound with the same object blurs the lines between players and brings questions of expressivity into a new light. In all cases, interaction and autonomy requirements should consider what machine performers are good at, what human performers are good at, and where there is overlap between the two. How will your design explore these abilities and relationships? A machine can augment or complement human performance; it can be a collaborator, an accompanist, or a muse. The idea of interacting with a robot to inspire and extend our activities is motivating (Albaono, 2016; Gordon, 2017).

References

Acoustical characteristics of musical instruments. (n.d.). DPA. Retrieved May 19, 2020, from https://www.dpamicrophones.com/mic-university/acoustical-characteristics-of-musical-instruments

Albaono, J. (2016, December 14). *Review: Polyend PERC PRO mechanical drummer.* Ask.Audio. http://ask.audio/articles/review-polyend-perc-pro-mechanical-drummer

Andy Cavatorta. (2018). *The Rain Drop Drum Machine.* http://andycavatorta.com/rddm.html

Baginsky, N. (n.d.). *The Three Sirens—A robotic rock band.* Retrieved June 5, 2020, from http://www.the-three-sirens.info/binfo.html

Barton, S. (2013). The human, the mechanical, and the spaces in between: Explorations in human-robotic musical improvisation. *Ninth Artificial Intelligence and Interactive Digital Entertainment Conference, 9*(5), 9–13.

Bretan, M., Gopinath, D., Mullins, P., & Weinberg, G. (2016). *A robotic prosthesis for an amputee drummer* (arXiv:1612.04391). arXiv. https://doi.org/10.48550/arXiv.1612.04391

Dahl, S., Grossbach, M., & Altenmüller, E. (2011). Effect of dynamic level in drumming: Measurement of striking velocity, force, and sound level. *Proceedings of the Forum Acusticum (European Acoustics Association)*, 621–624.

Felix Thorn. (n.d.). *Felix's machines.* Retrieved June 5, 2020, from http://felixsmachines.com/

Gordon, M. (2017, August). Polyend Perc Pro. *Sound on Sound*. https://www.soundonsound.com/reviews/polyend-perc-pro

Jim Murphy. (2014). *Nudge: A mechatronic drum player on Vimeo*. https://vimeo.com/85900549

Kapur, A. (2011). Multimodal techniques for human/robot interaction. In J. Solis & K. Ng (Eds.), *Musical robots and interactive multimodal systems* (pp. 215–232). Springer Berlin Heidelberg. http://link.springer.com.ezproxy.wpi.edu/chapter/10.1007/978-3-642-22291-7_13

Kapur, A., Darling, M., Murphy, J. W., Hochenbaum, J., Diakopoulos, D., & Trimpin. (2011). The KarmetiK NotomotoN: A new breed of musical robot for teaching and performance. *NIME*, 228–231.

Kapur, A., Murphy, J., Darling, M., Heep, E., Lott, B., & Morris, N. (2016). *MalletOTon and the Modulets: Modular and extensible musical robots*, 69–72. https://doi.org/10.5281/zenodo.1176050

Kapur, A., Singer, E., Benning, M. S., & Tzanetakis, G. (2007). Integrating hyperinstruments, musical robots & machine musicianship for North Indian classical music. *Proceedings of the 7th International Conference on New Interfaces for Musical Expression*, 238–241.

Liquid Percussion with Trimpin. (2008, February 25). [Video recording]. https://www.youtube.com/watch?v=VD7xg8IFcLQ

Logos Foundation—Automotons and Robots. (n.d.). Retrieved June 2, 2020, from https://www.logosfoundation.org/instrum_gwr/automatons.html

Long, J., Carnegie, D., & Kapur, A. (2016). The closed-loop robotic glockenspiel: Improving musical robots with embedded musical information retrieval. *Proceedings of the International Conference on New Interfaces for Musical Expression*, 2–7. https://doi.org/10.5281/zenodo.3964607

MechDrum. (n.d.). Retrieved June 24, 2020, from https://mechdrum.wordpress.com/

MechDrum Video/Pictures. (2017, October 20). https://mechdrum.wordpress.com/video/

Michael Irving. (2020, February 26). *Shimon the singer-songwriter robot to launch first album and tour*. New Atlas. https://newatlas.com/robotics/robot-singer-songwriter-shimon-album-tour/

Murphy, J., Carnegie, D. A., & Kapur, A. (2014). Little drummer bot: Building, testing, and interfacing with a new expressive mechatronic drum system. *Proceedings of the International Computer Music Conference. ICMC*, 663–670.

Perc Drumming Machine. (n.d.). Polyend. Retrieved June 1, 2020, from https://polyend.com/

Percussive Arts Society—Rudiments. (n.d.). Retrieved June 29, 2020, from https://www.pas.org/resources/rudiments

photonflood (Director). (2009, December 7). *Robotic Drum Kit, Drumoff* [Video recording]. https://www.youtube.com/watch?v=8zTz6TskNjA

Schedel, M., Smith, B., Cosgrove, R., & Hwang, N. (2021, June 1). RhumbLine: Plectrohyla exquisita—Spatial listening of zoomorphic musical robots. *International Conference on New Interfaces for Musical Expression*. NIME 2021. https://doi.org/10.21428/92fbeb44.9e1312b1

Schlueter, B. (2018, March 31). Jazz chops for rock drummers. *DRUM! Magazine*. https://drummagazine.com/jazz-chops-for-rock-drummers/

Segal, L., Rubin, J., & Talmudi, A. (2010). *Robotic drums | 2010*. https://www.liatsegal.com/2010/10/robotic-drums.html

Singer, E., Feddersen, J., Redmon, C., & Bowen, B. (2004). LEMUR's musical robots. *Proceedings of the 2004 Conference on New Interfaces for Musical Expression*, 181–184.

Snar: An automated snare drum by Godfried-Willem Raes. (n.d.). Retrieved October 18, 2022, from https://logosfoundation.org/instrum_gwr/snar.html

Van Rooyen, R., Schloss, A., & Tzanetakis, G. (2016). Snare drum performance motion analysis. https://www.researchgate.net/publication/303873184_Snare_Drum_Performance_Motion_Analysis

Van Rooyen, R., & Tzanetakis, G. (2015). Pragmatic drum motion capture system. *NIME*, 339–342.

Weinberg, G., & Driscoll, S. (2007). The interactive robotic percussionist: New developments in form, mechanics, perception and interaction design. *Proceedings of the ACM/IEEE International Conference on Human-Robot Interaction*, 97–104. https://doi.org/10.1145/1228716.1228730

Zareei, M. H., Carnegie, D. A., McKinnon, D., & Kapur, A. (2015). Rippler: A mechatronic sound-sculpture. *Journal of Comparative Media Arts*, *1*. https://millihertz.net/s/Zareei_Rippler_CMA.pdf

CHAPTER 5

PERCUSSION

Part 2: Design

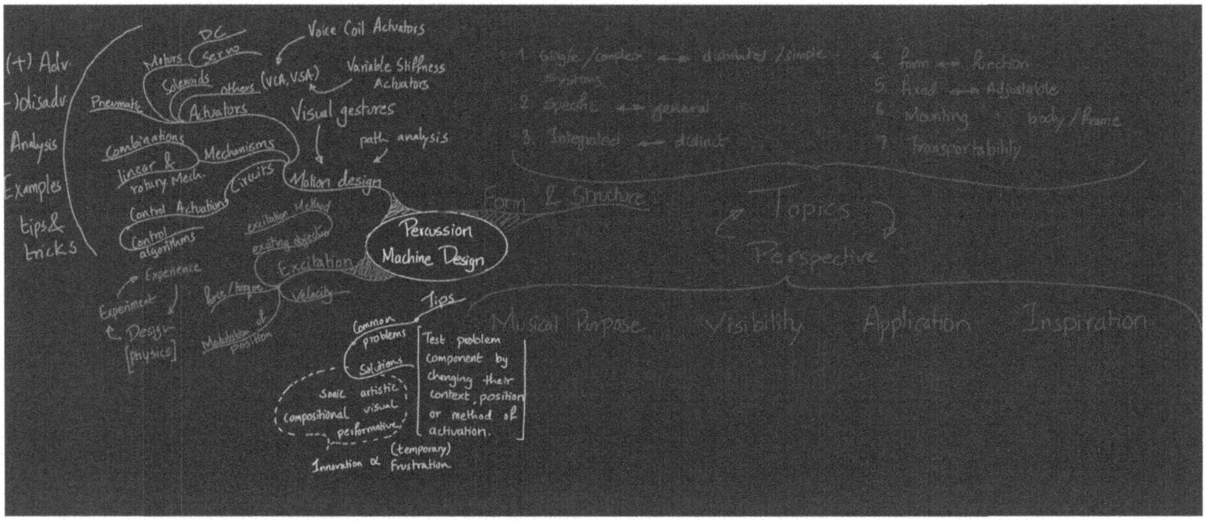

FIGURE 5.2.1. **Mindmap by Yash Garje.**

Sonic Objects

Requirements determine whether the machine will be *flexible* or *fixed*, and if the latter, what type of sound it will produce. If the machine is going to be *flexible* (e.g., a modular percussion machine), test it with a variety of sonic objects without optimizing for any one in particular. If it is *fixed*, acquire the sonic object that meets your instrumental and timbral requirements either by buying *off-the-shelf*, through *rummage*, or by building it *from scratch*.

Buying *off-the-shelf* may be the path of least resistance, given percussion's modular nature. One of the disadvantages of this approach, particularly if you are drawn to the convenience and selection of our internet supermarket behemoths, is that choices are typically made based on price or availability rather than sound quality.

The second option, finding sonic objects through *rummage*, is fruitful in the case of percussion. You can find percussion instruments almost anywhere. I have frequented rummage, home goods, and tile stores equipped with a wooden dowel to test any object that evinced the slightest musical potential (sometimes accompanied by quizzical looks from other customers and store employees). Be rigorous when testing: Two seemingly identical copper mule mugs on a *HomeGoods* shelf may have wildly different musical properties in terms of pitch, timbre, and resonance. Strike each object in multiple places (lightly if you don't yet own it), as the excitation point makes a big difference. Experiment with damping by placing the object on different supporting structures, which will affect the frequencies at which it resonates. My found treasures include metal bowls, candlestick holders, glass vases, and ceramic tiles.

The third option, making a sonorous object *from scratch*, is possible, but it requires expertise, effort, and patience. Creating *from scratch* provides the greatest control over a design, which is particularly useful when dealing with complex electromechanical systems. Start with materials that sound good. In this case, *from scratch* is harder than *rummage* because the raw materials available are likely either (1) not able to be tested (e.g., if ordered) or (2) in a form that doesn't reveal the sonic properties of the final form (e.g., a 4′ × 8′ rubber sheet). One strategy to overcome these obstacles is to choose materials known for their sonic properties, such as rosewood or padauk. It is unlikely that any old piece of wood from Home Depot will yield jaw-dropping sonic richness. Synthetic materials are durable, inexpensive, and resistant to environmental changes, unlike wood. Whatever you choose, think sustainably: there are many opportunities to reuse what has already been made! With materials in hand, the following sections detail fundamental design considerations for the most encountered percussion categories: *membranophones* and *idiophones*.

MEMBRANOPHONES

Membranophones are one of the most common types of percussion instruments, including drums, found across musical cultures and styles. Membranophones consist of a flexible material (*head*) that is drawn tightly and secured over a resonating *body*. Membranophones can be made *from scratch* or acquired *off-the-shelf*. While membranophones are simple in many ways, creating a durable instrument that holds its pitch, produces compelling and consistent timbres, and is capable of sufficient volumes is not trivial. For these reasons (and others), the best option for most is to purchase a professionally manufactured membranophone and devote your time to the actuation system. For those intrepid souls who seek to *make from scratch*, a primer on membranophone instrument design is provided on the companion website ▶.

IDIOPHONES

Idiophones are solid objects that are both rigid and flexible enough to produce sound when struck (Hornbostel & Sachs, 1961). Examples of idiophones include woodblocks, claves, bells, xylophones, and triangles. Two prominent categories of idiophones are *free-bar* and *vibrating-beam*. A free-bar idiophone is defined as a sonic object that can vibrate freely, such as in a marimba or xylophone. *Vibrating-beam* idiophones (*lamellophones*) are characterized by a *tine* (tongue, prong, lamella, rod) that is fixed at one end, including kalimbas, mbiras, thumb pianos, music boxes, and tongue drums. Fixing the beam at one end results in vibration patterns that are unique from those of free-bars. Vibrating-beam instruments typically produce overtones that are unevenly spaced and inharmonic (a non-integer relationship to the fundamental frequency). Spectral analyses of vibrating-beam instruments have been conducted, as illustrated in Table 5.2.1. These overtones are initially apparent and then quickly decay, contributing to the characteristic attack of the sound.

TABLE 5.2.1. **Frequency relationships of spectral components of free-bar instruments.**

Instrument	f_2/f_1	f_3/f_1
Mbira	~5 (4.2–5.8)	~13.6
Kalimba	6.5 (5.7–6.8)	13.7 (8.9–18.5)

Note: f_1 is the fundamental, f_2 is the second partial, and f_3 is the third partial (Chapman, 2012; McNeil & Mitran, 2008).

As with membranophones, commercial *off-the-shelf* idiophones are plentiful and are a fine choice for those who want to advance to excitation and structural design. For those pursuing the *from scratch* approach, instrument design primers are provided on the companion website, covering fundamental aspects such as tuning, mounting, and amplification ▶.

Excitation

From Chapter 4: The Process of Making a Musical Machine, sound production requirements include:

- excitation object (e.g., drumstick)
- force or torque
- velocity
- position (x, y, z) relative to the sonic object, modulation of position

With the sonic object chosen, what excitation methods are possible and what methods are desirable? Percussion instruments are typically excited by striking them, but they can also be excited through friction (e.g., scraping, bowing) and vibration (e.g., via a transducer). Experiment with your sonic object and vary your methods, materials, striking locations, and dynamics. Some approaches yield amazing sounds; others not so much. Focus on the ones that compel you and make them the focus of your excitation design.

The next step is to determine the kinetic quantities (e.g., velocities, distances) necessary to achieve your *musical* and *excitation requirements* (i.e., dynamic range, repetition rate, articulations). For example, you determine the ideal sound is produced by striking the drum one inch to the left of center with a thin wooden dowel. You review the excitation requirements that you previously set: an articulation rate of at least 100 ms and a dynamic range of at least 30 dB. It then becomes a matter of determining the velocities, displacements, and paths of motion required to achieve these goals. These can be calculated mathematically (see chapter 2 of the Companion Website for concepts and equations). An experimental setup could involve a human percussionist or a variable (electro)mechanical system. The movement and position of the striker could be tracked via a high-speed camera or a motion capture system. Strike velocity can be measured through video analysis or by using sensors such as a tachometer, laser surface velocimeter (LSV), linear velocity transducer (LVT), or piezoelectric sensor. Articulation rate can be determined through audio recordings, video analysis, or a combination of both. Dynamic range can be measured using a calibrated SPL meter. Contact force and duration can be calculated using strain gauges or conductive material attached to the drumstick tip (Dahl et al., 2011). Quantifying timbral qualities can be achieved using spectral analyzers, such as those found in *Audacity* or *Sonic Visualiser*. Discovering the quantities that achieve your musical and excitation requirements will help you develop your excitation design.

If some of these measurement methods are prohibitive, trial and error will be necessary. Gather available actuators, research their specifications, experiment with how they can excite the sonic object, and test their limits. This approach is more limited (there are only so many actuators available in one's workshop) and can be time-consuming and expensive, but it can eventually get you in the ballpark of your musical requirements. There are benefits to this approach when first starting out. It can help you build an actuator inventory for use in future projects. It is a way to learn what works by experiencing that which does not. Calculation can save you time, effort, and money, but experimentation is valuable too.

Motion

What motions are necessary and what motions are possible given your requirements? For percussion, primary considerations are the striker's distance from the sonic object, whether that distance is variable, how the striker can move to hit different parts of the sonic object (this becomes more complicated if the object isn't flat), and what visual gestures are desired. These considerations are dictated by the *properties and relations* that apply to your project. The simplest case is to have a 1-DOF actuating mechanism fixed to the sonic object. The next level of complexity involves varying the stick height by changing the angular position (if the striker is connected to a rotational motor), linear position (using an actuator with position control), or by having a second motor move the entire mechanism up or down (as in *Nudge*). Moving laterally or front-to-back requires another motor to move the entire mechanism (as in *Haile* and *Nudge*). An alternative is to build or utilize a multi-DOF "arm" to which the striking mechanism attaches. The paths of motion that the percussion machine travels depend on the sizes, shapes, and orientations of the sonic objects it will strike. One option is to design the motion around a known configuration, such as a traditional drum set. Another approach is to limit movements to specific locations where percussion instruments can be placed. A more complex possibility is enabling the machine to move about a space, requiring feedback and navigation capabilities. In all cases, remember visual gesture requirements. As motion paths suggest or require specific types of physical forms, some of the choices you make in this step will affect the structural design that will be articulated later.

Mechanisms, Actuators, and Circuits

The next step is to select mechanisms, actuators, and circuits to drive the desired motions. For those requiring a primer in these areas (or would benefit from a refresher), the Companion Website ▶ contains sections devoted to *The Mechanical*, covering physical principles, mechanisms, motion conversions, connections, and methods, and *Actuators, Circuits, and Code*, providing technical descriptions and projects to help you turn information into physical action. These pages provide a reference that you can keep side-by-side as you work through the design process.

A variety of mechanism designs can be used in a percussion machine. Levers are commonly implemented, particularly the third-class variety (e.g., EMMI's *CADI*, MPR Lab's *modular percussion*). Linkages are often necessary intermediaries, for example, when changing the vertical and horizontal position of a striking object (e.g., *Mech Drum*). The possibilities of four-bar configurations and treadles in percussion machines are vast and inspiring. Rotary systems can utilize gear trains or chain drives (LEMUR's *Modbots*), or a striking object, such as a drumstick, can be coupled directly to the shaft of a motor or gearbox (*Drums v.0.1* by Liat Segal). Cams are another possibility. In many cases, actuators strike the sonic object directly (Logos's *<Vibi>*), often using some kind of end-cap material to produce the desired timbre (Jazari's percussion machines).

The actuators used affect the mechanical design, musical capabilities, and visual impact of a percussion machine. Solenoids (linear, rotary, and soft shift), DC motors, servomotors, pneumatic cylinders, VSAs, and VCAs are possibilities. Solenoids are a popular choice because they are fast, durable, inexpensive (at least some are), available, and relatively narrow, which is useful when a design requires many of them in a tight space. They are driven by simple electrical circuits (see Project 1: Solenoid Percussion on the Companion Website ▶) and can be modulated to provide dynamic range. The disadvantages of linear solenoids for percussion are that the forces they produce are relatively small (at least for the more affordable ones), their *throw* is relatively short (which limits the scale of motion and visibility), they produce mechanical noise (when the armature reaches full extension by contacting the case), and they require the design of additional mechanisms to convert linear into rotational motion when using a stick or mallet. Linear solenoids are used in EMMI's *CADI*, Raes' *<Vibi>*, and LEMUR's *Modbots*.

Rotary solenoids are typically more expensive than linear solenoids, but they have an advantage in that they do not require conversion of linear to rotational motion. They can be unidirectional or bidirectional and can exert greater starting torque than linear solenoids. The disadvantages of rotary solenoids are that they are relatively expensive and their width is problematic in some implementations (Long et al., 2015). Rotary solenoids have been implemented in Murphy's *Nudge* (Murphy et al., 2014) and the KarmetiK *NotomotoN* (Kapur et al., 2011).

Pneumatic systems are an attractive choice for percussion because they are fast and powerful. Their disadvantages include the requirement of additional components, including air compressors, solenoid valves, and pneumatic cylinders. They are expensive, difficult to modulate, and generate noise from solenoid valve exhaust, air compressors, and the contact of the piston and the internal walls of the cylinder. Machinic percussion instruments that use pneumatic actuation include the MPR Lab's *Percussive Aerophone*, Ken Caulkin's installations, and Ashura of Z-Machines.

The aforementioned actuators (solenoids and pneumatic cylinders) are essentially on/off devices, making it challenging to control velocity, acceleration, and position. While some amount of velocity and motion control is possible through mechanical and electrical design, these limitations inhibit the realization of percussive nuances and techniques that are of interest to many musicians. Fortunately, some actuators can address these issues, specifically Soft Shift solenoids, DC motors, servomotors, and VCAs.

Soft Shift solenoids can retract quickly (as typical solenoids do), but their retraction velocity can be controlled by varying input current. They provide a starting force three to five times higher than standard solenoids, are capable of variable positioning, and prevent noise, wear, and vibration at positional extremes (*Ledex® Linear Soft Shift Solenoid | Johnson Electric*, n.d.).

DC motors are another possible solution. Because they rotate, they obviate the need for other mechanisms to translate linear to rotational motion. They are widely available, come in a variety of sizes, and are capable of high speeds and high torques, typically achieved through gear reduction. Brushless Direct Current (BLDC) motors are compelling as they are more efficient, last longer, offer better performance and controllability, and generate less heat than induction motors (*What's the difference between AC, DC, and EC motors?*, 2017). They replaced solenoids in the robotic marimbaist *Shimon*: the BLDC motors were faster (39.2 Hz vs. 8.3 Hz), had a wider and more stable dynamic range (57–83 dB vs. 73–83 dB), and received higher expressivity rankings in a subjective evaluation versus human performance (Yang et al., 2020). *Electronically Commutated (EC)* motors are BLDC motors controlled by integrated circuitry. EC motors were used in STAMPER's actuated bass drum pedal (Uriu et al., 2023).

Motor-based percussion systems typically require position sensing and control (though a zip tie or a string can produce interesting results if left to rotate unfettered). This means added components (e.g., an encoder) and motion control algorithms (e.g., Proportional-Integral-Derivative—PID—controller). Getting all of this work is not trivial, as the velocity, acceleration, inertia, position, direction of the striker, and backlash in the mechanical system must be accounted for. This is further complicated when attempting to produce different dynamics and percussion techniques. The preceding should not discourage you from going this route—the performance this complexity yields is worth the effort—but you should know what you are getting yourself into.

Servomotors are useful for percussion because they provide control of rotation angle, speed, and acceleration. Disadvantages of servomotors are that their control signals are more complex. "Hobbyist" servos, such as those used in remote-controlled vehicles, are the most common. Their advantages include being inexpensive, their position can be controlled precisely, and their rotational motion aligns naturally with a percussive striker. Their disadvantages are that they do not provide position feedback externally, are loud, and are less powerful than some of the other options mentioned. Direct-drive servomotors have been used in percussion machines due to their high speed and precise torque control (Karbasi et al., 2022). Servomotors have also been used in *Drums v.0.1* by Talmudi, Rubin, and Segal; *Drum Machine* by Sand and Strømme; *Nudge* by Murphy; and *Game*lan *Robot* by Long (Long et al., 2015).

VSAs enable control of force, position, and stiffness. When a drum is struck, the stick bounces off the drumhead. Human players modulate this bounce by applying forces with their arms, wrists, and fingers. Doing so enables techniques such as the bounced double stroke, which involves a single wrist movement (strike one) and subsequent finger flexion (strike two). By modulating actuator stiffness, VSAs can achieve both single and double strokes and may be capable of more drum rudiments (Kim et al., 2014). A disadvantage of custom VSAs is the design time and expertise required, as well as the procurement of components (such as motors, sensors, and springs) and performance testing.

VCAs are one of the most promising choices for a machine percussionist. They are powerful, fast, and quiet. They are direct-drive motors and, thus, are capable of high acceleration with no backlash and minimal hysteresis, which allows precise positioning when paired with a sensor. They are capable of displacements of up to five inches, can move bi-directionally, and can produce forces of up to hundreds of pounds. Their disadvantages are that they require complex control systems, including a position sensor and a motion control method (such as a PID controller), and they are expensive. Their advantages are evident in *MechDrum*'s impressive capabilities (Van Rooyen et al., 2017).

Voice coils can drive loudspeakers and other physical objects (the term "voice coil" is used to describe speaker drivers, actuators, and motors because the operating principles are similar, although the design and performance of each device vary). A signal generator drives the voice coil, exciting the resonances of the attached object. The signal's amplitude can be modulated, enabling gradual attacks and variable sustain. They have been coupled to sheet metal and a panoply of everyday objects, as David Tudor did in *Rainforest*. Electromagnetic actuators have also been incorporated into a Rhodes piano, whose sonic objects are metal tines coupled to tone bars (Shear & Wright, 2011).

Choose Your Path

How do you choose one of these options? The answer depends on purpose, relevant *properties and relations*, requirements, and the excitation and motion designs. A design that requires rapid articulation, a limited dynamic range, small movements, and numerous actuators may be best realized by linear solenoids. A design that prioritizes high dynamic levels and more visible linear motion may be a good match for a rotary motor. A design that seeks to produce a variety of nuanced percussion rudiments may be a good fit with VSAs or VCAs. Cost, logistics, and experience will also influence these choices. For most budgets, a VCA could be a choice for a single percussion instrument but not for an ensemble of 28. A percussion machine that will be transported in a small car to perform acoustic chamber music in intimate venues would not be particularly well-suited for a pneumatic system powered by an industrial air compressor. If a performance is booked in three months that will include a yet-to-be-designed percussion machine, then the time is probably not now to start experimenting with custom

VSAs. Aspire to ideals in the design process, but at the same time be aware of factors that may delay or prevent musical activity.

PERCUSSION EXCITATION SYSTEMS

With a sense of available actuators and mechanisms, how can we put them together? One option is to design from the "bottom-up": start with the actuator, incorporate it into an excitation mechanism(s), and then design the larger form and structure that comprises it. Alternatively, designs can be "top-down": start with the larger idea for the machine's form and structure and then work down to the level of individual actuators. Each approach has its advantages, depending on the project's purpose and vision. Several resources articulate how percussion mechanisms can be designed (Kapur et al., 2007; Long et al., 2015), some of which are summarized below. If you are working "top-down," feel free to skip to Form and Structure and then return to this topic.

Direct Linear Striker

An actuator can be positioned close to the sonic object, requiring only a frame to hold it in place (e.g., the instruments of *Jazari* and *Logos*). A key factor is the distance between the striker and the sonic object, which affects the instrument's timbre and dynamic range. Being able to adjust the actuator's height to achieve the correct position is a welcome convenience. Gravity is a concern: actuators that are mounted so that they extend against gravity's force may require modifications. Orientation has other issues: the armature of some solenoids will simply fall out if turned upside down. The structure holding the actuator must be designed. It could attach to a sonic object (e.g., the rim of a snare drum), a stand (e.g., a mic stand), a custom frame, or a multi-DOF arm (e.g., a humanoid robot). An example of a direct linear striker is shown in Figure 5.2.2.

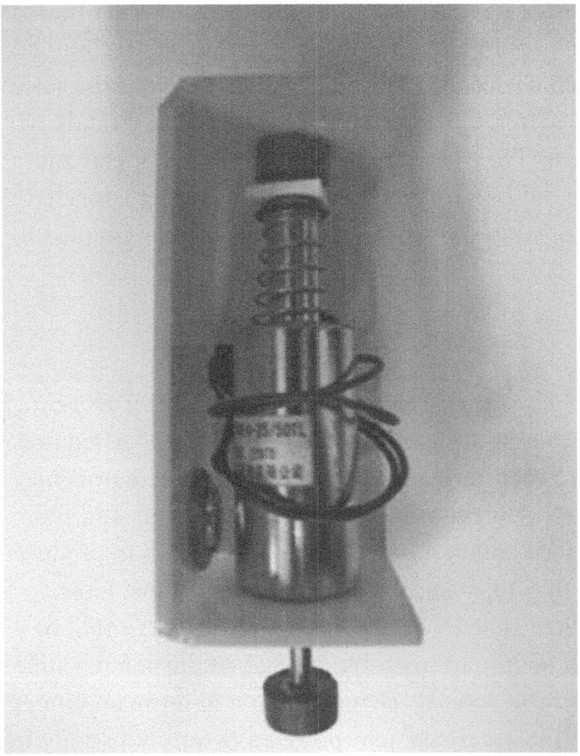

FIGURE 5.2.2. **A direct linear striker (Long et al., 2015).**

Linear Striker with Pivot

A *linear striker with a pivot* consists of a linear solenoid (usually of the "pull" type) connected to a shaft that rotates around a pivot attached to the mechanism's structure. When the solenoid is activated, the shaft is pulled down, contacting the drum, and returning to its initial position via the force of a spring. A consideration in this design is how the shaft is connected to the actuator. In EMMI's *CADI* and the MPR Lab's *Modular Percussion v.2*, a cap is glued or press-fitted onto the armature of a pull solenoid (in *Modular Percussion v.3*, there is no cap). A linkage with an axle on each side then connects the two parts, allowing linear motion to be converted into rotational motion (Figure 5.2.3).

FIGURE 5.2.3. **Designs that feature linkages to attach actuators to shafts. From Left to right: EMMI's CADI and the MPR Lab's modular percussion v.2 and v.3.**

The *Trimpin Hammer* (Kapur et al., 2007; Long et al., 2015) features a ball joint attached to the solenoid armature, another ball joint fixed to the structure securing the actuator, and a rod that extends through both (secured by shaft collars), serving as the striking mechanism (Figure 5.2.4). When the pull solenoid retracts, the ball joints rotate, allowing the striking rod to contact the sonic object. Ball joints are a compelling solution as they allow both vertical and horizontal motion.

FIGURE 5.2.4. **Variants of the Trimpin Hammer. Left from Long et al. (2015), right from Kapur et al. (2007).**

Linear Striker with Attachment

Linear actuators can also be used by attaching a striking object directly to them. The MPR Lab's *Percussive Aerophone* features foam discs secured directly to the shafts of pneumatic cylinders (Figure 5.2.5). The armatures or pistons of many linear actuators, such as solenoids and pneumatic cylinders, can rotate freely; therefore, a structure is required to constrain horizontal movement to strike the sonic object in the desired location. In the MPR Lab's *Percussive Aerophone*, acrylic fence posts play this role.

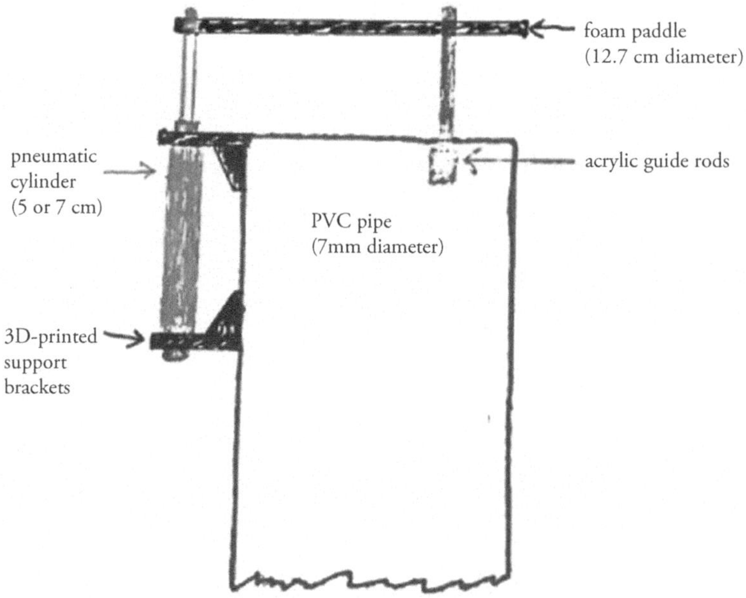

FIGURE 5.2.5. **Linear striker with an attachment: the MPR Lab's Percussive Aerophone (Sundberg et al., 2018).**

Rotary Strikers

Rotary strikers are advantageous in percussion machine designs because they eliminate the need for mechanisms that convert linear to rotational motion. With the motor selected, primary mechanical design considerations include coupling the striking object to the actuator and securing the actuator-striker to a supporting structure. The striking object can be connected directly to the motor (RC Servo) or may require a custom coupler (MPR Lab's *Whamola*), as shown in Figure 5.2.6. Couplers can be made more flexible, for example, by incorporating springs to accommodate rebounding forces involved in high-speed drumming (Karbasi et al., 2022). Regarding supporting structures, brackets can be custom-made, but commercially available ones often work just fine (the brackets for the *Whamola* were purchased from a hardware store).

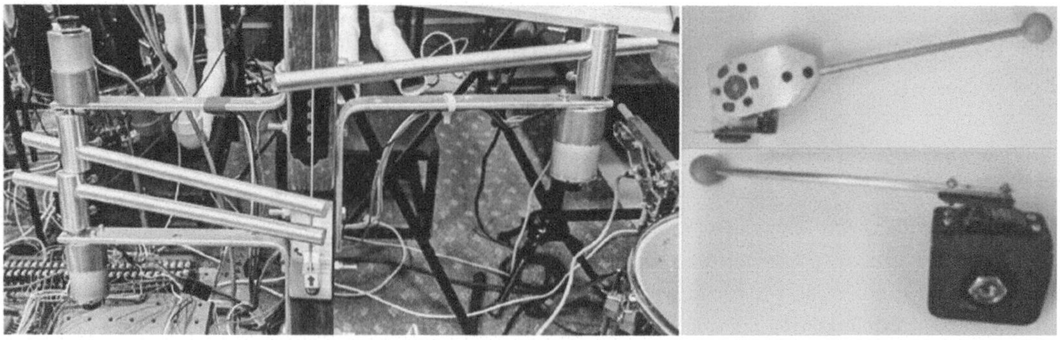

FIGURE 5.2.6. **Rotary strikers. Whamola—MPR Lab (Left), RC Servo (top right), and Rotary (bottom right) striking mechanisms—Long et al. (2015).**

Some of these mechanism-actuator combinations have been tested in experimental settings to determine maximum repetition rate, dynamic consistency, dynamic range, latency, and maximum loudness (Kapur et al., 2007; Long et al., 2015). The results of these experiments reveal the pros and cons of each system across the parameters mentioned. There is no single best solution.

Distributed Systems

Actuators can be separated from other structural and mechanistic parts of the excitation system. A striking mechanism on a drum could be linked via Bowden cables to an actuator situated in an independent enclosure. A bank of pneumatic cylinders seated in a carriage could slide along a marimba's surface, connected by air hoses to solenoid valves and an air compressor in an adjacent room. These *distributed* systems are beneficial when mass is an issue (e.g., for a movable carriage) or the aesthetic goal is to minimize the visibility of the actuators.

MAKE ONE MODULE

As detailed in Chapter 4, once you have developed a design, create one module, test it, and revise the design as necessary. Once you have a module that meets your requirements dependably, start thinking about the larger structure that will house it (and its replications).

Form and Structure

Structural design is contingent upon the *properties and relations* that pertain to your project:

- single / complex ⇔ distributed / simple
- specific ⇔ general
- integrated ⇔ distinct
- form ⇔ function
- fixed ⇔ adjustable

At this point in the process, you likely have a good idea about where your design locates on these dimensions, for example, whether the machine will be a *single* entity capable of *complex* movement or contain *simpler, distributed* components. There are many structural questions still to address. For an *integrated* system, what constitutes the structure's shape, dimensions, and materials? For a *general* system, how will excitation mechanisms interact with different sonic objects? Where will electronics be enclosed and situated? How will the structure convey artistic interpretations of *form*? To navigate these questions, start with sound, considering how actuators can be positioned relative to sonic objects.

MOUNTING

One possibility is to mount actuators directly to the sonic object (e.g., a drum counter hoop). If optimized for a *specific* instrument, this method can help realize an *integrated* system. If connections allow actuators to be moved and configured among diverse sonic objects, mounting facilitates a *general* approach. You can create your own mounting system, but this can be more challenging than it seems. The good news is that commercial hardware specifically designed for percussion (e.g., from companies such as Latin Percussion) is abundant. These products—including clamps, mounts, and rods—can seem expensive, but they are usually durable and work seamlessly with *off-the-shelf* percussion instruments. General-purpose hardware (e.g., for home construction) is equally valuable in these endeavors. The MPR Lab's *Modular Percussion v.2* utilizes LP claws, which attach directly to the counter hoop of a drum, and standard U-bolts to position the actuating machines (see Figure 5.2.3).

FRAMES

Fixing actuators onto a frame that also supports a *specific* instrument facilitates the development of an *integrated* system. The structure conforms to the qualities and relationships of the instrument–actuator combination. In EMMI's *MADI* (see Chapter 5.1), the structure was built around the snare drum. In the MPR Lab's *Percussive Aerophone* (Figure 5.2.7), the aesthetic goal was for the structure to be inconspicuous, so aluminum channels were minimized to only those that would support the tubes and the components' enclosures were located behind the tubes. Fixed-frame integrated systems expand sonic potential as actuators are tuned to the properties of sonic objects. This requires fastidious calculation and adjustment in the structural design.

FIGURE 5.2.7. The MPR Lab's *Percussive Aerophone*.

Structural pieces can also be *integrated* in a *general* way, allowing actuators and instruments to conform to the structure's qualities. As an example of the latter, I used a garden lattice as a frame to hold a variety of percussion actuators built by the MPR Lab and instruments in an installation entitled *Machinic Tides* (Figure 5.2.8).

FIGURE 5.2.8. *Machinic Tides* at the Worcester PopUP (July 2019).

A motivation for *integrating* actuators and sonic bodies is the possibility of creating *forms* that are as much machine as they are sonic object, blurring the line between player and instrument. This arena is home to physical audiovisual art that synthesizes technological and cultural artifacts, exemplified in the works of Trimpin and Felix Thorn. The heterogeneous and ubiquitous nature of percussion lends particularly well to this form of art.

STANDS

Structure may consist of excitation mechanisms attached to stands positioned near sonic objects. This approach makes player and instrument more *distinct* and is also conducive to *general* applications, given the ease of reconfiguring stands. In practice, a threaded flange or tapped hole in the excitation system allows it to be mounted on a mic or camera stand (EMMI's *CADI*, MPR Lab's *Modular Percussion v.3*, *MechDrum*). This approach utilizes components that already exist and are widely available, saving time and often money. Mic stands are great because many venues already have them: all you need to bring are the actuators!

MULTI-DOF ARMS

Multi-DOF arms make player and instrument more *distinct*, requiring stabilizing structures as simple as a base or as complex as an anthropomorphic body. Depending on the arm's capabilities and programming, these systems can be tailored to a *specific* instrument or used more *generally*. There may be one arm that can move to and sound objects in varied locations (e.g., the ABB robotic arm). There may be many arms of assorted sizes and complexity, each responsible for striking a set of instruments in a region (e.g., the anthropomorphic robot rock drummers mentioned). If the machine is stationary, its form will be affected by the expanse of the percussion instrument ensemble. The ability to ambulate introduces additional complexities but affords more *general* applications. The size, shape, configuration, and materials of the design may fall anywhere on the *emulative* ⇔ *novel* continuum, as determined by the project's aesthetic and artistic vision.

FORM AND FUNCTION

Percussion machines that are *distinct* from the sonic objects they sound reinforce traditional notions, appearing as *players*. Their forms evoke notions of drummers and lend themselves particularly well to established musical genres. In terms of structure and form, these *distinct* machines can be anthropomorphic (e.g., Stickboy, Ashura) or not (e.g., *MechDrum*). The difference between them is often a matter of *form* and *function*. To date, most examples of *machinic percussion players* that gravitate toward the *form* pole have followed the humanoid route. Possibilities of *form* that explore the *emulative* ⇔ *novel* dimension more deeply abound. Imagine multi-legged insect players, Tetris configurations of geometric shapes, or atomic units distributed and connected as if on the molecular level. Other *machinic percussion players* have focused more on *function*, resulting in structures and forms that support the machine's musical performance abilities (e.g., *MechDrum*).

ADJUSTABILITY

The *function* of a machine depends on the extent to which its components are *fixed* or *adjustable*. Some kind of *adjustability* is necessary to maximize musical performance. Changing the vertical displacement of an actuator by a fraction of an inch meaningfully affects the speed and dynamic range it can produce. Vibration from playing, idiosyncrasies of the sonic object, jostling during transport or setup, actuator variance, imperfect measurement or fabrication, friction, and deformation of parts can derail what was perfect on paper or even once humming along smoothly in practice. The fix to such issues is the ability to make fine adjustments. This is a feature of many acoustic instruments and is particularly important for excitation systems. A system can be *fixed* in place if relevant factors are known and can be concretized, but the realities of use make *adjustability* a necessity. Whether the physical system is *fixed* or *adjustable*, establish a calibration routine to maximize musical performance capabilities. This may be software-based or involve changing the physical positions of actuators (this is a case where actuator choices can affect structural design).

Structural pieces must not damp the sonic object's vibrations. Locate supports at vibrational nodes that allow desired partials to be heard and *fix* the supports in place. Depending on the sonic object, some *adjustability* may be useful in determining the optimal structural configuration. My piece, *Water, Rhythm, and Light* (2016), features modular percussion machines playing ceramic tiles, whose resonance was significantly damped by some mounting solutions. After some experimentation, the best option proved to be suspending them using strings.

STABILITY

The structure of a percussion machine must be stable, requiring *that* some components be *fixed* solidly in place. Percussion's nature involves collisions at high velocities and with large forces, which can crack plastic, break adhesive bonds, and loosen nuts. Figure out which pieces are bearing the brunt of that stress and which are not. Choose strong materials, such as metal, for the force-bearing parts and reserve plastic for junctures subjected to lower stress. These issues were encountered during the design of the bracket for the *Percussive Aerophone,* which holds the pneumatic cylinder in place. The requirements were that the bracket be low-profile, preserve the sonic characteristics of the tube, and remain stable despite thousands of cycles of pneumatic striking. Powerful adhesives failed for some of the brackets, which was addressed by securing the brackets to the tubes with bolts. For other brackets, the adhesives remained intact. This was fine until the original pneumatic cylinders were replaced with cushioned ones to reduce mechanical noise. The problem was that the new cylinders were a different size than the original ones, which meant that the pertinacious brackets had to be laboriously removed from the tubes. Another key lesson was learned:

> Only adhere pieces when you are sure that they will not be adjusted later. Envision what might need adjusting in the future. Parts secured with bolts and nuts are usually easier to (re)move than those secured with adhesives.

ELECTRONICS

Electronics and connectors must be factored into a structural design. Systems that are *distributed / simple* often consolidate electronics (power supply, microcontroller, motor controllers, solenoid circuits, etc.) in a separate box that connects to the various percussion actuators via cables. The benefits of this design are that the box keeps electronics safe from the outside world (and the outside world safe from the electronics), it is configurable (if you want the actuator further away, use a longer cable; if you want another actuator, just plug it into an available channel), and it consolidates components (as opposed to each actuator having its own microcontroller and electronic circuitry). These points are particularly relevant in the case of percussion, which often requires different instrument configurations. The downsides to this arrangement are that it involves a lot of cabling, which can look messy, and it needs time to set up and take down. Conversely, each percussion actuator could be completely self-contained with its own microcontroller, control circuitry, and power supply. If the program was stored on the microcontroller or the latter communicated with a host computer wirelessly, there would be no need for a data cable. If a battery were used as the power supply, then no cables would be needed at all. This would increase the cost and size of the percussion actuators but would provide a visually clean option that would be easy to set up and take down. For percussion machines that tend toward the *single / complex* pole, it makes more sense to locate the electronics systems in the same structure as the actuators. This reduces the need for cabling and connections to separate enclosures (a benefit for those who seek a cleaner look or are looking to make technology more invisible), it reduces wire runs, and sometimes it is just nicer to have everything in one place (e.g., when moving or setting up machines, especially when space is limited).

TRANSPORT

Transport, setup, and takedown time are affected by the extent to which structures are fixed or flexible. With flexible percussion systems, each actuator must be set up, positioned relative to each instrument, and connected to the electrical system. This takes time that is sometimes unavailable amid the frenzy of pre-show setup. *Fixed* actuators can significantly reduce setup time. At the same time, *fixed* systems add size that must be accounted for with custom cases. Percussion instruments occupy a considerable amount of space on their own, and this space requirement increases even further when they are equipped with mechanical actuators. Conversely, *flexible* systems often can be broken down and stored in compact cases. This could be the difference between being able to use your own car and having to rent a truck. This is another reason that smaller, modular forms have been popular. When venues can supply the instruments, there is even less to bring.

In all cases, remember form and visibility when designing a structure. It is easy for functional requirements to dictate the form of the machines, but there are so many fantastic artistic, philosophical, and cultural possibilities when thinking beyond *function*. A worthwhile approach is to study the examples given in Chapter 5.1 as well as works of art that have nothing to do with musical machines for inspiration.

References

Chapman, D. M. (2012). The tones of the kalimba (African thumb piano). *The Journal of the Acoustical Society of America, 131*(1), 945–950.

Dahl, S., Grossbach, M., & Altenmüller, E. (2011). Effect of dynamic level in drumming: Measurement of striking velocity, force, and sound level. *Proceedings of the Forum Acusticum (European Acoustics Association)*, 621–624.

Hornbostel, E. M. von, & Sachs, C. (1961). Classification of musical instruments (A. Baines & Wachsmann, Klaus P., Trans.). *The Galpin Society Journal, 14*, 3–29.

Kapur, A., Darling, M., Murphy, J. W., Hochenbaum, J., Diakopoulos, D., & Trimpin. (2011). The KarmetiK NotomotoN: A new breed of musical robot for teaching and performance. *NIME*, 228–231.

Kapur, A., Trimpin, E. S., Suleman, A., & Tzanetakis, G. (2007). A comparison of solenoid-based strategies for robotic drumming. *ICMC, Copenhagen, Denmark,* 393–396.

Karbasi, S. M., Jensenius, A. R., Godøy, R. I., & Tørresen, J. (2022). A robotic drummer with a flexible joint: The effect of passive impedance on drumming. *Proceedings of the 19th Sound and Music Computing Conference*, 232–237. https://doi.org/10.5281/zenodo.6797833

Kim, Y. G., Garabini, M., Park, J., & Bicchi, A. (2014). Drum stroke variation using variable stiffness actuators. *2014 IEEE/RSJ International Conference on Intelligent Robots and Systems*, 3892–3897.

Ledex® Linear Soft Shift Solenoid | Johnson Electric. (n.d.). Retrieved May 27, 2025, from https://www.johnsonelectric.com/en/products/solenoids/soft-shift-linear-solenoid#faq

Long, J., Murphy, J. W., Kapur, A., & Carnegie, D. A. (2015). A comparative evaluation of percussion mechanisms for musical robotics applications. *2015 6th International Conference on Automation, Robotics and Applications (ICARA)*, 173–178.

McNeil, L. E., & Mitran, S. (2008). Vibrational frequencies and tuning of the African mbira. *The Journal of the Acoustical Society of America, 123*(2), 1169–1178.

Murphy, J., Carnegie, D. A., & Kapur, A. (2014). Little drummer bot: Building, testing, and interfacing with a new expressive mechatronic drum system. *Proceedings of the International Computer Music Conference. ICMC*, 663–670.

Shear, G., & Wright, M. (2011). The electromagnetically sustained Rhodes piano. *NIME Proceedings, Oslo, Norway*. New Interfaces for Musical Expression, Oslo, Norway.

Sundberg, K., Barton, S., Walter, A., Sane, T., Baker, L., & O'Brien, A. (2018). A robotic percussive aerophone. *Proceedings of the 18th International Conference on New Interfaces for Musical Expression*, 409.

Uriu, D., Iiyama, S., Okada, T., Handa, T., Maekawa, A., & Inami, M. (2023). STAMPER: Human-machine integrated drumming. *Extended Abstracts of the 2023 CHI Conference on Human Factors in Computing Systems*, 1–5. https://doi.org/10.1145/3544549.3585619

Van Rooyen, R., Schloss, A., & Tzanetakis, G. (2017). Voice Coil Actuators for Percussion Robotics. *Proceedings of the International Conference on New Interfaces for Musical Expression*, 1–6.

Van Rooyen, R., Schloss, A., & Tzanetakis, G. (2017). *Voice coil actuators for percussion robotics. What's the difference between AC, DC, and EC motors?* (2017, August 24). Machine Design. https://www.machinedesign.com/mechanical-motion-systems/article/21835874/whats-the-difference-between-ac-dc-and-ec-motors

Yang, N., Savery, R., Sankaranarayanan, R., Zahray, L., & Weinberg, G. (2020). *Mechatronics-driven musical expressivity for robotic percussionists* (arXiv:2007.14850). arXiv. https://doi.org/10.48550/arXiv.2007.14850

CHAPTER 5

PERCUSSION

Part 3: Making

Realization

BOM, FABRICATION, AND COST

The BOM and fabrication considerations associated with making a machinic percussion instrument are similar to those of other kinds of projects, so revisit the general descriptions of those topics in Chapter 4: The Process of Making a Musical Machine. Specific to percussion, the materials that you choose should accommodate the forces output by the system. If you are creating a drum-thumping robot that can fill an arena with sound, 3D-printed plastic is unlikely to be the best choice for parts that will be subjected to significant strain. Regarding cost, the price of actuators can make a difference. More powerful solenoids and quieter motors are usually more expensive. That said, it is possible to build a percussion machine for a relatively small amount of money, albeit with some compromises in performance. Make the most of what you have, which will reveal what you need (and can save for).

Control and Feedback

Many of the design considerations mentioned in the Control and Feedback section of Chapter 4 are not specific to percussion instruments and should be considered here. These include choosing

- whether to generate musical instructions from a host computer or store a musical program on the microcontroller.
- a communications protocol if using a host computer.
- if the percussion machine can sense the output of other musicians. If it can, how will it retain, interpret, and respond to input information?

For those requiring a primer (or who would benefit from a refresher), the Companion Website ▶ contains sections devoted to *Sensing, Communications, and Control,* which include technical descriptions and projects that can help you collect and interpret information, as well as connect and control system components.

Percussion has proven to be fertile ground for exploring control and feedback in human–machine interaction. The musical robot *Haile* interprets the playing of human percussionists, including note onsets, pitch, amplitude, and spectra, and generates responses using different modes of interaction performed via two robotic arms (Weinberg et al., 2006; Weinberg & Driscoll, 2006). *Shimon*, an anthropomorphic marimba-playing robot, utilizes interaction modules, including call-and-response, opportunistic overlay improvisation, and rhythmic phrase-matching improvisation, in human–robot jazz improvisations (Hoffman & Weinberg, 2011). Sheffield and Gurevich (2015) explored *cooperative* human–machine interaction using two found-object percussion instruments that could be simultaneously played by both human performers and mechatronic actuators (solenoids, motorized faders). As the human performer plays the instrument, the sound is transduced into a control signal, driving the other instrument (and vice versa). The performer can turn the actuation on or off via a capacitive strip on the instrument, leading to experimentation and unique performance behaviors by the participating human musicians. Bretan et al. (2016) made a robotic prosthesis for an amputee drummer. The design features brushless gimbal motors and a timing belt coupled to a standard drumstick, all of which are secured to a socket worn by the drummer. Electromyography sensors determine muscle activity, generating control signals for the motors. The system is *cooperative* as the drummer controls initial onsets through arm motion while bounces are generated by the robotic system. STAMPER (Uriu et al., 2023) is a human–machine integrated drumming system where actuated pedals (APs) are connected to multiple bass drums. In a three-AP system, a human playing AP1 can control the output of AP2 and AP3 by selecting the mode (metronome or delay) and moving their knees. This movement, when interpreted by a motion capture system, modulates musical parameters, including tempo, volume, and the number of repetitions.

At the actuator level, solenoid-based systems, which include solenoid valves used to control pneumatic actuators, can function without feedback and are optimized through mechanical design, spring selection, adjustable positioning, and careful listening. With that said, feedback makes the optimization process easier and more precise. If the position and sonic output of a striker (e.g., via a proximity sensor, camera, encoder, etc.) is measured, the mapping between input (e.g., actuator ontime, pulse width modulation—PWM—duty cycle) and output (sound level) can be determined automatically to establish the instrument's dynamic range. The Robotic Glockenspiel (Long et al., 2016) does just this by utilizing electromagnetic pickups to measure the onset and amplitude of bar strikes. This information is used to autonomously calibrate the instrument's dynamic response and latency, as well as to synchronize with external sequencers, signal processing devices, and human players.

Rotational motors and VCAs used in percussion systems typically require feedback. The striker's position must be known to enable tight control of direction, acceleration, velocity, and timing. Without such information, it is easy to miss notes, play them at the wrong volume, or extend past boundaries, which can damage components. Positional feedback can be provided by limit switches, potentiometers, or encoders. This data is then integrated into a control scheme (e.g., PID controller), which attempts to reconcile the actual and ideal position of the striker.

Evaluation

The methods used to evaluate percussion machines depend on the project objectives, but two areas must be addressed in most cases. The first is dynamic range. Testing dynamic range requires a calibrated SPL meter or a real-time analyzer (RTA). Microphone gain and position must be consistent across trials. The sound's intensity and the perception of loudness that results are a function of the distance between the source and receiver. Determine the relevant distance in your situation. For example, if the percussion machine is meant to play in a small ensemble, then you want to know the loudness at the positions of other ensemble members. Document your setup and maintain consistency across all experiments. Loudness measurements at one foot in a small room are going to be difficult to reconcile with those at ten feet in a concert hall.

Second, the temporal capabilities of percussion machines, particularly in terms of speed and precision, should be evaluated. Timing performance can be analyzed using audio and video recordings. Programs such as Audacity allow you to record audio and precisely measure the time intervals between events at the millisecond level. If you are using video to analyze the performance, the higher the resolution (e.g., frames per second or fps), the better. Analyzing video can reveal travel time and distance from a striker's first movement to its contact with a surface. From this information, the maximum playing rate can be determined, and acceleration and velocity can be calculated (these quantities could also be measured using an Inertial measurement unit—IMU). This information can be reconciled with sound-intensity measurements to better understand the relationship between velocity, distance, time, and dynamic range.

The other aspects of your machine that need to be evaluated depend on the requirements you specified. Follow the guidelines given in Chapter 4 to ensure that the methods you use are appropriate, rigorous, and repeatable.

Example

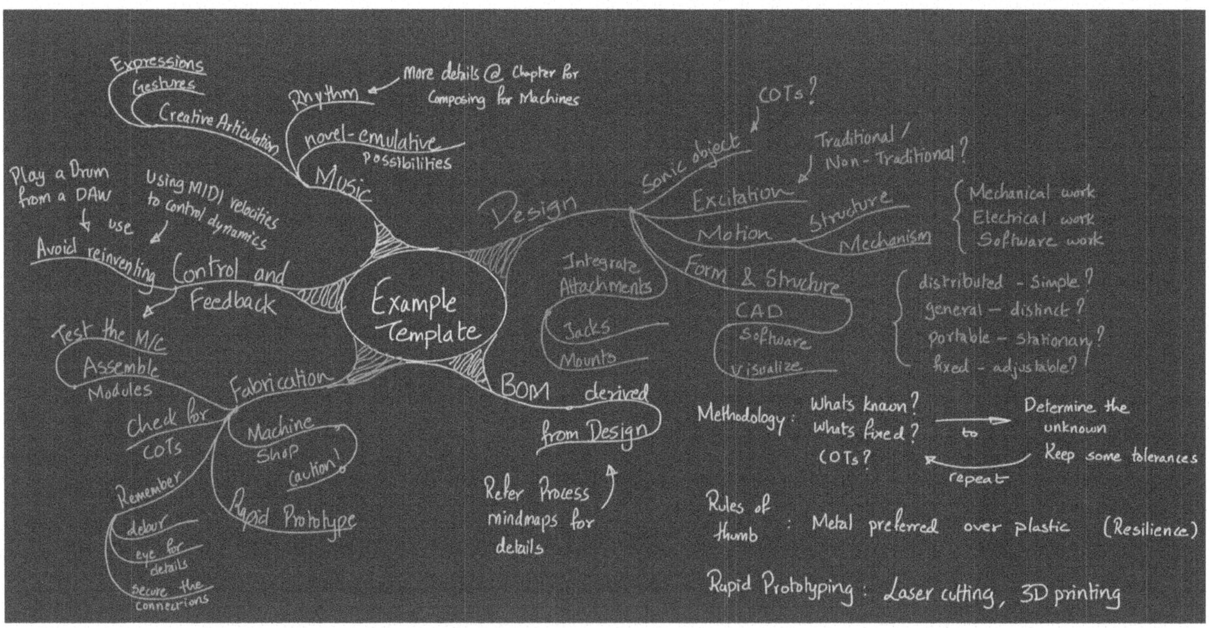

FIGURE 5.3.1. **Mindmap by Yash Garje.**

IDEATION

In this section, the design process described in this chapter will be put into practice. We begin by establishing **Identity, Purpose, and Usage**. Here, the goal is to design a percussion machine that can play a drum set. Drum sets are a collection of diverse objects whose contents and positions can change, so the system should be *simple*, *distributed*, *general*, and *adjustable*. It will be used primarily for composition, rehearsal, and recording with small acoustic ensembles, so it should be *transportable*. We aim to create a system that is affordable, constructed from readily available parts, and easy to build and program. The flexible percussion examples in the **Prior Art** section, including LEMUR's *ModBots*, EMMI's *CADI*, and Polyend's *Perc*, provide references for inspiration. Regarding **Musical Performance** requirements, the desired timbres are those of a traditional drum set, strikers should be able to play at rates of 100 ms or less, latency should be less than 50 ms, and the dynamic range should be from 40 to 70 dB SPL at minimum. The articulation rates and dynamic range should enable the performance of standard drum rudiments. **Visually**, we seek a non-anthropomorphic, minimalist *form*, limiting structural and mechanical aspects to those needed to support the actuators (thus, the design will lean toward *function*). There should be some visible movement to create associations with traditional drumming (without being excessively *emulative*) to engage audiences and human performers in collaborative scenarios. Regarding **Interaction and Autonomy**, the machine will play back sequences programmed in a DAW, such as Ableton Live, via MIDI.

DESIGN

The **Sonic Objects** are the components of a drum set, which will be acquired *off-the-shelf.* The drums will be **excited** using traditional drumsticks; it is assumed that the velocities, forces, and positions required have been experimentally determined. As these are 1-DOF actuating mechanisms, their **motion** will be determined by the stroke of the actuator and the distance of the striking object from the drumhead. Considering the possibilities for the mechanism design, a **linear striker with a pivot** meets our criteria. A linear solenoid will be used, given that it is fast, durable, easy to work with, and inexpensive. To control the solenoid, we will utilize the MOSFET-based circuit featured in Project 5.1: Solenoid Percussion on the Companion Website ⊙.

The excitation system design must include the following:

- a solenoid and a drumstick
- a way to convert the linear motion of the solenoid into the rotational motion of the striker (because of the pivot)
- structural pieces that stabilize and connect the solenoid and drumstick
- electrical connections.

As this will be a *simple-distributed* system, the primary structural considerations are:

- a mounting system that will allow the mechanism to be *adjustable*
- an enclosure for the microcontroller, power supply, and electrical connections

Guided by the aforementioned criteria, sketches of the basic idea for the mechanism are seen in Figure 5.3.2.

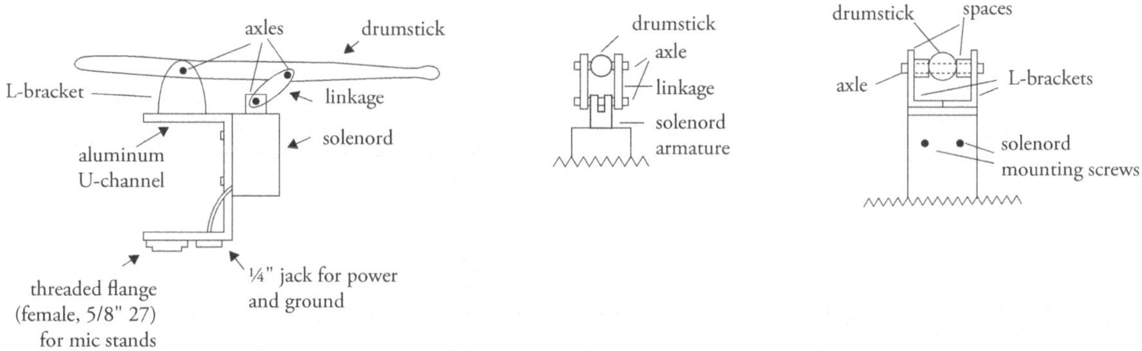

FIGURE 5.3.2. **A design for a flexible percussion machine.**

Structure and Excitation Mechanism

The *primary structure* consists of an aluminum U-channel, to which the solenoid, drumstick brackets, electrical connections, and mounting point attach. A linkage connects the solenoid's armature to the drumstick, converting linear to rotational motion. On the front, one axle runs through the solenoid armature, and a second axle runs through the drumstick. On the back, an axle (pivot) secures the drumstick to the *pivot structure* (two L-brackets facing each other or a U-channel). The *pivot structure* attaches to the top of the *primary structure* with bolts and nuts. Ball bearings can be recessed into the drumstick at these locations to minimize friction and ensure smooth movement. Spacers may be needed on the axles to prevent lateral movement, depending on the dimensions. The axles can be held in place using either e-clips that fit into cut grooves, end caps, or end brackets. The power and ground wires of the solenoid terminate in a ¼″ TS jack mounted on the bottom of the U-channel. Next to the jack is a threaded flange (female, 5/8″ 27) that can connect to a standard US mic stand.

The next step is to figure out the exact dimensions of the components. CAD software is helpful at this point, as it can help ensure that components fit together correctly and the files it generates can be utilized by fabrication machines. If such software is unavailable or intimidating, this process can also be done manually. In either case, we begin with the predetermined dimensions, which, in this instance, include the solenoid and the drumstick. From there, we can dimension the other structural components.

The dimensions of commercially produced components, such as a solenoid, can be found on the product's datasheet (locate the model number and conduct an internet search). In our case, we have chosen a solenoid from Adafruit, which helpfully provided this information on their product page (Figure 5.3.3).

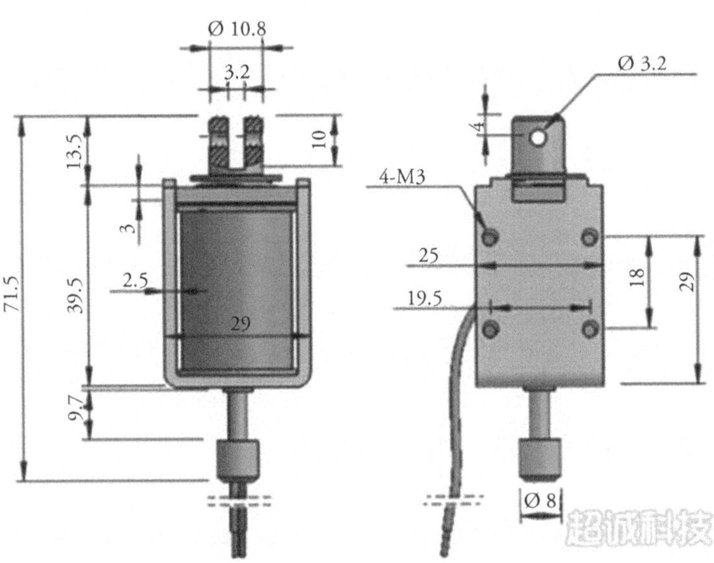

FIGURE 5.3.3. **Dimensions of a solenoid (Adafruit.com, n.d.).**

The width and height of the case from the rear (the side that will attach to the primary structure) are 25 mm and 39.5 mm, respectively. With these dimensions, we can specify the *primary structure* (note that standard U-channel dimensions treat the "bottom" of the "U" as the width and the "sides" of the "U" as the height). Given our minimalist aspirations, the "U" does not need to be much bigger than the solenoid case, so a width (which corresponds to the height of the solenoid) of 2″ and a depth (which corresponds to the width of the solenoid) of 1″ should be sufficient. The next task is to figure out what height (of the "U") is necessary. The L-brackets can be designed as needed, so the primary restriction is the size of the mounting flange and electrical connection. For the latter, we will use a ¼″ two-conductor (TS, or tip-sleeve) jack, as shown in Figure 5.3.4.

FIGURE 5.3.4. **1/4″ TS Jack connected to a ¼″ male cable.**

For the mounting flange, we will use a 5/8″-27 female to ¼″-20 male adapter, as shown in Figure 5.3.5.

FIGURE 5.3.5. **5/8″-27 female to ¼″-20 male mic adapter.**

The width of each of these components is less than an inch, so 2″ of "height" in the U-channel should be sufficient. This brings up another key lesson:

> **Lesson**: Leave some "padding" around your measurements as the extra space is often useful if not necessary. This is a good idea in many domains.

Now that we know the dimensions needed for the *primary structure*, the next step is to determine its materials. One option would be to make the object *from scratch* by 3D printing it out of plastic. This could work, but a more robust material for the primary structural component is preferable. Another issue is that the mic adapter has ¼″-20 male threading, which would ideally screw directly into the *primary structure*. We could use a ¼″-20 nut, but the shaft of the adapter is likely not long enough to secure the part. It is possible to tap plastic (a *tap* is a tool that creates threads in a solid material), but the threading is not resilient: cycles of inserting and removing a bolt will weaken the coupling. Metal is a better material to use in this case, which brings another key lesson:

> **Lesson**: Tap into metal (and not plastic) to ensure resilient connections, particularly when significant forces are involved.

Aluminum is a great material: it is strong, lightweight, affordable, and easy to work with. The U-channel could be made *from scratch* out of flat bar stock, but aluminum U-channel is widely available *off-the-shelf*. Vendors such as McMaster-Carr have a variety of options, including the 2″ W × 2″ H stock that we need (it comes in 4 and 8-foot lengths, which can then be cut down to the 1″ depth required). In terms of thickness, 1/8″ could work, but the extra stability of the ¼″ option helps one sleep better at night.

We can dimension the *pivot structure* in a similar way. One option would be to design it in CAD and 3D print it, which would beneficially allow making different shapes (such as the arch in the diagram) with no tapping necessary (the parts would use bolts and nuts to secure to the *primary structure*). Another possibility would be to buy aluminum L-channel stock (1/2″ width, 1″ height, 1/8″ thickness should work) and cut two pieces configured facing each other. Another option is to buy more aluminum U-channel (1″ width, 1″ height). With the aluminum parts, the arch shapes can be made using machining tools (which require skills and access to a machine shop) or by hand. The option you choose depends on fabrication resources and aesthetic preferences.

There are at least two possible designs for the linkage. The first (not pictured) would be to use a single linkage that fits both in the slot of the solenoid armature and in a slot cut in the middle of the drumstick. The armature slot is 3.2 mm, so the linkage thickness would have to be 3 mm or less (which is quite thin). Plastic, frailty, and rapid movements are not a great combination. Another issue is cutting the slot in the drumstick, which would require woodworking tools and skills. Alternatively, linkages can be placed on the outside of the armature and drumstick for more stability, as shown in Figure 5.3.2. The primary considerations for the linkages are materials and dimensions. These parts could be made from metal or plastic. This is essentially a 2D part, so it can be laser cut instead of 3D printed, which is faster and allows us to use clear acrylic (aligned with our transparent machine aesthetic). A thickness of 1/8″ should be fine, but 1/4″ is also an option. The length of the linkage will be determined by the position of the drum, the throw of the armature, and the length of the spring. These parameters can be measured, but some trial and error will also likely be necessary.

The length of the axles and the width of the spacers depend on the diameter of the drumstick (here, we will use a diameter of 0.5″). The axles will be made from 0.25″ steel rod, cut to the different sizes needed.

- The **rear axle** should extend beyond the width of the pivot structure so that the end caps can be press-fit into place (which are available from vendors such as McMaster-Carr or can be 3D printed). If the pivot structure is 1″ wide and the two end caps are each 0.25″, a 1.5″ rear axle should be sufficient. The interior of the pivot structure is 0.75″ (assuming a thickness of 0.125″), so the width of the drumstick (0.5″) leaves 0.25″ of space. To prevent any lateral movement, two **spacers** (each 0.125″) will be placed on either side of the drumstick on the inside of the L-brackets (as shown in Figure 5.3.2). Ball bearings (with an inside diameter equal to the axle diameter) will be recessed into the drumstick to minimize friction as the drumstick rotates.
- The **front axles** need to be the diameter of the drumstick (0.5″) + the width of the acrylic linkages (0.125″ + 0.125″ = 0.25″) + the two end caps (0.25″ total) + extra space to prevent friction (0.25″) = 1.25″ length for each axle. For the front top axle, small spacers (0.125″ or less) can be used to prevent lateral movement. For the bottom front axle, the width of the armature is 10.8 mm, which leaves about 30 mm of space, which can be offset with the same small spacers (2 of them at 0.125″ each).

Electronics

The electrical control system needs to translate computer instructions into microcontroller signals that result in the appropriate electrical voltage and current being delivered to the solenoids. Build the circuit and write the code following Project 1: Solenoid Percussion and Project 2: Solenoid Percussion with Variables and Arrays in the *Actuators, Circuits, and Code* sections of the Companion Website ⊕. During testing, you can connect the solenoids directly to the circuit. Later, you will use jacks, wires, and cables to connect components, as depicted in Figure 5.3.6.

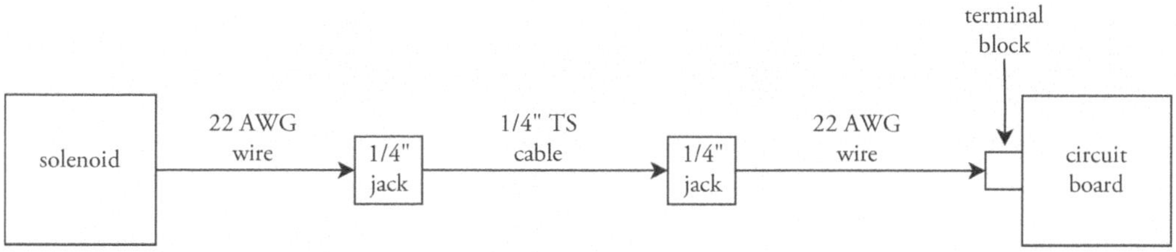

FIGURE 5.3.6. **Block diagram of system connections.**

When everything is working correctly, the electronic components must be physically configured and secured. The system is *distributed*, so the electronics will be housed in a separate enclosure, the specifications of which depend on the system's capacity (i.e., the number of strikers used). The circuit should fit within the boundaries of a small breadboard. The enclosure must accommodate ¼″ TS jacks (one per striker), a 2.1 mm jack to connect to the power supply, the Arduino's USB jack to connect to the host computer, the circuit board, and the wiring that connects the components. Make a model of the enclosure by following these steps:

1. Determine the dimensions of components either by looking them up on the internet or physically measuring them (calipers provide precise measurements).
2. Make scaled models of the length and width of each component in the form of rectangles using software (e.g., Fusion, Adobe Illustrator, Google Drawings) or by cutting them out of a piece of paper.
3. Arrange the components in a way that conserves space but also allows for wiring and ventilation.
4. Draw a rectangle that encompasses the components, incorporating modest buffers. Measure the length and width of this rectangle and scale to the actual size: these are the dimensions of the enclosure that you need.
5. Determine height by measuring the distance between the enclosure bottom and the tallest component. Account for standoffs needed to elevate components (e.g., the microcontroller) above the enclosure. Include some buffer space.

A top-view layout for an electronics enclosure of a three-striker percussion system is illustrated in Figure 5.3.7:

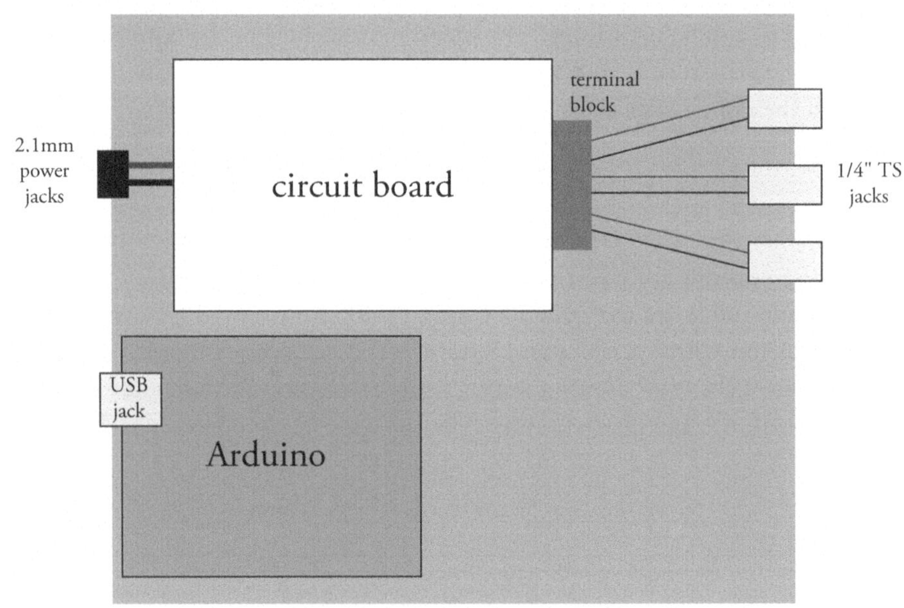

FIGURE 5.3.7. **Electronics enclosure for a three-striker percussion system.**

BOM

With the necessary parts identified, create a BOM that helps organize your procurement efforts and provides documentation for others who contribute to or replicate your project. A partial BOM is given as a sample in Table 5.3.1. A comprehensive BOM for the project, including links to components, is available on the Companion Website ⊙.

TABLE 5.3.1. **Sample BOM for a percussion machine.**

Part name	Description	Manufacturer	Product #	Specifications	Quantity
Striker	Drumstick	Vic Firth	TMB1	.5″ OD, 15.5″ L	1
Actuator	Solenoid	Adafruit	413	12 V DC	1
Primary structure	Aluminum U-channel	McMaster-Carr	9001K99	2″ W, 2″ H, .25″ T, 4′ L	1
Primary structure	Screws	n/a	n/a	M3, 15 mm L	8
Primary structure	Washers, nuts	n/a	n/a	M3	4
Pivot structure	Aluminum U-channel	McMaster-Carr	9001K52	1″ W, 1″ H, .125″ T, 4′ L	1
Linkage	Acrylic sheet	McMaster-Carr	8560K239	12″ × 12″	1
Electrical connector	USB C to A jack	Adafruit	4259	30 mm diameter, 41 mm length	1
Electrical connector	1/4″ TS jack	Switchcraft	11PKG	.75″ W, 1 A, 24 V DC	1
Mounting system	5/8″-27 to 1/4″-20 adapter	On-Stage	MA-125, INV#12531		1
Axle	1/4″ steel rod	McMaster-Carr	8927K18	.25″ OD, .5′ L	1
Axle	Ball bearings	McMaster-Carr	57155K356	.25″ ID, .5″ OD	2
Axle	End caps	custom	N/A	.25″ ID, .25″ H	6
Axle	Spacers	McMaster-Carr	90176A153	.125″ L, .25″ ID	6

FABRICATION AND ASSEMBLY

The aluminum parts can be fabricated in a machine shop or a workshop with a manual hacksaw.

1. Cut, deburr, and sand the parts so that they are smooth.
2. Measure and drill holes to attach the solenoid and the pivot structure to the primary structure using screws, washers, and nuts.
3. Drill a hole in the bottom of the primary structure for the ¼″ jack by measuring the width of the threaded shaft using calipers and cut a hole just large enough for the shaft to fit through (but not the base of the jack).

4. Drill and tap a ¼″-20 hole in the bottom of the primary structure and screw the mounting adapter in. Ensure that there is sufficient space for both connectors to fit.
5. The linkage will be laser cut. Design the part in software (e.g., Adobe Illustrator, Inkscape, Fusion) and save the file in an appropriate format (e.g., .pdf, .svg, .dxf: check the requirements of the laser cutter). The protocols for using the laser cutter should be available from the manufacturing facility you are using (often, there is a safety course that you must pass to use the machine). If you don't have access to a laser cutter, you can send the file to an on-demand manufacturing service (e.g., SendCutSend, Xometry), or the linkage can be made manually.
6. Drill two holes in the drumstick (one for the pivot, one for the linkage) that are large enough for the axle to fit through comfortably. Make sure that each of these holes is centered relative to the diameter of the stick. Drill a larger hole "on top" of these smaller holes that is the diameter of the ball bearing, but stop when you are halfway through the stick. This will allow the bearing to be recessed. If the hole is sufficiently tight, the bearing can be press-fit into place. If not, it can be glued.
7. Put the axles through the brackets, spacers, and drumstick and then secure them with the caps.

Electronics

8. Solder wires (22 AWG is usually sufficient; check the current requirements of your system) from the solenoids to the ¼″ TS jacks on the strikers. Insert the jack through the primary structure and secure it using the included nut.
9. While the circuit will work in its breadboard form, it is not advisable to keep it this way, as wires can come loose and can potentially short against other connections. To make circuit connections more robust, solder components into a protoboard or design a PCB (printed circuit board), which can be manufactured by companies such as JLCPCB (PCB design is beyond the scope of this book).
10. The electronics enclosure could be made *from scratch*, acquired through *rummage*, or purchased *off-the-shelf*. Ensure it has the correct dimensions and ventilation (the circuit will generate heat) and drill the necessary holes (2.1 mm power, ¼″ TS, and USB jacks).
11. Solder wires to the ¼″ jacks and connect them to the circuit board via the terminal blocks. Insert the ¼″ jacks into the enclosure holes and secure them with the included nuts.
12. Insert and secure the 2.1 mm barrel jack to the enclosure, connecting the power and ground wires to the circuit board. Fasten the circuit board to the enclosure by using tapped standoffs.

Putting It All Together

With the excitation mechanism and the electronics enclosure assembled, the next step is to connect and configure the components. Mount the striker on a mic stand and position it near a drum. Plug one end of a ¼″ TS cable (ensure the wire gauge is rated for the current the solenoid will draw) into the striker and the other into the electronics enclosure. Connect the host computer to the Arduino using a USB cable and plug the power supply into the 2.1 mm barrel jack on the enclosure.

Completing fabrication and assembly offers a moment to reflect on the initial vision and requirements. The resulting device is *general* (it can play various percussion instruments), *functional*, minimal in its materials (which makes it cheaper and simpler to build), *adjustable* (via the mic stand), easy to *transport*, and evokes associations with traditional drumming without being exceedingly emulative. You are ready to make music!

CONTROL AND FEEDBACK

We will write music in a DAW and play it with the percussion machine. To translate the instructions from the computer into physical actions performed by the percussion machine, use the code introduced in Project 9: Play a Drum from a DAW on the Companion Website ⓟ. To ensure the instrument produces a range of dynamic levels that correspond to MIDI velocities, we will also incorporate the code from Project 10: Using MIDI Velocities to Control Dynamics, available on the Companion Website ⓟ. This design does not incorporate low-level feedback, as modulating solenoid ontimes will allow us to achieve the required dynamic range. High-level feedback is also not incorporated, as the system's purpose is to play back sequences made in a DAW for the composition, rehearsal, and recording.

MUSIC

The work and dedication required to make the machine should not distract from the reason we do such things in the first place: to make music! Thinking about the musical possibilities of a percussion machine brings us back to the *emulative* ⇔ *novel* dimension. We may choose to follow the path paved by human percussionists. Human musicians possess extraordinary motor abilities honed over thousands of hours of practice. Research on this topic (Dahl, 2004, 2011; Dahl et al., 2011) can be helpful to those trying to emulate such abilities. This has been part of Robert van Rooyen's approach (Van Rooyen et al., 2016; Van Rooyen & Tzanetakis, 2015), the benefits of which are evident in the success of *MechDrum*. Part of the equation involves understanding what human percussionists do when they play, part is designing and building a system that can physically produce those motions (which does not necessarily have to be anthropomorphic), and part is developing algorithms that can control the machine as needed. Given the variation and nuance of human performance, this is not an easy task.

As a composer, I am interested in the *novel* possibilities of machinic percussion in rhythm, time, and articulation. Machine percussionists are uniquely able in speed, rate contours, tuplets, polyrhythms, polytempo, complex meters, polymeter, augmentation, diminution, expressive timing, and *machine rhythms*. Concepts and creative possibilities in these rhythmic domains are described in more detail in Chapter 9: Composing with and for Machines.

Machinic percussion allows composers to explore articulation. For percussion instruments, variation in articulation is achieved by modulating a striker's height, velocity, position, and timing. Figure 5.3.8 illustrates the variation in elevation that occurs when an expert percussionist plays a drum rudiment such as a double-stroke open roll (Van Rooyen et al., 2016).

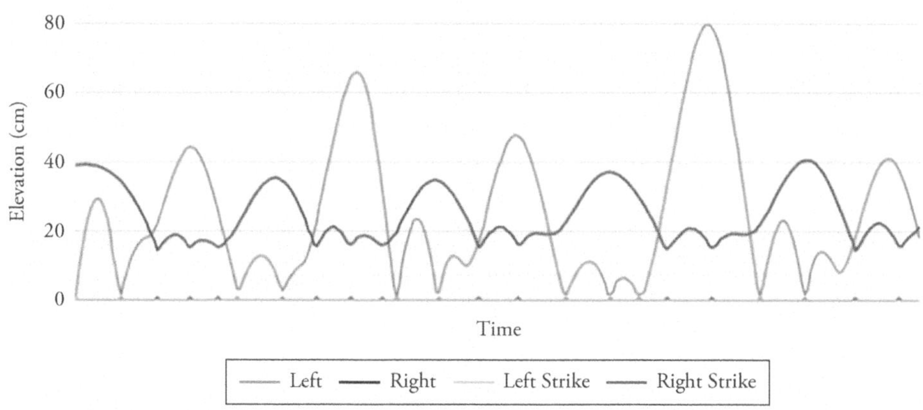

FIGURE 5.3.8. **Analysis of a double-stroke open roll as played by a human percussionist where the *y*-axis is elevation and the *x*-axis is time (Van Rooyen et al., 2016).**

The differences in elevation between the left and right hands are attributed to hand dominance (the left hand is dominant in this example), which contributes to expressivity and the nuance we hear in human performance. The goal of a machine percussionist could be to control these parameters precisely. Such a machine could be capable of new kinds of articulations by varying elevation, timing, position, and velocity both within and between "hands." The palette widens with the addition of more strikers. Such abilities could produce new kinds of sonic gestures, expressions, and even music.

Summary

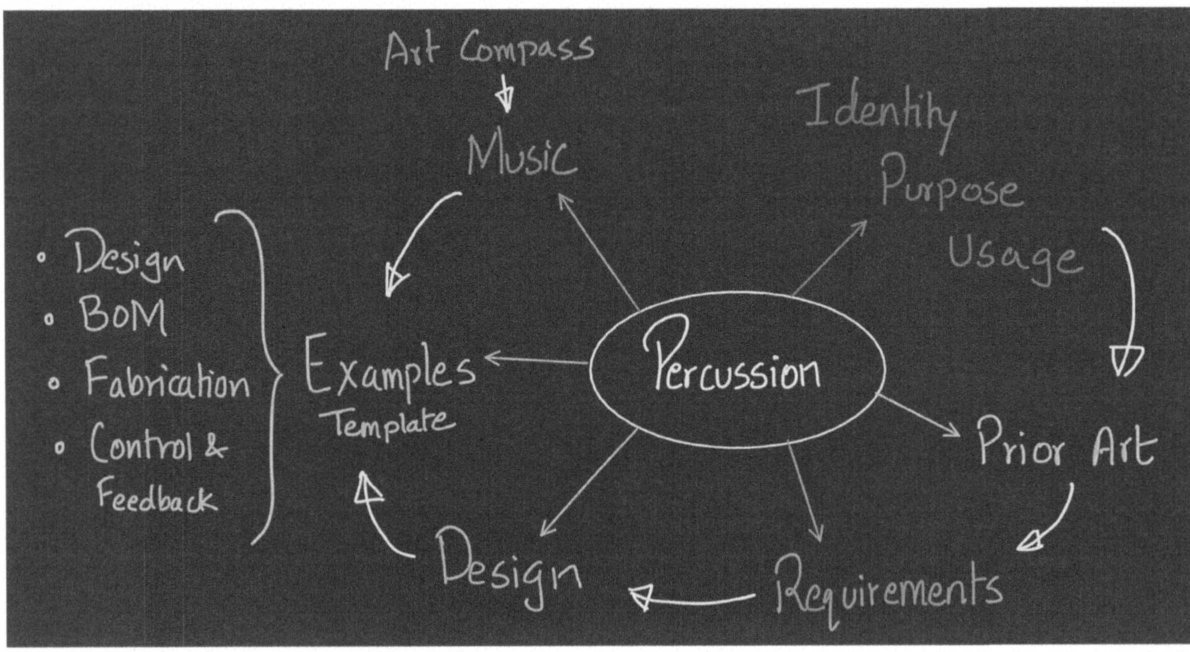

FIGURE 5.3.9. **Mindmap by Yash Garje.**

In this chapter, the design and realization template developed in Chapter 4: The Process of Making a Musical Machine was applied to a percussion machine. The identity, purpose, and usage of percussion vary widely in music, influencing the design of the abilities, form, and movements of a machinic percussionist. Many examples of machinic percussionists exist, which can be grouped into a smaller number of categories, including *flexible ⇔ fixed*, *form ⇔ function*, *complexity of motion*, *emulation ⇔ novelty*. The requirements for a percussion machine specify its musical (time, dynamic range, and articulation are particularly important), visual, and interactive characteristics. In the design, the sonic object, which can be acquired *off-the-shelf*, through *rummage*, or *from scratch*, will in part determine the mechanical design, so it should be considered from the start (even if the machine's purpose is to play different types of sonic objects). Percussion objects include **membranophones** (heads and bodies connected by tensioning hardware) and **idiophones** (bars, tines, and resonators), which can be designed and modified to suit specific needs. The requirements and design of an excitation system involve specifying masses, velocities, accelerations, forces, displacements, and paths of motion to achieve musical and visual objectives. Mechanisms used in percussion machines include levers, linkages, and gear trains. Actuators play a crucial role in mechanism design, influencing the machine's musical and kinetic capabilities.

Actuators featured in machinic percussionists include solenoids, DC motors, servomotors, pneumatic cylinders, VSAs, and VCAs. Actuators and mechanisms combine to form **excitation systems**; examples include the **direct linear striker, linear striker with pivot, Trimpin Hammer, linear striker with attachment**, and **rotary strikers**. When the design for an excitation system is complete, make one module, test it, and revise as necessary.

The structure that supports the excitation system (and its replications) is determined by musical, visual, and practical goals. One option is to have small, simple mechanisms that can be *distributed* and configured *flexibly*. At the other end of the continuum, a *single*, more *complex* system can create a variety of musical and visual gestures (and will also have different constraints in doing so). While contemplating possibilities, determine whether the machine is designed to play a *specific* instrument or will be more *general*, capable of playing many. Find the balance between *form* (size, shape, angles, color, materials, visual aesthetics) and *function* (musical capabilities, ability to give gestural cues to other performers). Determine to what extent instruments and actuators will be *integrated* or *distinct*. These choices will influence which aspects are *fixed* and which are *adjustable*, which impacts pragmatic (e.g., transportability) and musical matters (e.g., tuning).

The control of percussion machines is dictated by artistic goals. They can be played from predetermined sequences stored on a microcontroller, a DAW on a connected computer, or the directives of an AI. While it is possible to make a machine without feedback, information about the velocity, acceleration, and position of the striking object is often useful, particularly in extending dynamic range and ensuring precise timing. Audio feedback can enable exploration of timbral possibilities.

The fabrication and cost of building a percussion machine depend on the resources available. While expensive actuators and materials can improve a system's performance, it is possible to build a percussion machine for a relatively low cost. Make the most of what you have, which will show you what you need (and can save up for).

The methods used to evaluate a percussion machine depend on the requirements, but dynamic range and temporal performance (speed, precision) apply in almost all cases. Define your methods, perform standardized tests multiple times in different sequences, and document your results thoroughly. These practices can be tedious, but they can give you (and others who continue your work) tremendous insights about the next steps to take.

References

Adafruit Industries. (n.d.). *Large push-pull solenoid*. Adafruit. Retrieved July 7, 2020, from https://www.adafruit.com/product/413

Bretan, M., Gopinath, D., Mullins, P., & Weinberg, G. (2016). *A robotic prosthesis for an amputee drummer* (arXiv:1612.04391). arXiv. https://doi.org/10.48550/arXiv.1612.04391

Dahl, S. (2004). Playing the accent-comparing striking velocity and timing in an ostinato rhythm performed by four drummers. *Acta Acustica United with Acustica*, *90*(4), 762–776.

Dahl, S. (2011). Striking movements: A survey of motion analysis of percussionists. *Acoustical Science and Technology*, *32*(5), 168–173.

Dahl, S., Grossbach, M., & Altenmüller, E. (2011). Effect of dynamic level in drumming: Measurement of striking velocity, force, and sound level. *Proceedings of the Forum Acusticum (European Acoustics Association)*, 621–624.

Hoffman, G., & Weinberg, G. (2011). Interactive improvisation with a robotic marimba player. *Autonomous Robots*, *31*(2–3), 133–153.

Long, J., Carnegie, D., & Kapur, A. (2016). The closed-loop robotic glockenspiel: Improving musical robots with embedded musical information retrieval. *Proceedings of the International Conference on New Interfaces for Musical Expression*, 2–7. https://doi.org/10.5281/zenodo.3964607

Sheffield, E., & Gurevich, M. (2015). Distributed mechanical actuation of percussion instruments. *Proceedings of the International Conference on New Interfaces for Musical Expression*, 11–15. https://nime2015.lsu.edu/proceedings/183/0183-paper.pdf

Uriu, D., Iiyama, S., Okada, T., Handa, T., Maekawa, A., & Inami, M. (2023). STAMPER: Human-machine integrated drumming. *Extended Abstracts of the 2023 CHI Conference on Human Factors in Computing Systems*, 1–5. https://doi.org/10.1145/3544549.3585619

Van Rooyen, R., Schloss, A., & Tzanetakis, G. (2016). Snare drum performance motion analysis. https://www.researchgate.net/publication/303873184_Snare_Drum_Performance_Motion_Analysis

Van Rooyen, R., & Tzanetakis, G. (2015). Pragmatic drum motion capture system. *NIME*, 339–342.

Weinberg, G., & Driscoll, S. (2006). Toward robotic musicianship. *Computer Music Journal*, *30*(4), 28–45.

Weinberg, G., Driscoll, S., & Thatcher, T. (2006). Jam'aa—A Middle Eastern percussion ensemble for human and robotic players. *International Computer Music Conference*, 464–467.

CHAPTER 6

STRINGS

Part 1: Context and Ideation

The essence of a string instrument is a cord and a structure that stretches it into a pitch-producing range. Strings are musically flexible, capable of an expansive range of pitches, diverse timbres, virtuosic monophonic lines, chordal accompaniment, and acoustic or electric amplification. This chapter will explore the world of string machines by situating prior examples in terms of the *properties* and *relations* described in Chapter 4: The Process of Making a Musical Machine. The design of novel string instruments involves establishing musical performance, visibility, and interaction requirements, which influence excitation, pitch-producing, and damping systems. Structural possibilities are shaped by aesthetic vision and pragmatic considerations, such as actuator mounts, wire routing, and transportation. The chapter concludes with an example that integrates these ideas into a machine that you can build.

String Basics

A musical string is a cord, thread, or fiber that is stretched between two fixed points at a tension that produces sound when set into motion. Strings are typically long, thin, and cylindrical. They need to be flexible enough to oscillate freely but also durable enough not to stretch (which would change the pitch they produce) or break. They can be made from a variety of materials, including metal, nylon, and rubber.

A string's movement influences the sound that it produces. An essential aspect of string movement is illustrated by the concept of simple harmonic motion (SHM). In SHM, the net force acting on an object is proportional to its displacement from equilibrium in the opposite direction of that displacement. Put differently, the greater the displacement, the greater the force acting to return the object to its resting position. The maximum displacement of an object relative to its equilibrium position is its **amplitude**. In an ideal system with no **damping** (loss of amplitude and energy due to friction, air resistance, or internal forces), the net force acting on an object that is displaced to a position ($x = A$) would cause it to move back toward and through the equilibrium position ($x = 0$) due to inertia and then to a position of equal amplitude opposite the starting position ($x = -A$). The net force would then act in the opposite direction, moving the object back through equilibrium and to the original position of displacement ($x = A$) (see Figure 6.1.1).

Musical Machines. Scott Barton, Oxford University Press. © Oxford University Press 2026. DOI: 10.1093/oso/9780197599891.003.0006

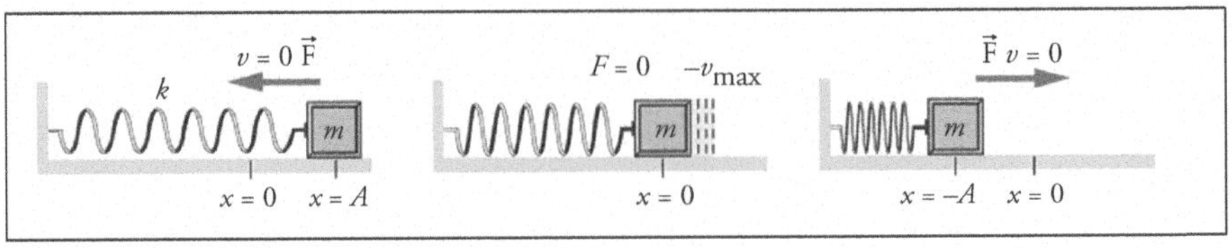

FIGURE 6.1.1. **Simple harmonic motion (image adapted from Moebs et al., 2016; cc by 4.0).**

With no damping, this process would repeat perpetually, resulting in oscillation with an amplitude of A. In the case of a string, an external object (such as a pick) applies the initial displacing force, thus **exciting** it into motion. Damping is a factor in the real world, so the string eventually loses energy and will return to its equilibrium position.

Oscillatory movements produce modes of vibration. The fundamental mode of vibration has a wavelength of twice the length of the string. A complete cycle of a sinusoidal waveform has a "positive" and a "negative" portion, which correspond to the alternating compressions and rarefactions of air molecules that are necessary to produce sound (see Figure 6.1.2). The entire length of the string moves to the "positive" and "negative" positions to create a cycle; thus, the wavelength (λ) of the resultant waveform is twice that of the length of the string.

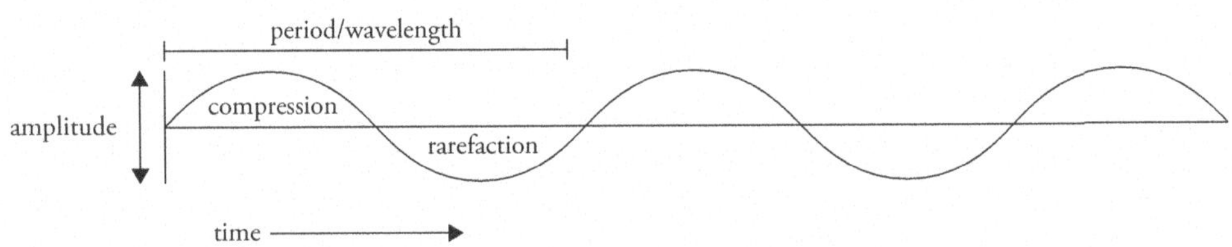

FIGURE 6.1.2. **A sine wave showing period, compression, and rarefaction.**

The frequency (f) of the string's vibration is a function of its length (L), tension (T), and linear density (D), which is its mass per unit of length (linear density roughly corresponds to string diameter):

$$f = \frac{\sqrt{\frac{T}{D}}}{2L}$$

This equation illuminates the variables that can be adjusted to modify frequency. If the string length and linear density are fixed, then the only way to produce different pitches is by adjusting string tension. If tension and linear density are fixed, then the only way to change pitch is by adjusting the effective vibrating length. Linear density can be adjusted when selecting materials but is more challenging to change during a musical performance.

The above equation allows us to determine the fundamental frequency of the string, which corresponds to the pitch that we perceive. In reality, a string's movement is more complex than a single fundamental frequency value implies, containing multiple simultaneous **modes of vibration** (Figure 6.1.3). Different modes of vibration occur at different frequencies, each one related to the fundamental frequency by an integer ratio (e.g., 2:1, 3:1), forming a harmonic series (Media Example 6.1.1 ▶). These harmonics coalesce in a complex waveform that shapes the timbral character of the sound produced.

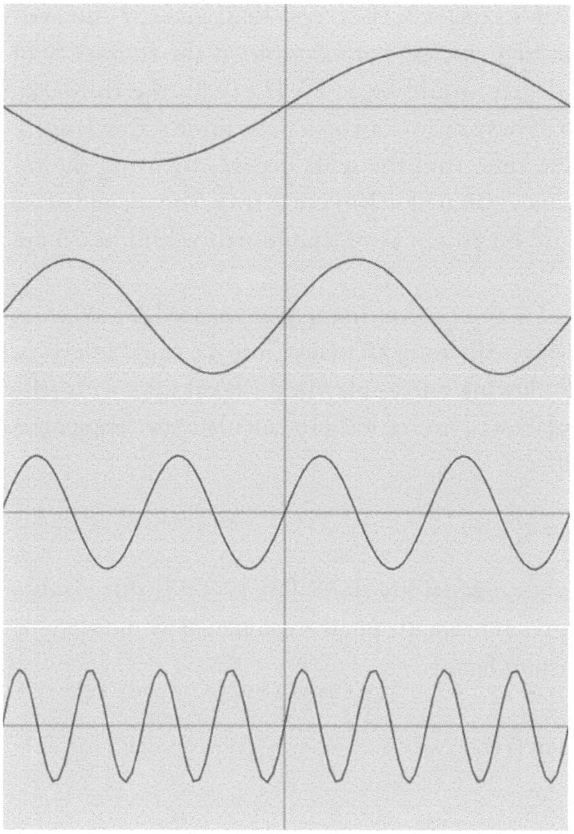

FIGURE 6.1.3. **Modes of vibration of a string, where periods are inversely proportional to frequency.**

The frequency spectrum that a string produces depends on its physical properties. Matching the physical properties of a string to a particular musical usage is a process as string **scaling**. As with many things in life, balance brings the best results: too much tension and the string could break; too little tension and the pitch is inconsistent. The length, mass, and rigidity must find a middle ground to produce a balanced frequency spectrum and drive the soundboard appropriately. These parameters vary depending on the register and range of pitches desired. Experimentation is required: try commercially produced strings (guitar, piano, bass, etc.) as well as materials that might yield interesting results (wire, nylon filament, etc.). With an idea of what strings you are going to use, you can think about the structure that will secure them and amplify their vibrations.

TUNING

Pitch can be varied on a tensioned string by modifying its vibrating length, which is accomplished by **stopping** it at different positions. In some cases, these locations are contained in a player's memory (human or machine), who navigates to them using a motoric system. Alternatively, positions can be specified in the physical characteristics of the instrument (e.g., frets on a guitar). In either case, the tuning system must be determined in accordance with the musical performance requirements. There are limitless tuning possibilities, but the most common are **just intonation** and **equal temperament**.

Just intonation is a system where pitch intervals are formed according to low-integer ratios relative to a reference or tonic pitch. For example, a just tuning that approximates the major scale is formed by the ratios 1:1, 9:8, 5:4, 4:3, 3:2, 5:3, 15:8, and 2:1 (octave). These ratios can be used to determine scale frequencies. For example, if the frequency of the tonic is 100 Hz (1:1), the second scale degree would be 112.5 Hz (9:8), the third 125 Hz (5:4), the fourth 133 Hz (4:3), and so on. These ratios can also determine string length. Start with a string length for the tonic pitch, then find the scale degrees by using the inverse of the ratios. For example, if the tonic necessitated a 100 cm string, the second scale degree would be 89 cm (8:9), the third would be 80 cm (4:5), the fourth would be 75 cm (3:4), and so on.

Twelve-tone equal temperament (12-TET), the predominant system used by Western instruments since the eighteenth century, involves dividing an octave into 12 equal intervals. This division offers the enormous benefit of allowing one to play in different keys with relatively minimal compromises in intonation (relative to just ratios). To calculate the frequencies of the 12-TET scale, use the following formula:

$$C = \sqrt[n]{2}$$

where C is the constant factor, and the scale has n equal divisions. For 12-TET, this yields a factor of 1.0595. The frequency of the next highest chromatic pitch is calculated by multiplying the frequency of the current pitch by the constant factor:

Pitch	Frequency (Hz)
A2	110
A#2	116.54 (110 * 1.0595)
B2	123.47 (116.54 * 1.0595)
C3	130.81 (123.47 * 1.0595)

As with string length, the position of frets is determined by the inverse of the constant factor, which in the case of 12-TET is 0.9438 (1/1.0595). The fret position for the next highest chromatic pitch is calculated by multiplying the current fret position by the inverse of the constant factor. For a 100 cm string, this would produce the following fret positions:

Fret	Position from bridge (cm)
1	94.38 (100 * .9438)
2	89.08 (94.38 * .9438)
3	84.07 (89.08 * .9438)
4	79.35 (84.07 * .9438)

These positions can be used in a string machine that mechatronically changes pitch via a movable tangent (e.g., via a track).

The preceding are only two of a panoply of tuning possibilities. Tunings expand the harmonic palette, inspiring creativity and facilitating musical exploration. Machinic precision and flexibility bring previously challenging or esoteric tuning possibilities within reach. Adhering to conventional models is fine if they align with your musical vision, but consider some of these alternatives as well; they may serve as a source of inspiration.

Prior Art

The history of string-playing automata spans centuries, populated by examples varying in form, function, and usage. To help identify similarities, differences, essential characteristics, and trends among this sundry group, we will use the *Properties and Relations* described in Chapter 4. Of these, the relations between *players* and *instruments* are particularly definitive. *Players* ⇔ *Instruments* can be segmented into four subgroups: (1) *independent players*, (2) *attachments* to existing instruments, (3) novel instruments that *integrate* mechatronic actuation with sonic objects, and (4) *cooperative* instruments that require human–machine interaction. These categories have ramifications on the form, motion, visibility, musical abilities, and psychological associations of string-playing machines; thus, many of the other properties and relations mentioned (such as *emulation* ⇔ *novelty*) are considered within the relation of *players* and *instruments*.

The first group includes string machines that are *independent players*: they are whole, complete, and distinct relative to the instrument they interact with. Independent players often take forms most familiar to us: the clearest examples of this category are anthropomorphic robots, including the violin-playing Toyota Robot (Media Example 6.1.6 ⓟ), *Mach* of Z-Machines, and *Fingers* of Compressorhead (Figure 6.1.4).

FIGURE 6.1.4. From left to right: Mach of Z-Machines (Image by Hiroshi Ishii, https://www.flickr.com/photos/isinoid/10671544034/in/photostream/, cc by 2.0), Compressorhead (image By Torsten Maue from Magdeburg, Deutschland—Musikmesse Frankfurt, CC BY-SA 2.0, https://commons.wikimedia.org/w/index.php?curid=32858466).

Emulating a human string player answers design questions about form, motion, how strings are excited, and how pitches are produced. It provides a way to address visibility issues.

The movement of a human finger in stopping a string is small (on the scale of centimeters), imperceptible at viewing distances typical of many concert venues. The human solution to this problem is to engage more of the body in grander and more complex gestures, augmenting expression. Rock guitarists jump around the stage; folk guitarists tell stories with their facial expressions; violinists gesticulate to make tension, resolution, and phrasing clear. Machinic movements that follow these more prominent and complex paths can similarly enhance visual experience.

Machines that emulate human musicians evoke associations in the minds of viewers and listeners. The designers of these machines engage with identities and performance practice (e.g., *bands*, *soloists*), inserting their creations into a rich world of imagery and gestures that have developed in branches of human culture. These associations give listeners and viewers a context to understand the purpose of the machines and their expressions. They invite direct comparisons with human performers so that similarities and differences can be distinguished. Such associations are useful when trying to produce music in a particular style, such as jazz, country, or rock.

Emulating human string players has its downsides. Complexity is introduced into the design that is not required to achieve musical objectives. A human hand is miraculous in its ability to manipulate multitudinous objects to accomplish a wide variety of outcomes. Not all of this functionality is necessary to pluck or stop a string. Emulation has ramifications on the speed at which larger pitch intervals can be produced. A human hand has only five fingers that span a relatively short distance. A primary way to change a string's pitch is to alter its effective vibrating length. The range of pitches over the length of a string exceeds the span of the fingers, which requires another system (i.e., the arm) to reach distant locations. The size of a registral jump is proportional to the linear distance traveled, where more considerable distances require more time (given the same speed).

An emulative machine may evoke associations and comparisons with human players that limit or even detract from the perceived value of what the machine is or what it can do. This is particularly evident when considering the nuanced articulations and expressive techniques that human string players are capable of, such as the way a human guitarist minutely varies the angle and depth of a pick in combination with the thumb when plucking a string. Replicating this nuance mechatronically is challenging. Shortcomings in domains like this do not mean that the machine is comprehensively inferior; rather, it begs the question of what it does best. The problem may not be faulty design or manufacture; rather, it is the frames of reference used when making comparisons.

Machinic *players* can interact with sonic objects that include, but are not limited to, *off-the-shelf* instruments built for humans. The latter are common in machinic systems for good reasons: They are readily available, well-established, and thoroughly understood. They provide a guide for the development process. They allow researchers to study and better understand human capabilities. With that said, a machinic player could interface with any kind of sonic object, including those acquired through rummage or built from scratch.

The second category of *player* ⇔ *instrument* examples are defined by an actuating system built around and *attached* to an extant string instrument. While the instrument in these situations could be a novel design, more commonly it is well-established and originally intended for human players (i.e., *off-the-shelf*). Instruments featured in popular and folk musical genres are typical, such as guitars, basses, banjos, and pianos, speaking to the musical traditions and interests of builders. Examples include systems built around acoustic guitars, seen in Figure 6.1.5, and Logos Foundation's <Player Piano> I & II (Raes, 2023).

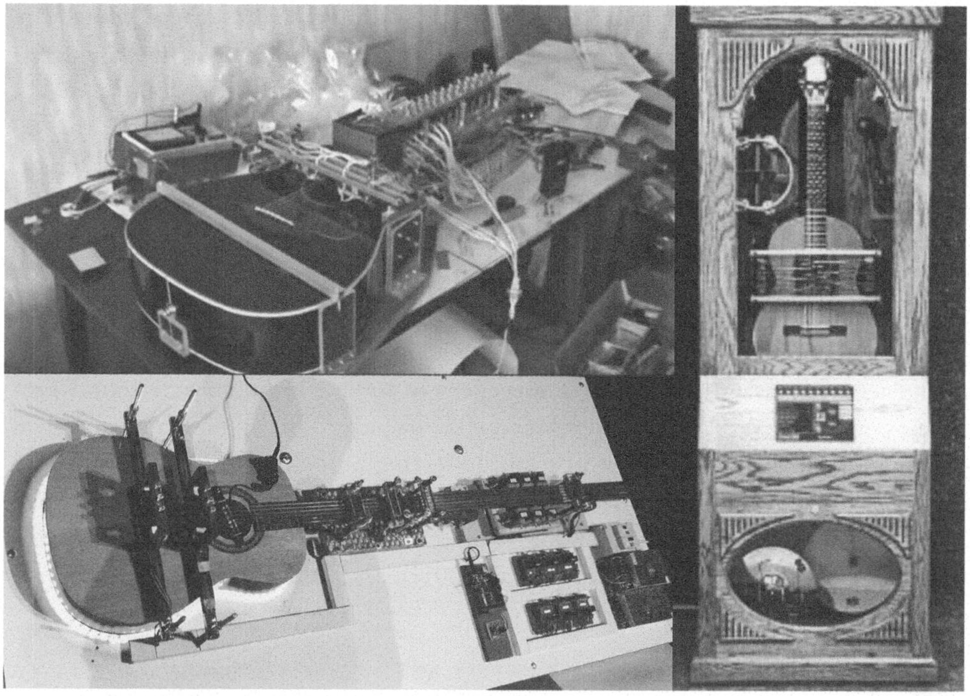

FIGURE 6.1.5. **Top left: Robot guitar by the m@P Group (Vinsmoke G., 2015), bottom left: Robot Guitar (lexademin, n.d.), right: Automated guitar from Ragtime WEST (ragtimewest.com).**

String instruments from the orchestra have also been equipped with automatic actuation systems such as the Hupfeld Phonoliszt-Violina (invented in 1907 and debuted at the World's Fair in 1910), RO-BOW by kinetic sculptor Seth Goldstein, and Haathani, which was designed to play Carnatic music (Sankaranarayanan & Weinberg, 2021) (Figure 6.1.6).

FIGURE 6.1.6. **Left: Hupfeld Phonoliszt-Violina (image © Jorge Royan / http://www.royan.com.ar, CC BY-SA 3.0, https://commons.wikimedia.org/w/index.php?curid=23403774); top right: Ro-Bow, the self-playing violin (image provided by Seth Goldstein); bottom right: Haathani (image from Sankaranarayanan & Weinberg, 2021, CC By 4.0).**

Using an existing string instrument as a basis is advantageous because it provides a reference point for design. The number, configuration, and tensioning of strings are determined, as is the structure that comprises them. Varied actuation possibilities abound, but at least some of the choices have already been made. A disadvantage of this approach is that an instrument designed for human players may not readily accommodate machinic actuation. The spacing of strings may be too tight for some actuators. The associations with familiar string instruments may or may not be desirable, depending on the builder's goals. The actuation systems of *attachments* are often conspicuous, clarifying the mechanical causality of sound production.

The third category of *player-instrument* relations includes systems that *integrate* the sonic object and the actuating system. Without a precedent to follow, sonic objects, sound production systems, and structures can be developed concurrently. This approach has resulted in string-playing machines that are sleek in form, reflecting a predilection for *function* and minimalism, and modular, driven by a desire to expand pitch range and polyphony. Examples include LEMUR's *GuitarBot* (Singer et al., 2003), *Swivel* (Murphy et al., 2013), EMMI's *PAM* (initial design 2007–2008), *MechBass* (McVay et al., 2011), and *Strumbot* (Richard Vindriis, 2015), pictured in Figure 6.1.7.

FIGURE 6.1.7. **Top left to right: GuitarBot by Lemur (by joce01_y—Flickr: LEMUR3.jpg, CC BY 2.0, https://commons.wikimedia.org/w/index.php?curid=17883121); Swivel (Murphy, McVay, et al., 2015); PAM by EMMI. Bottom from left to right: Mechbass (image from https://www.researchgate.net/publication/314126205_An_Inclusive_Musical_Mechatronics_Course/figures?lo=1), strumbot (Vindriis & Carnegie, 2016).**

These examples are *emulative* in some ways and *novel* in others. The names *GuitarBot* and *MechBass* are derived from those of traditional instruments, creating explicit connections to performance conventions and musical styles. The physical and sonic characteristics of the machines make these connections less clear. *MechBass* has four strings that are excited by guitar picks, but it bears little physical resemblance to a bass. Its sleek aluminum modules are distinct and stacked vertically. There is no fretboard or finger or hand-like assemblage. Instead, pitches are created by a sliding carriage holding a solenoid-driven clamping device. Picks are used to pluck the strings, but they are positioned around a motor-driven wheel. Musically, many of the recorded performances of the machine involve playing along with a human-produced track (Media Example 6.1.2 ▶), which strengthens associations with human practices. This is not to say that such examples are the limit of the machine's capabilities; rather, they are evidence of the interests, influences, and philosophies of the builders.

GuitarBot is less like a guitar than the name suggests. Guitar picks pluck strings, and the instrument's range is similar to that of a guitar, but that is pretty much where the similarities end. *GuitarBot* consists of four rectangular aluminum modules placed side-by-side, with the picking mechanisms located at the bottom of the structure and the strings extending vertically upward. It is angular and industrial in stark contrast to the warm, colored, wooden curves that delineate an electric guitar's body. The modules are supported by a stand, allowing them to bounce and shift while they play. The kinetic results are of a vibrating, oscillatory character, compelling in performance but unequivocally machinic. There is no fretboard: pitches are produced by a bridge that moves along a track. Unlike *MechBass*, the bridge is always in contact with the string, producing characteristic portamenti as it moves from note to note. The result is a process of pitch production whose looks and sounds differ considerably from those of a guitar. The interactions in Mari Kimura's *Guitarbotana* reveal *GuitarBot*'s distinct personality as decidedly string-like but not guitar-like. A similar appraisal can be made of the instrument's use in Pat Metheny's *Orchestrion*.

PAM is connected to a broader family of string instruments, although its ties to specific, culturally situated examples are less clear. It is a monochord with no frets, 15 tangents to produce different pitches, a damper, and a rotating pickwheel secured in a lattice-like aluminum frame. The instrument's structure is composed of a T-slotted aluminum channel and a wooden bar that is painted black. The tangents are made from clear acrylic, as are the plates that hold the circuit board and connections in plain sight. Together, the instrument looks nothing like a guitar. Sonically, two strings of different gauges are tuned to the same pitch, creating intonational idiosyncrasies. The sound of the tangents contacting the strings is neither like the sound of legato nor picked articulation. The ability to sound rapid sequences across large pitch intervals results in sequences that are not idiomatic to human guitar playing. In sum, the instrument sounds little like a guitar (for a direct comparison, listen to my composition *Rise of a City* (Media Example 6.1.3 ▶), which is a duet between the two instruments).

Despite these differences, audience members commonly refer to PAM as a guitar. Why are these connections being made? One possibility is that some listeners have not yet developed a comprehensive mental lexicon of musical instruments, so they forge connections with familiar examples. A more intriguing possibility is that connections exist in less obvious domains, such as amplification, timbral processing, or the structure and role of the monophonic lead lines. The lesson is that associations with the familiar can be evoked even when a builder attempts to make something novel.

Other examples of *integrated* machines present as kinetic musical sculptures. One example is the Gravity Harps by Andy Cavatorta (Figure 6.1.8).

FIGURE 6.1.8. **Gravity Harps by Andy Cavatorta (https://andycavatorta.com/gravityharps.html).**

Each harp is a pendulum that contains multiple strings, each tuned to a different pitch. As the pendulum swings back and forth, the inside string (relative to the supporting structure) meets a fixed pick, producing a note. The harp's cylinders rotate, allowing other strings to sound. The instrument was a collaboration for Björk's Biophilia Project, inspired by forces of nature (in this case, gravity) that could be used to create music. The instrument is striking not only because of its size but also because of its hypnotic oscillations and segmented rotations. Each harp has a weight that is visually and sonically palpable, evident in the structure's distribution of mass and volume and expressed in the rhythms of the music. It would be possible to pluck variably tuned strings with simpler mechanisms, and the oscillations of the pendulum could be mathematically modeled to produce analogous rhythms, but the experience would not be the same. The Gravity Harps are a beautiful example of attention to both visual and sonic aspects, as *form* and *function* are deeply and inseparably integrated.

The animated kinetic sculptures of *Animusic* provide similar insights into how physical form and movement affect musical expression. *Resonant Chamber* features an ensemble of string instruments whose necks protrude from a unified body constituted by an amalgam of resonators (Media Example 6.1.4 ▶). The mechanisms that excite and stop the strings are large, following complex paths of motion (relative to typical human production). The excitation mechanisms are bird-like, pecking away at the strings, while the stopping mechanisms appear as spider legs, flowing outward and snapping back into position. *Resonant Chamber* illustrates the artistic and expressive potential of attending to form, movement, and visibility beyond what is functionally required to pluck or stop a string. The instrument is boldly *novel* in concept and realization, yet its abstractions are inspired by familiar musical instruments and animal species; it is also *emulative*. The example is a video and not a physical realization (a design of this complexity would require significant resources to actualize), but it nonetheless provides a concept to aspire to.

The artistic potential of the *integrated* approach is illustrated by the music made by the machines mentioned. *GuitarBot*'s idiosyncratic character is defined by portamenti, heard on Mari Kimura's *Guitarbotana* and Pat Metheny's *Orchestrion*. *Swivel* is also capable of portamenti, yet the timbres and gestures it produces make it slide-guitar-like in ways that *GuitarBot* is not. PAM's timbral character, partly a function of its two strings, acrylic tangents, and wood-aluminum frame, evokes connections with string instruments around the world, which combine with its unequivocal machinic identity.

The fourth subcategory of *player-instrument* relations involves *cooperative* instruments, in which humans and machines interact using the same physical medium to produce music. Early examples of cooperative instruments include the pianola (pianos can be considered string instruments), whose automatically produced pitches and rhythms were shaped by human input in the realms of dynamics, tempo, and texture. More recently, cooperative systems such as *The Continuator* (Pachet, 2003) extend a Markov model that learns input patterns and produces contextually appropriate responses, enabling human–machine call-and-response possible via automated pianos such as the Disklavier (Media Example 6.1.5 ▶). Wang et al. (2024) employed an RNN to generate accompanying chord progressions based on human melody input, performed on a shared keyboard interface using an anthropomorphic robotic arm.

The MPR Lab's *Cyther* (Barton et al., 2017) is a zither that can be played simultaneously by a human and mechatronic actuators. Playing the instrument has revealed a world of musical expression previously unavailable. Working with it has clarified the similarities and differences between my performative strengths and those of the machine, inviting me to reconsider the performative roles that I play. For example, when *Cyther* plays rapid passages that coalesce into dense and unwieldy harmonic textures, I become a physical filter, muting and unmuting the 10 strings with my hands, fingers, arms, and other objects to transform the sonic stream into diverse shapes. This contrasts with a conception of a performer's role as being primarily responsible for sounding specific pitches at specific times. With the machine tending to "low" level matters involving individual notes, my attention can move to "higher" planes, such as the shaping of phrases, dynamic contours, and form. While the machine is capable of some degree of articulatory variety, it is limited; therefore, my focus is on ways to convey nuanced, emotional, and dynamic expressions. These interactions have produced music that I have neither made nor heard before, resulting from a shared *physical* medium within which human and machine interact. They signal to me great promise in future *cooperative* machines.

Others have seen the potential of *cooperative* string instruments. In some examples, the machinic part is responsible for string excitation via plucking, solenoid hammering, electromagnetic "bowing," tremolo-producing vibration, and damping (Leigh et al., 2019; Leigh & Lee, 2021; Ogata & Weinberg, 2017). In others, the machine frets the strings. *Robo-Tar* is an array of actuators that attaches to the neck of a guitar, allowing it to create chord shapes and individual notes (within a 2.5-octave range) via a foot controller while the human player picks the strings (*Robo-Tar*, n.d.). Instead of replacing the functionality of the fretting hand, *Guitar Machine* (Leigh & Maes, 2018) features mechatronic "fingers" that allow a human and the machine to fret notes simultaneously on the same guitar neck. The interaction between human and machine enhances learning and expressive possibilities, enabling simultaneous melody and accompaniment, expanded natural harmonics, and more complex phrases that would otherwise be difficult to realize given the physiology of the human hand.

Requirements

Equipped with knowledge of prior art and the *properties* and *relations* that characterize string machines, we can specify what kinds of music the machine should be able to produce, what it will look like, and how it will interact with others.

MUSICAL PERFORMANCE

Start with *the sound*. This may involve an existing instrument; if so, procure it so that timbre becomes part of the process during the design and testing of the excitation systems. When building *from scratch*, determine whether the timbre will be produced acoustically, electrically, or a combination of both. If the instrument is going to be acoustic, discover what is necessary to deliver the timbre that makes you happy. It is risky to assume that an excitation system designed for a timbrally underwhelming test rig will seamlessly integrate with a future instrument that has different dimensions and materials. Get your sound first and then design around it. This is also a good idea for an electric instrument, though its timbral character also depends on the pickups, processing units, amplifiers, and speakers used. Establishing the instrument's sound will help you later dial the excitation mechanisms precisely.

How can the instrument's sound change? Timbral transformations can be effected by varying excitation location, angle, velocity, force, objects, and signal processing. The modes of excitation dictate electromechanical requirements in terms of the number of actuators, DOFs, size, and materials. It will also affect the system's electrical design. Perhaps audio processing happens in the machine's circuitry; perhaps it is sent out to external processors and amplifiers. While it is impossible to imagine all timbral possibilities, try to establish a range that is realizable.

In some cases, the noise of actuators contributes to the attack of a note in unique and pleasing ways. In other instances, noise occludes musical outputs in ways that range from mildly distracting to turn-it-off annoying. The question of how much noise is permissible in machinic string instruments partly depends on whether the projection will be acoustic or electric. If acoustic, actuator noise will likely be an issue, given the relatively constrained dynamic range of most string instruments. If the output of the string instrument is going to be amplified, mechanical noise could be less of a problem. In either case, steps can be taken to minimize actuator noise, which will be covered in Part 2 of this chapter. Electric instruments have their own noise problems, with potential sources including pickup design, grounding, cables, and other nearby sources of electromagnetic energy. Establish a permissible noise floor (in dB SPL) for the expected performance contexts (e.g., recording studio, concert hall).

Musical time requirements for string instruments are subject to the same guidelines as other instruments in some ways and are unique in others. The ranges in the general categories of IOI, temporal accuracy, and latency, as discussed in Chapter 4, are applicable here: <100 ms for IOI and <50 ms for temporal accuracy and latency. Specific to strings, pitch-producing, excitation, and damping systems have different capabilities. Solenoid tangents positioned close to a string will be able to stop and release a string very quickly (likely near 50 ms IOI, depending on the system), while a stepper-motor-based pickwheel system or a servo-based damper may not achieve these rates. As a result, *legato* passages may be possible at fast speeds, while articulations involving the picker and / or damper will be slower. Determine what kinds of articulations are important to you and use those to help establish temporal requirements.

The pitch requirements of a string instrument are consequential for its physical structure and electromechanical design. As with other instruments, you could start with pitch range or pitch set first. A one-octave range is usually not enough; two octaves is far better. Regarding the pitch set, the chromatic scale is a popular choice given its flexibility, but there are hundreds of other possibilities. Think about the musical contexts in which your instrument will be used and the affordances and constraints that the pitch choices you make will present. If it can only play the *C Major* scale and a collaborator wants to play in *A Major*, what will you do? What if you want to play non- or micro-tonal music when the instrument is fixed to a scale associated with a particular tonal system (or vice versa)? Retuning is a possibility, but one that can become cumbersome, particularly if the instrument takes time to hold tuning changes. I learned this lesson with *Cyther v.2*, which has 10 strings that span over two octaves. When a fellow composer or collaborator wanted to change the tuning, it required time and effort to accommodate their harmonic vision. One solution to this is to make the tuning process automatic, which can occur before a performance (e.g., the *Gibson Robot Guitar*) or during one (e.g., *Cyther v.3*). The dynamic tuning option allows for pitch modulation, including portamento and glissando.

> Experiment with pitch sets to see which ones inspire you creatively. At the same time, think about tuning not only in terms of what you need now, but also what others may need later. Establish what aspects of tuning are going to be fixed (e.g., frets) and what aspects will be adjustable (e.g., string tension). For the elements that are adjustable, do they need to be set before a performance, or can they be changed while the instrument is being used?

A subset of pitches is not necessarily a disadvantage. The constraints of *Cyther*'s 10 pitches have been creatively inspiring. It precludes the combinatoric possibilities of two octaves of the chromatic scale, but fewer options make the consideration of each more thorough. I combine subsets of different scales, allowing me to allude to the scales through intervallic sequences or combinations without presenting or committing to them in their entirety. This puts primacy on the pitch *intervals* that are possible. It also compels one to look beyond pitch as the only source of musical interest, instead exploring rhythm, alternative means of articulation, and preparing the instrument with different materials.

What kinds of *modulations* within the established pitch set and range will be possible? Continuous pitch changes, such as portamento, require dynamically changing either the scale or tension of the string. In both cases, producing larger pitch intervals takes longer than producing smaller ones. Discrete pitch movements are often achieved by independent actuators positioned at different locations, enabling rapid production of large pitch intervals. The downside is that fixed configurations do not allow continuous movements. These choices are not right or wrong, but they impact the instrument's structural and electromechanical design, so ponder the possibilities and set your requirements at the outset.

Pitch accuracy is a requirement common to all instruments; we can use the psychologically based guidelines given in Chapter 4 of ±12 cents here. In the case of string machines, the pitch production method will impact accuracy. If your pitch accuracy requirements are tighter or looser than the standards given above, this is the time to establish an acceptable range.

As with other instruments, the dynamic range required by a string machine depends on the musical contexts in which it is used. Will the instrument appear in acoustic chamber settings or rock venues? Will it blend with other strings or need to compete with louder instruments such as percussion and brass to be heard? Answering these questions reintroduces the acoustic / electric divide. As with timbre, the dynamic capabilities of an electric instrument will depend on the equipment that processes the output signal, the most important of which is the amplifier and speaker. For acoustic instruments, a 30 dB dynamic range is standard for violins and violas, providing an ambitious goal to work toward (you will likely want to start with a more modest range). Keep in mind that dynamic range is a product of both instrument and player, so setting and working toward this requirement is necessary whether using an instrument *off-the-shelf* or built *from scratch*.

What articulations is your string machine going to be capable of? Distinctions between *legato* and *staccato* passages are essential, as is the ability to produce different kinds of *accents*. Setting these requirements will determine the capabilities and relationships between the excitation, pitch-producing, and damping systems.

VISIBILITY

Some *visibility* questions are answered by thinking through the *properties* and *relations* described in Chapter 4. If you seek to build an anthropomorphic robot that plays a guitar, you already have a model of form and movement. Possibilities expand and become more amorphous as a design aspires toward *novelty*.

Visibility and musical performance are often at odds in string machines. The movements required to excite, stop, and damp a string are small, particularly in comparison to those needed to play other instruments, such as percussion. When speed is desired, the distance of movements becomes even smaller, which is why many designs feature actuators fixed in position centimeters away from a string. All of this is great for improving the musical performance of the machine, but it often results in a system whose motion is only perceptible when viewed at close range. Larger and more complex physical gestures, such as the kind seen in Animusic's *Resonant Chamber*, are more challenging to design, are more costly, and will usually result in diminished performance abilities. The question then becomes, how important is it that the movements of the machine are seen? Are they necessary for communication with other musicians? Do they meaningfully enhance or change the experience of the audience? Do they result in new kinds of sono-kinetic gestures? You will need to make compromises, which may result in other requirements. For example, making the picking mechanism on a string instrument more visible may result in slower performance, which could be addressed by introducing a second picking mechanism to work in conjunction with the first.

Visibility requirements of a string instrument include the scale of components and how they are distributed in space. String instruments are often compact because they are designed according to the span of human fingers. Mechanically actuated instruments have no such obligation: strings could stretch to the horizon and could be separated by multiple inches or even feet. They do not need to be oriented in the same direction or on the same plane. The possibilities of form and movement multiply exponentially when the constraints of a human player are relaxed.

INTERACTION AND AUTONOMY

Many aspects of interaction and autonomy requirements, as described in Chapter 4, are general and not specific to strings. These include who the machine will interact with; whether the system's output is predetermined, results from human actions, or is algorithmically generated; how it will sense and interpret external information; and how it will respond to collaborators.

Some aspects of *external* interaction and autonomy are specific to string machines. The first involves visual communication. If string playing is small in scale, other elements of the system (such as a robotic arm, images or text on a screen, or a set of lights) are required to communicate intention, desires, interpretations, timing, and cues to other collaborators. Consider the role that string instruments typically play in musical contexts. If the machine is to be a rock-like Guitar Hero, then its relationships to others might be different than if it is part of a group that creates textural background pads. Aspects of these relations include who leads and who follows, who is featured and when (e.g., solo), and who is responsible for establishing musical qualities such as tempo, meter, or key.

Within the *internal* systems of a string machine, interaction and autonomy require sensing and feedback. This involves transducing the string's movements into an analogous electric signal, which is converted to the digital domain as a string of numbers. This information is processed to determine fundamental frequencies, spectral distributions, temporal synchronization, or rhythmic groupings. Sensing may also involve positional information, such as that of a tuning machine (string tension), a moving bridge (vibrating length), or a picking mechanism. These subsystems interact with each other in sensing and communicating their states, which are interpreted by algorithms on the microcontroller (or connected computer). These interactions determine the ways in which the machine can function autonomously.

When establishing interaction and autonomy requirements, start at a higher level that concerns musical aspects. Figuring these out will determine which lower-level requirements are necessary.

References

Barton, S., Prihar, E., & Carvalho, P. (2017). Cyther: A human-playable, self-tuning robotic zither. *Proceedings of the International Conference on New Interfaces for Musical Expression*, 319–324. https://doi.org/10.5281/zenodo.1176266

Leigh, S., Jain, A., & Maes, P. (2019). Exploring human-machine synergy and interaction on a robotic instrument. In M. Queiroz & A. X. Sedó (Eds.), *Proceedings of the International Conference on New Interfaces for Musical Expression* (pp. 437–442). UFRGS. https://doi.org/10.5281/zenodo.3673027

Leigh, S., & Lee, J. (Jonna). (2021, June). A study on learning advanced skills on co-playable robotic instruments. *Proceedings of the International Conference on New Interfaces for Musical Expression*. https://doi.org/10.21428/92fbeb44.002be215

Leigh, S., & Maes, P. (2018). Guitar machine: Robotic fretting augmentation for hybrid human-machine guitar play. In T. M. Luke Dahl Douglas Bowman (Ed.), *Proceedings of the International Conference on New Interfaces for Musical Expression* (pp. 403–408). Virginia Tech. https://doi.org/10.5281/zenodo.1302643

lexademin (Director). (n.d.). *Russian robot guitar. Tico-Tico* [Video recording]. Retrieved July 20, 2021, from https://www.youtube.com/watch?v=B40XymKPicA&list=RDB40XymKPicA&index=2

McVay, J., Carnegie, D. A., Murphy, J. W., & Kapur, A. (2011). Mechbass: A systems overview of a new four-stringed robotic bass guitar. *Proceedings of the 19th Electronics New Zealand Conference (ENZCon), Dunedin, New Zealand*, 145–150.

Moebs, W., Ling, S. J., & Sanny, J. (2016). 15.1 Simple harmonic motion. In *University Physics Volume 1*. https://openstax.org/books/university-physics-volume-1/pages/15-1-simple-harmonic-motion

Murphy, J., McVay, J., Mathews, P., Carnegie, D. A., & Kapur, A. (2015). Expressive robotic guitars: Developments in musical robotics for chordophones. *Computer Music Journal*, 39(1), 59–73. https://doi.org/10.1162/COMJ_a_00285

Murphy, J., Kapur, A., & Carnegie, D. (2013). Swivel: Analysis and system overview of a new robotic guitar. *Proceedings of the 2013 International Computer Music Conference*. International Computer Music Conference.

Ogata, T., & Weinberg, G. (2017). Robotically augmented electric guitar for shared control. *Proceedings of the International Conference on New Interfaces for Musical Expression*, 487–488. https://doi.org/10.5281/zenodo.1176326

Pachet, F. (2003). The continuator: Musical interaction with style. *Journal of New Music Research, 32*(3), 333–341. https://doi.org/10.1076/jnmr.32.3.333.16861

Raes, G.-W. (2023). *Player Piano: Automated piano Vorsetzers by Dr. Godfried-Willem Raes.* Logosfoundation.Org. https://www.logosfoundation.org/instrum_gwr/playerpiano.html

Richard Vindriis (Director). (2015, November 18). *StrumBot presentation* [Video recording]. https://www.youtube.com/watch?v=lbwNu_Vg5ic

Robo-Tar. (n.d.). Robo-Tar. Retrieved March 6, 2022, from http://www.robo-tar.com/

Sankaranarayanan, R., & Weinberg, G. (2021, April 29). Design of Hathaani—A robotic violinist for Carnatic music. *International Conference on New Interfaces for Musical Expression.* NIME 2021. https://doi.org/10.21428/92fbeb44.0ad83109

Seth Goldstein (Director). (n.d.). *RO-BOW—A kinetic sculpture that plays a violin by Seth Goldstein* [Video recording]. Retrieved July 21, 2021, from https://www.youtube.com/watch?v=EPTUM2_bxnQ&t=1s

Singer, E., Larke, K., & Bianciardi, D. (2003). LEMUR GuitarBot: MIDI robotic string instrument. *Proceedings of the International Conference on New Interfaces for Musical Expression*, 188–191.

Vindriis, R., & Carnegie, D. (2016). StrumBot—An overview of a strumming guitar robot. *Proceedings of the International Conference on New Interfaces for Musical Expression*, 146–151. https://doi.org/10.5281/zenodo.1176135

Vinsmoke G. (Director). (2015). *Robot Guitar—Arduino—Aria sulla quarta corda—Italian Makers* [Video recording]. https://www.youtube.com/watch?v=eEeQKCgACzk&list=RDB40XymKPicA&index=3

Wang, H., Zhang, X., & Iida, F. (2024). Human-robot cooperative piano playing with learning-based real-time music accompaniment. *IEEE Transactions on Robotics*. https://ieeexplore.ieee.org/abstract/document/10726887/?casa_token=4iVKOvgBJjUAAAAA:G6JfrBKr6CXAGCLGEIuuVZhbAlTsEwPIwA-qxk6GRnJgdLti7mpHJqJdXZsXPGqd8WLA_QtqTA

CHAPTER 6

STRINGS

Part 2: Design

Sonic Objects

The sonic object of a string-playing machine consists of a string and a structure that allows it to be tensioned. These sonic objects can be acquired *off-the-shelf*, via *rummage*, or built *from scratch*. The advantages and disadvantages of these options depend on the object's design, fabrication process, sonic properties, and the associations it evokes. Whether a feature is an advantage or not depends on the goals of the builder, composer, or performer.

ACQUIRING

Acquiring any instrument, whether *off-the-shelf* or through *rummage*, has the general advantages expressed in Chapter 4: The Process of Making a Musical Machine. For string instruments, certain aspects should be settled, including fret spacing and alignment, a structure that can withstand the significant tension exerted by multiple strings, setting the string action, and intonation. For electric instruments, convenient predeterminations include pickup selection and placement, internal wiring, and interface elements (e.g., pickup switches, knobs, and output jacks). The form and configuration of strings, frets, pickups, necks, and bodies to which the actuation system must be designed has been codified, meaning fewer choices and potentially less design time.

The choice to acquire a string instrument depends on whether it is acoustic or electric. Acoustic string instruments are best acquired as they are challenging to make. Electric instruments tell a different story. Sonic properties are still a function of the materials used, the physical design and manufacture of the instrument, the selection of pickups, and so on; therefore, many advantages of acquiring *off-the-shelf* instruments apply. With that said, the body of an electric instrument does not play the same role in producing the resultant sound as in an acoustic instrument, given the effects of amplification and signal processing. Electric string instrument bodies of various shapes, sizes, and materials can achieve compelling timbral results, which makes the *from scratch* case compelling.

As with *players*, existing instruments evoke *associations* with musical conventions that should be considered when choosing to acquire. In the world of strings, the violin, viola, and cello are connected to string quartets, orchestras, and the compositions of the European Art Music tradition. Steel-string acoustic guitars accompany the musings of folk singer-songwriters. Electric guitars established the identity of rock with blaring open chords and soaring virtuosic solos. Whether aware of it or not, using an existing instrument in the design of a musical machine engages with rich histories and cultural traditions. These associations can influence how music is composed, performed, and heard by the people involved in a musical interaction. Some builders make this kind of engagement a central part of their work. Ajay Kapur draws upon the traditions of North Indian classical music in the design of hyperinstruments such as the *ESitar* (Kapur et al., 2007). *Hathaani* was designed to perform *gamakas* (a pitch ornamentation), fundamental to Carnatic musical performance practice (Sankaranarayanan & Weinberg, 2021). Z-Machines, One Love Machine Band, and Compressorhead use human-playable guitars and basses with specific cultural associations to form strong bonds with rock music traditions.

Associations can be deeply rooted in our cultural consciousness, which can be cumbersome for those seeking new stylistic territory. Combinations of instruments from different worlds, such as the electric guitar and the orchestra, lug the baggage of the genres in which they came to prominence (this issue applies particularly to the electric guitar, whose identity was forged in rock music). Part of the issue is what the instruments look like, part is what they sound like, and part is how they are played. If you use an established instrument in a new musical machine, you will have to negotiate associations, which can be deeply rooted. With that said, known objects can be recontextualized in interesting and successful ways. Doing so is how the identities of instruments develop, though the rate of such change can be glacial.

FROM SCRATCH

Sonic mavericks may be best served by making a string instrument *from scratch*. The *from scratch* case is less intimidating if you view string instruments as variations on the same basic idea: tension a string and couple it to a larger structure so that the former drives the latter when excited. Experimentation with this idea, particularly in combination with mechatronic actuation, may produce novel instruments. For those who seek to brave this path, ideas and lessons are provided on the companion website ▶.

IDIOMATIC DESIGN

Whether acquiring a sonic object or making it from scratch, following principles of *idiomatic design* when imagining the playing system is imperative. What are the object's affordances and constraints? What techniques have been developed to play it? Using *off-the-shelf* electric guitars opens the door to a world of amplifiers and signal processors that have been specifically designed for them. For acoustic string instruments, techniques idiomatic to human players, such as *sul ponticello* or *pizzicato*, are part of a rich lexicon developed over hundreds of years. What techniques are *idiomatic* to machinic actuation?

The answer to this question partly depends on how *emulative* or *novel* the design purports to be. One approach is to use a traditional instrument and emulate the physical structure and motions of the human who would play it. These anthropomorphic excitation systems can emulate human gestures (Shibuya, 2011), or their machinic capabilities can explore sounds and gestures humans cannot produce. Another option is to use the sonic products of human techniques as goals to achieve through alternative physical means. For example, bowing can be emulated by applying a rotating disc to a string. One could forgo emulation entirely, finding inspiration by moving away from established models in sonic and physical realms to find novel sounds and techniques.

Idiomatic design has challenges when working across modalities (e.g., making machines play instruments designed for humans). For example, configuring multiple actuators in a space designed for human fingers can be challenging, particularly when trying to make the actuators' motions visible. Strings on human instruments are sometimes configured in an arc (e.g., viola), allowing players to isolate a string and perform techniques such as arpeggios while bowing. A machine tasked with playing the same human instrument might feature a fixed bow around which the body of the instrument rotates (e.g., the Hupfeld Phonoliszt-Violina; see Crandall, 2013), but this makes it harder to produce passages that jump between non-adjacent strings. Solutions may involve greater DOFs, but these are more complex and costly. Alternatively, separating the strings and actuating them independently (a configuration not idiomatic to humans) may be a more effective way to mechanically realize familiar and novel musical techniques. This approach has been popular, as seen in the multiple one-string instruments of the Phonoliszt Violina (c. 1908) and the modules of *GuitarBot* and *MechBass* a hundred years later. The moral of this story is that designing a machine around a string instrument that is idiomatic to human hands can be like trying to fit a square peg into a round hole. It is imperative to consider what is idiomatic *to the machinic player*, which is simpler and less able than its human counterparts.

Excitation

With an instrument in hand, producing sound first requires a method for exciting the string. Strings can be excited directly through plucking, bowing, hammering, and scraping / sliding or indirectly through air currents, sympathetic vibration, and electromagnetic manipulation. A more extensive palette of excitation methods expands timbral and articulatory possibilities. The following sections describe these methods and the mechanisms and motors that can be used to achieve them. For those requiring a primer in mechanisms and actuators (or those that would benefit from a refresher), the Companion Website ▶ contains sections devoted to *The Mechanical*, covering physical principles, mechanisms, motion conversions, connections, and methods, and *Actuators, Circuits, and Code*, providing technical descriptions and projects to help you turn information into physical action. These pages provide a reference as you work through the design process.

PLUCKING

Plucking is a popular method of exciting a string (e.g., guitars). Force and velocity requirements must be specified when designing a plucking mechanism. As a starting point, studies have shown that plucking a string requires ~5–7 N of force (Li et al., 2019). Common plucking mechanisms include the rotating pickwheel, bidirectional servo, linear actuator, and harpsichord jack.

Rotational Picking

A rotating pickwheel consists of a motor coupled to a wheel around which picks (or plectrums) are oriented. Examples that feature a rotating pickwheel include LEMUR's *GuitarBot*, EMMI's PAM (Figure 6.2.1), *MechBass*, the Logos Foundation's *<Synchrochord>* (Raes, 2012), and *BassBot* (Richard Vindriis, 2011).

FIGURE 6.2.1. **Pickwheel mechanism of PAM by EMMI / MPR LAB.**

The rate of rhythmic intervals (the time between note onsets) is a function of the speed of the wheel and the number of picks it contains:

$$rhythmic\ \text{int}erval = \frac{rotational\ speed\ of\ wheel}{\#\ of\ \ picks}$$

For example, if the wheel completes a full rotation in one second (1,000 ms) and the pickwheel has eight evenly spaced picks, the mechanism can produce notes at 125-ms intervals. The primary factors in designing a pickwheel are rotational speed, torque, and the number of picks. Balancing these factors enables the motor to run at maximum efficiency while simultaneously producing the desired note rate and quality. Gear reduction (e.g., via a gearbox) is often required to produce the necessary torque.

The wheel design requires specifying the position of the picks and how they will be secured. Picks are usually evenly spaced to ensure consistent rhythmic intervals for a constant motor speed. The angle of the picks is crucial because it affects the torque required to move the pick past the string, which in turn influences the timbre and dynamic level of the produced sound. Picks can be fixed in place, for example, by using screws that fasten directly to the wheel (e.g., *GuitarBot*). Doing so requires measurment, calculation, experiment, and experience. It also requires that the other aspects of the system remain stable, such as the string's height. An adjustable system, such as clamps that secure the picks in place (PAM, *MechBass*), allows you to address performance (e.g., inability to pluck the string) or sonic issues (e.g., dynamic levels that are too soft) and to swap picks if they wear out or have an undesirable thickness. Disadvantages of adjustable systems include picks shifting during performance and the challenge of identically orienting the picks. McVay et al. (2011) illustrated such variation in the development of *MechBass* (Figure 6.2.2).

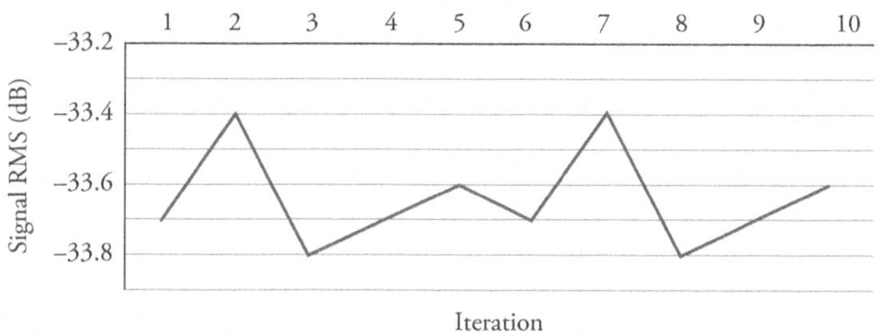

FIGURE 6.2.2. **Dynamic levels produced by two complete revolutions of clamping picking mechanism used in MechBass (McVay et al., 2011).**

Different types of motors can propel pickwheels. One option is a stepper motor, exemplified in EMMI / MPR Lab's *PAM* and *MechBass*. Steppers can be controlled precisely by indicating the number of steps the motor will travel. Figuring out the number of steps per pick is then a matter of dividing the number of steps per rotation by the number of (evenly spaced) picks:

$$steps\ per\ pick = \frac{steps\ per\ rotation}{\#\,of\ \ picks}$$

For example, if the stepper motor takes 400 steps per rotation and has eight picks, it would require 50 steps to pick the string. The advantages of stepper motors for picking mechanisms include precise control, high torque at low speeds, low cost, uniform mounting options, and simplicity (no external feedback is required). A disadvantage of steppers is that the number of steps instructed may not match the number of steps moved due to interactions between the pick, string, and motor, and a lack of feedback. Further, steppers are relatively slow compared to DC and servo motors, have less torque at high speeds than low speeds (see Figure 6.2.3), constantly draw current, and are less efficient.

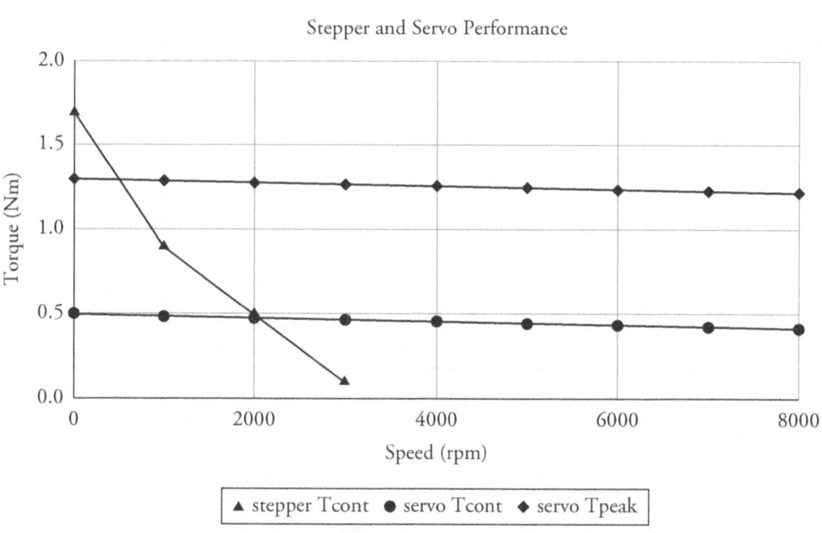

FIGURE 6.2.3. **Speed–torque curve for stepper and servomotors where Tpeak is peak torque and Tcont is continuous torque (data from Gill, 2021).**

Servomotors can be incorporated into pickwheel systems. *GuitarBot* (Singer et al., 2003) features a block with four picks connected to a shaft, driven by a continuous rotation DC servomotor via a belt and pulley (Figure 6.2.4). An encoder is mounted on the shaft, providing feedback on its position. The closed-loop system allows for precise positioning of the picks, and varying the motor's speed enables a range of rhythmic durations.

FIGURE 6.2.4. **GuitarBot pickwheel (image from Singer et al., 2003).**

Servomotors are capable of higher speeds than steppers, and their output torque is more consistent over their speed range (see Figure 6.2.3). Additionally, they provide momentary peak torque to compensate for load variations. Their disadvantages are that they typically cost more and are more complex to implement due to their closed-loop control system.

A servomotor is usually a better choice for a picking mechanism, given its higher speeds, more consistent torque, and ability to adjust to variable loads (the load varies depending on if the pick is in contact with the string). A stepper-based mechanism is a suitable choice for those seeking a lower-cost, simpler-to-implement motor that is functional and reliable (see Table 6.2.1).

TABLE 6.2.1. **Advantages of stepper and servo motors.**

Stepper	Servo
Precise control	Higher speeds
Simple to implement	More consistent torque
Low cost	Variable loads

Oscillating Picking

Instead of continuously rotating a pickwheel, an alternative approach is to use oscillating motion to move a pick back and forth along an arc whose path is bisected by a string. This method is implemented in the picking mechanisms of Z-machines's *Mach* (Figure 6.2.5), Trimpin's guitar-playing systems, and *Swivel*. *Swivel* uses a single RC servo whose driveshaft is extended to secure a pick. A picking event produces an arc of motion whose length is proportional to the thickness of the pick: thicker picks deform less and thus can function with shorter arcs (Murphy et al., 2015).

FIGURE 6.2.5. **Left: oscillating picking mechanism design. Right: picking mechanism of Z-Machines's Mach (image from Daito Manabe, 2013).**

A benefit of this approach is that it maximizes efficiency by reducing extraneous movement: the pick moves only the distance needed to excite the string. A downside of this economy is that it results in less visible movements.

A variety of motors can drive an oscillating picking mechanism. A motor capable of continuous rotation with position sensing (e.g., an encoder) is a possibility. In this case, the inertia of the pick needs to be overcome when at rest, accelerated to a requisite velocity while applying sufficient torque to sound the string, and then slowed down and stopped at a position at the same distance from the string as the starting point. This requires implementing and tuning a motion control algorithm such as PID, which can be complex. A mechanism, such as a treadle or a cam, simplifies the process, producing oscillating motion from continuous motor rotation (see Media Example 6.2.1 ▶). A stepper motor is another possibility, which can be programmed to move a specified number of steps in one direction and then move the same number of steps in the opposite direction on the next command. Servomotors are another option, featuring closed-loop feedback for precise control of position and speed. Even cheaper servos with only position sensing could provide reasonable results.

Reciprocating Picking

Picking machines can also utilize reciprocating motion. The design typically features two solenoids that face each other. The armatures of the solenoids are joined by a coupler that holds a pick. When solenoid A is activated, the armature retracts, pulling the pick across the string. Activating solenoid B pulls the armature in the opposite direction, picking the string again. This design is featured in Trimpin's *JackBox*, *KrautKontrol*, and the first version of EMMI's PAM (Figure 6.2.6).

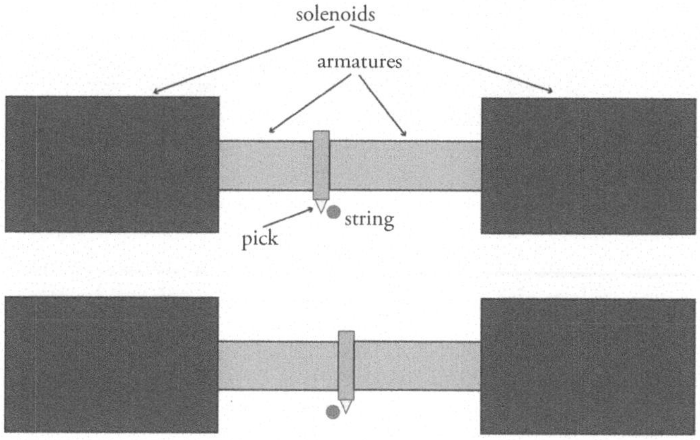

FIGURE 6.2.6. **Reciprocating picking mechanism.**

Reciprocal picking can also move along a vertical dimension. A harpsichord jack is an ingenious device that consists of a jack, tongue, plectrum, spring, and damper. When a key is pressed on a harpsichord, the jack moves up so that the plectrum plucks the string. When the key is released, the jack travels down and the tongue (to which the plectrum is attached) rotates so that the plectrum slides past the string without re-activating it. When the jack stops, the damper attenuates the string's vibrations, and an internal spring rotates the tongue and plectrum back to their initial positions (see Figure 6.2.7).

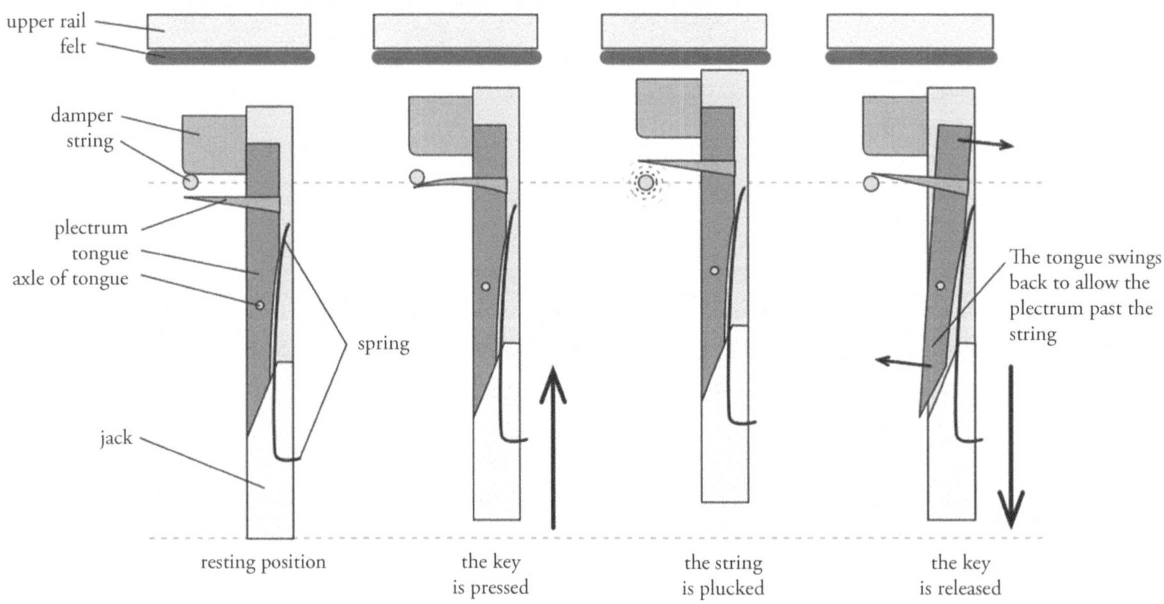

FIGURE 6.2.7. **The action of a harpsichord jack (image by Njohan, English captions and other edits by Jeff Dahl—Own work, CC BY-SA 4.0, https://commons.wikimedia.org/w/index.php?curid=3140667).**

Reciprocating motion can be combined with rotary motion to create more complex plucking paths. In the low-noise guitar robot (Li et al., 2019), a solenoid retracts, drawing a plectrum across a string in a linear motion. When the "finger" is fully retracted, a pair of magnets is aligned, providing an upward force that routes the "finger" along a semicircular path up and over the string (without contacting it) and back to its original position (Figure 6.2.8).

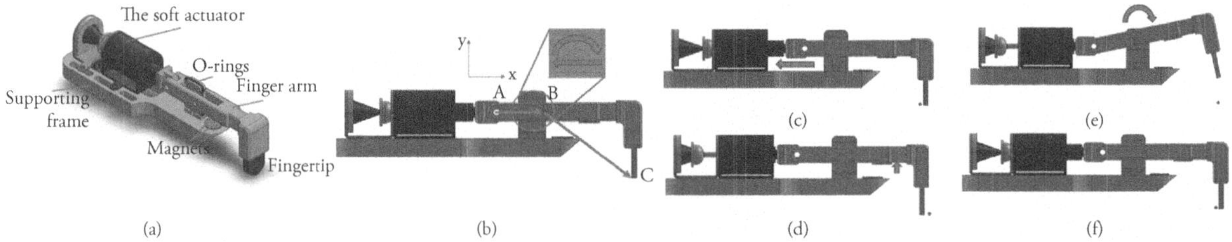

FIGURE 6.2.8. **Reciprocating and rotary motion in a picking mechanism (image from Li et al., 2019).**

Because a plucking action only occurs when the solenoid retracts, varying the voltage to the solenoid can increase dynamic range. Turning the actuator off and using magnets on the return path reduces power consumption (and thus heat). This design is capable of rapid actuation of up to 12 notes per second (nps) at 30 and 28 V (Li et al., 2019).

Strumming

Picking gestures need not be restricted to one string. Strumming and sweep picking are standard in the lexicon of string players, which can be mechanically reproduced either through separate picking mechanisms on individual strings or a mechanism that sweeps a plectrum over groups of strings. Baginsky's *Aglaopheme* uses a linear motor to sweep a plectrum across multiple strings. This picking assembly is coupled to a solenoid that pulls the plectrum away from particular strings as desired (N. A. Baginsky, n.d.). *StrumBot* features a parallel selective compliant articulated robotic arm (pSCARA) driven by two servomotors (Figure 6.2.9). The design allows the picks to be positioned at different points along a string to modulate timbre (the sound tends to be thinner and sharper towards the fixed end and warmer and fuller away from it).

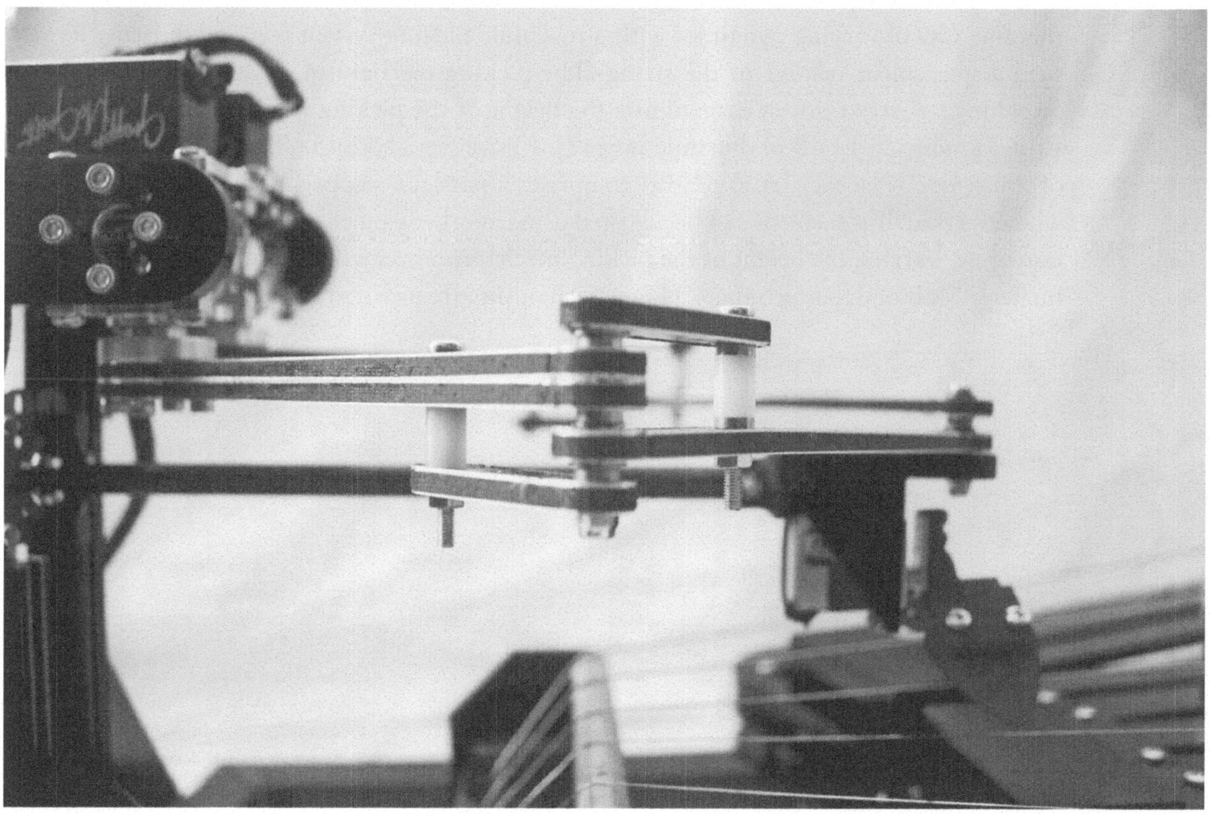

FIGURE 6.2.9. **The strumming arm of strumbot (image from R. Vindriis & Carnegie, 2016).**

Two picks are connected to a servomotor on the arm's end effector. Rotating the servomotor changes the angle of the picks, allowing them to engage and disengage with the strings as the arm moves laterally. The angle of the picks is controllable, allowing dynamic variation. *Strumuké* is a stepper motor-driven electromechanical system that performs strumming actions on a ukulele (Moy et al., 2021). Its pick holder is spring-loaded, providing different effects depending on the direction of the strumming. In each of these examples, varying the motor's speed and position produces individual notes, connected arpeggios, full chords, and everything in between.

Dynamic and Timbral Variation

Actuation force and velocity are essential in determining a plucking system's dynamic range (the range between the loudest and quietest sound possible). These factors can be manipulated by varying the voltage supplied to an actuator. Li et al. (2019) were able to produce up to ~10 dB of dynamic range by changing the input voltage (26 V–30 V) to a solenoid picker (as compared to ~20 dB produced by a human player).

Factors other than force and velocity are also in play, including the vertical displacement of the plectrum relative to the string. Most picking mechanisms are mounted "above" the strings (i.e., perpendicular to the length of the string), which is logical, given there is usually little space between the strings and the instrument's body. The vertical position of a plectrum relative to a string influences dynamics: the smaller tip of the pick results in softer sounds, while the thicker part of the pick (toward its middle) generates louder outputs. It follows that one way of varying dynamics with a machinic picking system is by modulating its vertical displacement relative to the string. The picking mechanism of *MechBass* is mounted on a hinge. A servo-driven cam adjusts the height of the picking mechanism relative to the string, producing 12 dB of dynamic range (J. Murphy et al., 2015). The picking mechanism of *Protochord* (Placencia et al., 2019) comprises a pancake stepper motor-driven pickwheel, whose vertical displacement can be adjusted using another pancake stepper motor coupled to a leadscrew. Varying the height of the picking mechanism over a 2 mm range produces approximately 17 dB of dynamic range. These mechanisms are pictured in Figure 6.2.10.

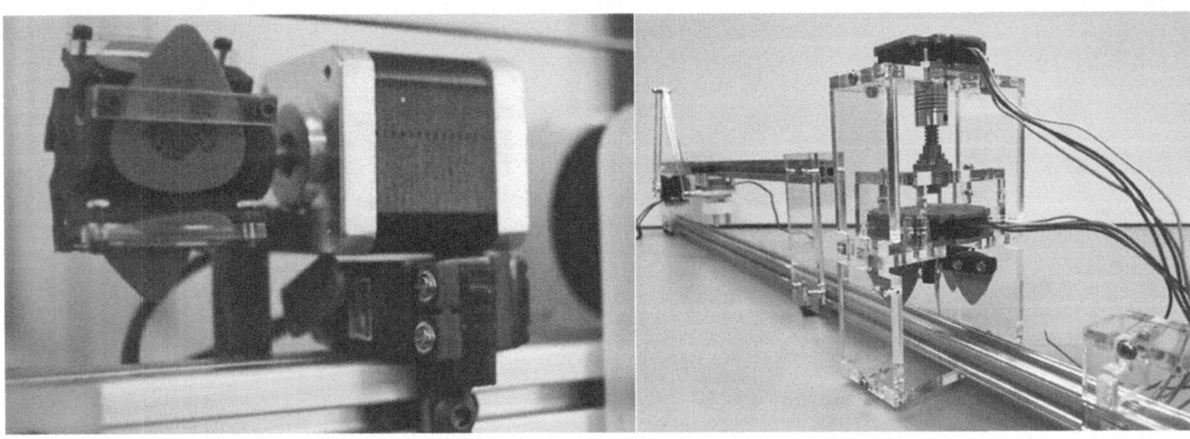

FIGURE 6.2.10. **Height-adjustable picking mechanisms. Left: MechBass (McVay et al., 2011); right: Protochord (Placencia et al., 2019).**

Height-adjustable picking mechanisms also help compensate for string displacement that results from stopping and fretting. The distance of this displacement is relatively small to the naked eye, but it is often significant in picking, where millimeters impact sonic results. In the worst cases, displacement is so substantial that the picker misses the string entirely. In more modest cases, variations in dynamics, articulations, phrasing and rhythm occur. The degree of displacement depends on the instrument's design. A neck (or equivalent) with frets can mitigate this issue, while tall bridges can be more problematic. Displacement also depends on where the string is stopped (closer to the fixed ends produce smaller displacements) and the amount of force applied to the string, which is a function of the specifications and placement of the string-stopping actuators. The height of a picking mechanism can be automatically adjusted to compensate for vertical displacements that result from string stopping, providing more uniform excitation.

More DOFs, materials, and control schemes enhance dynamic, timbral, and articulatory effects. A plectrum's angle relative to the string affects the sound produced. By changing the pick angle, *StrumBot* can produce a dynamic range of 20 dB (R. Vindriis & Carnegie, 2016). Swapping different kinds of picks (e.g., thickness, material) produces sonic variation. Trimpin's *If VI were IX*, for example, alternates between a plastic pick and a plectrum made from silicone. The firmware architecture of the chordophone *Azure Talos* enables tremolo, muted picking, and ghost notes (Tait, 2023). Journeying further down this road reveals complex yet exciting challenges, which are fertile ground for future exploration.

Noise

Most actuators produce noise, such as from gear teeth meshing or solenoid armatures contacting their metal cases, which can affect an instrument's dynamic range. At its best, noise can help articulate notes and define the instrument's idiosyncratic character. At its worst, noise masks musical sounds to the point that the latter are barely distinguishable. Methods to attenuate noise include the *Kapur Fingers*, which modify a push solenoid with foam and grommets to make it quieter (Kapur et al., 2007). Another approach (Li et al., 2019) draws inspiration from soft robotics, utilizing a custom-designed elastomeric cone tuned to balance the non-linear forces generated by an electromagnetic solenoid. This design produced significantly less noise than other common actuators, including a stepper motor, pneumatic actuator, solenoid, and servomotor (see Figure 6.2.11).

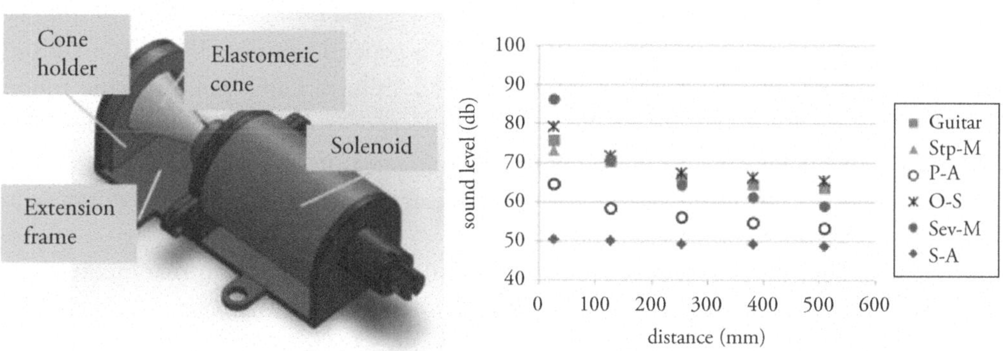

FIGURE 6.2.11. **Left: Elastomeric actuator; right: noise tests where step-m is a stepper motor, p-A is a pneumatic actuator, o-S is an ordinary solenoid, Sev-M is a servomotor, and s-A is the silent actuator. The *Y*-axis represents the sound level (dB), and the *X*-axis represents the distance between the sound meter and the device being tested (images from Li et al., 2019).**

The developers of *StrumBot* were sensitive to noise problems, integrating motors and parts (such as rubber O-rings on wheels) to reduce noise emissions below a target of 60 dBA at 1 m. Tests showed noise levels of 56 dBA at 1 m for movements involved in a typical song for *StrumBot*, compared to 60 dBA for *Swivel* and 69 dBA for *MechBass* (R. Vindriis & Carnegie, 2016). These results demonstrate that incorporating noise reduction into the design process can yield significant improvements; make it part of your requirements.

Choosing a Picking Mechanism

What is the best design for a picking mechanism? Vindriis et al. (2011) compared three systems—a small stepper (based on *GuitarBot*), a large stepper (with controllable height), and a solenoid system (similar to EMMI's PAM)—according to maximum speed and consistency of sound produced (amplitude). The small stepper (with four picks) was the slowest, producing 12 picks per second (*pps*), but it was the most consistent in sound production. The solenoid system was faster (20 pps) but produced inconsistent amplitudes. The large stepper system (with eight picks) could produce 25 pps, also with amplitude variability. In a separate study, the rotary stepper-motor-driven pickwheels of *MechBass* and *Swivel* were capable of 9 pps and 8 pps, respectively (J. Murphy, McVay, et al., 2013). The five-arm pickwheel of *Protochord* could produce 32 pps across all vertical positions tested (Placencia et al., 2019). Dynamics, timbre, and sustain are also considerations, which can be modified using the techniques previously discussed.

In all cases, choosing a picking mechanism design depends on your musical requirements. Will the machine play a single string or strum groups of strings at once? How fast does the picking need to occur? What dynamic range is satisfactory? (The answer to this question will also depend on whether the instrument is electric, acoustic, or a hybrid.) What sorts of articulations should be possible? Answering these questions will enable you to select and combine the approaches detailed in this section. A picking mechanism that synthesizes these approaches—variable velocity of the plectrum across the strings, precise control over plectrum distance and angle relative to the strings, and variability of plectrum stiffness—is compelling indeed.

BOWING

A bow grips and releases a string in rapid alternation. When designing a bowing mechanism, the first consideration is whether to use an instrument designed for human players. Doing so requires confronting idiomatic issues, which can be addressed through anthropomorphic emulation (i.e., making a robotic arm) or creating a novel device.

The *emulative* approach is exemplified in Toyota's *Violin-Playing Robot* (*Toyota unveils personal-transport, violin-playing robots*, 2007), which is a general-purpose domestic assistant that can also play the violin (Media Example 6.2.2, ▶). Shibuya Labs' violin-playing robot (Shibuya, 2011) purported to study both the physical aspects of violin playing and explore the role of *kansei*, which is connected to "sensibility," "feeling," and "mood" in musical expression (Shibuya, 2011, p. 179). The result was a 7-DOF robotic arm that can play sustained notes on individual strings (Media Example 6.2.3, ▶).

Other builders have developed novel, non-anthropomorphic mechanisms that incorporate bows designed for human performers. Such a structure is realized in *automated composition for violin bow, toy piano and tape recorder* (Wouter van Veldhoven, 2014; Media Example 6.2.4 ▶), where a violin bow is fastened to a rod that is connected to the rocker of a four-bar linkage. As the crank rotates, the rocker oscillates back and forth, bowing the string in alternate directions (Figure 6.2.12).

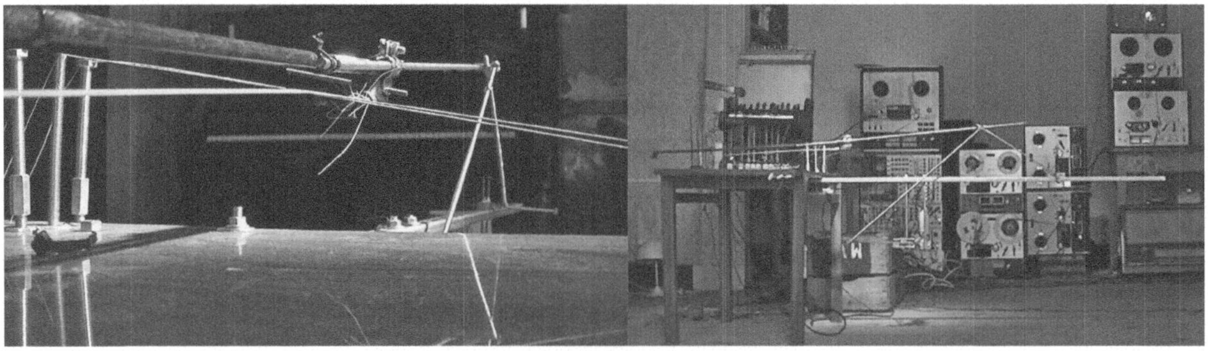

FIGURE 6.2.12. **Four-bar bowing mechanism (Wouter van Veldhoven, 2014).**

The musical machine *Ghost Play* controls bow speed and pressure to produce *spiccato* and *tremolo* techniques (Kamatani et al., 2022). The bowing mechanism comprises three electromagnetic linear shaft drives that enable the bow to move along multiple DOFs. A sensor is attached to the bow to measure bowing pressure, used for calibration and to calculate bow displacement. Examples of the machine's performance can be seen on the companion website (Media Example 6.2.5 ▶).

A rotating wheel applied to a string can also produce a bowing effect. The hurdy-gurdy features such a wheel, attached to a hand crank powered by a human musician to make sound (see Figure 6.2.13).

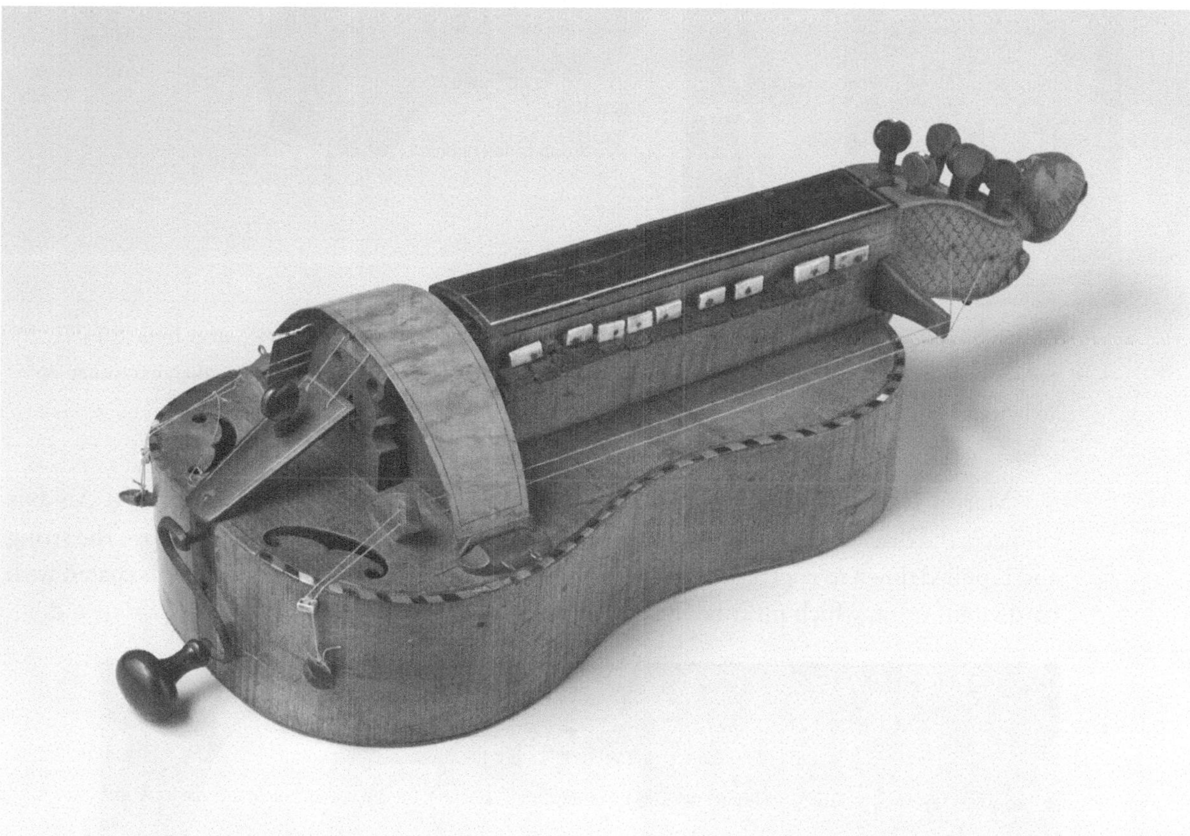

FIGURE 6.2.13. **A hurdy-gurdy.**

In the mechatronic world, bowing mechanisms have been developed using cables wrapped around motor-driven pulleys. Examples include the Logos Foundation's <Hurdy> and *Hathaani* (Figure 6.2.14). Timbral, dynamic, and articulatory variation is produced by modifying the speed and direction of bowing (via the motor attached to the drive wheel), pressure exerted on the strings (via the servos in *Hathaani*'s design), position of the bow (which is modulated by a servo-driven slider-crank mechanism in *Hathaani*), and tightness of the bow material.

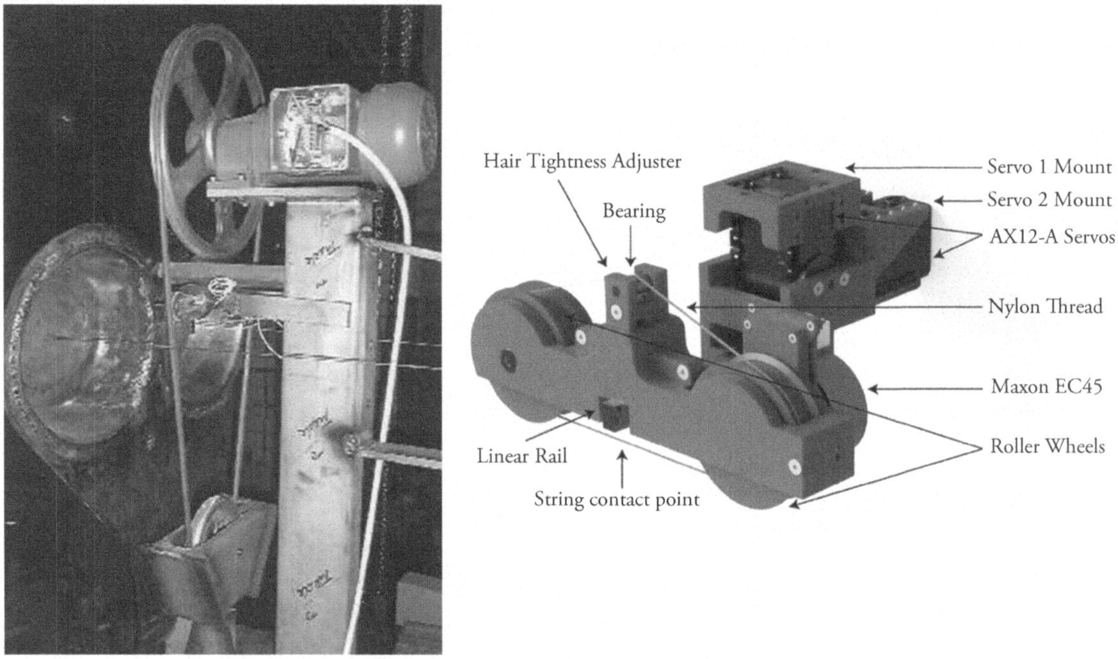

FIGURE 6.2.14. **Cable-pulley bowing mechanisms. Left: Logos Foundation's <Hurdy> (image from https://www.logosfoundation.org/instrum_gwr/hurdygurdy.html); right: Hathaani (image from Sankaranarayanan & Weinberg, 2021, CC by 4.0).**

Martin Riches's *String Thing* uses a cambered plexiglass disc to bow the string. An axle connects the disc to a motor that rotates at 100 rpm. The disc sits about 1 mm above the string and is pulled down to contact the string by a solenoid (see Figure 6.2.15). The disc is coated with hard violin rosin, which must be re-applied every 10 minutes (*More about StringThing*, n.d.).

FIGURE 6.2.15. **Bowing mechanism of string thing (image from *More about StringThing*, n.d., photo: Martin Riches).**

Strings can be bowed by other strings made of materials such as nylon filament. Pioneers of this method include composers C. Curtis Smith and Stephen Scott, the latter of whom founded the Bowed Piano Ensemble in 1977. Works such as Scott's *Entrada* (Media Example 6.2.6 ▶) show a group of human musicians using strings (and other sundry materials) to bow the strings of a grand piano. The sonic effects of the technique are striking as they simultaneously evoke notions of bowing as commonly encountered and unique timbres. One can imagine mechatronic systems that replicate these techniques.

HAMMERING

Hammering is a common method for exciting strings. The piano is a prime example of this method. The mechanical process that enables this action is complex, as seen in Figure 6.2.16.

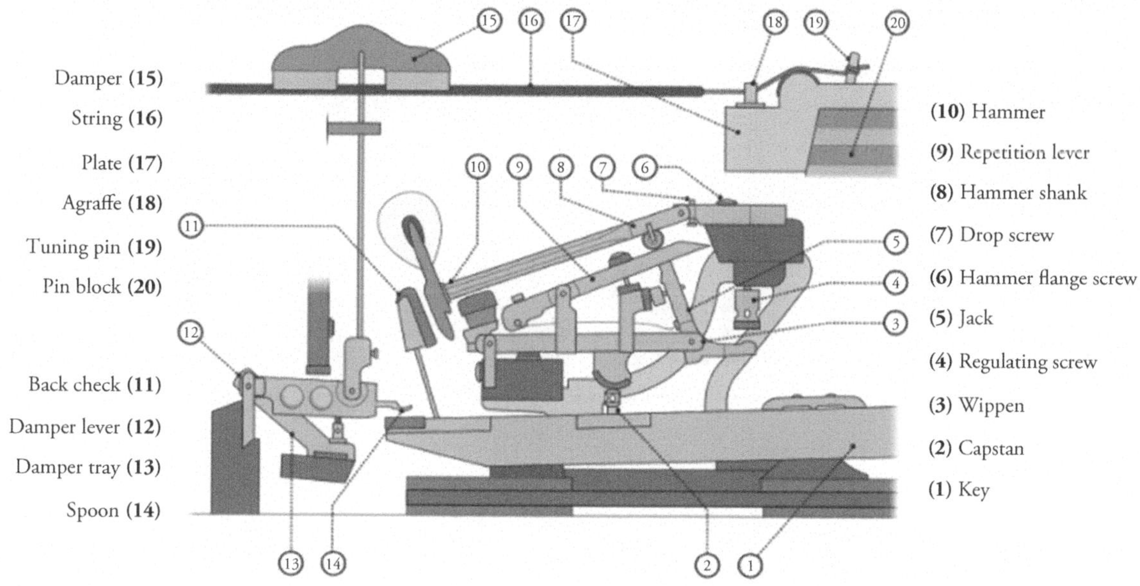

FIGURE 6.2.16. **Piano mechanism. Image by Olek Remesz (wiki-pl: Orem, commons: Orem) / Made in cooperation with User:Bechstein.—Own work, CC BY-SA 3.0, https://commons.wikimedia.org/w/index.php?curid=1104318.**

Why does a seemingly simple action require such a complicated device? The reasons are numerous, but primary among them are the need to translate the forces and velocities of input actions (e.g., pressing keys with fingers) into forces and velocities of output actions (e.g., hammering strings) that produce sound with a clear and consistent attack, sustain, duration control, a wide dynamic range, and little noticeable latency between input and output. As the list of desired sonic features grows, the complexity of the mechanism follows.

Mechatronic string-hammering systems can use human-playable physical interfaces or not. One approach is to place actuators where human fingers typically go, for example, by placing rows of solenoids directly over a piano keyboard. The Logos Foundation's <*Player Piano I*> illustrates this method (Figure 6.2.17).

FIGURE 6.2.17. **Logos Foundation's <Player Piano> (image from https://logosfoundation.org/instrum_gwr/playerpiano.html).**

By utilizing technology (here, the piano) that is already well-developed, this approach avoids wheel-reinventing that accompanies a *from-scratch* attempt. It also provides a compelling visual experience as keys are depressed and released without direct human intervention, continuing the tradition of twentieth-century player pianos. A disadvantage of the *piano attachment* is unnecessary mechanical complexity. Actuators press interface devices (keys) designed for humans that activate other mechanisms to hammer a string. These intermediary stages could be reduced by an actuating system that directly hammers the strings.

Direct mechatronic string-hammering systems vary in complexity. One of the simplest designs consists of a linear actuator (such as a solenoid) positioned directly under or over a string. The type of actuator used will determine its orientation. A push solenoid can be placed below a string so that when activated, the armature extends to contact and excite the string (e.g., the MPR Lab's *Cyther*, see Figure 6.2.18).

FIGURE 6.2.18. **Cyther's hammering and damping actuators.**

Conversely, a pull (or push-pull) solenoid whose armature extends through its body could be positioned above a string so that when activated, the armature retracts to engage the string (Figure 6.2.19).

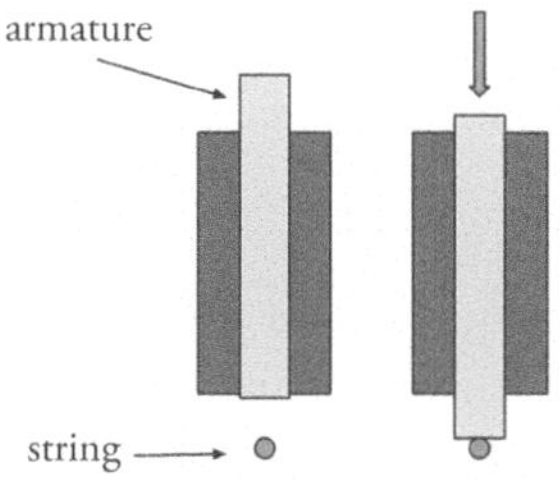

FIGURE 6.2.19. **Pull solenoid as a string hammer.**

The armature can contact the string directly; however, experimenting with armature cap materials facilitates timbral and articulatory effects. The solenoids on *Cyther* contain 3D-printed Polylactic acid—PLA—caps and foam pads that increase surface area (to make sure the string is contacted properly), make the hammering actions more visible, and affect the produced timbre (Figure 6.2.18). More complicated systems may utilize an actuator to drive a mechanism, such as a lever, linkage, or cam (see *The Mechanical* section of the Companion Website ▶) that engages the string.

SCRAPING / SLIDING

A string can be excited by moving an object (*slide* or *scraper*) along its length. The position of the *slide* or *scraper* affects the vibrating length of the string, changing the pitch produced. The material of the *slide* or *scraper* affects the sustain, timbre, and volume of the resultant sound: hard, smooth surfaces such as metal or glass are popular choices for *slides*, while *scrapers* are typically made of metal or wood (which could be smooth or rough). The shape of the slide or scraper determines the surface area that will contact the string. Less contact surface area means less friction (with all other things being equal), which means the slide or scraper will be easier to move. A cylindrical shape minimizes the contact surface area, allowing the cylinder to become tangent to the string when the two are joined; therefore, rods and hollow tubes are common choices for *slides*. Scrapers can also be cylindrical, though more friction may be sonically desirable when *scraping*, so different shapes are worth experimenting with.

Slides are common in the world of human-played instruments. They are used in blues, rock, and country genres and were a significant factor in the development of the Hawaiian *steel guitar*. The slide concept has also been incorporated into mechatronic instruments. *Swivel* utilizes a fretter arm connected to a stepper motor. As the stepper motor rotates, the fretter arm slides along the string, changing the pitch (Figure 6.2.20).

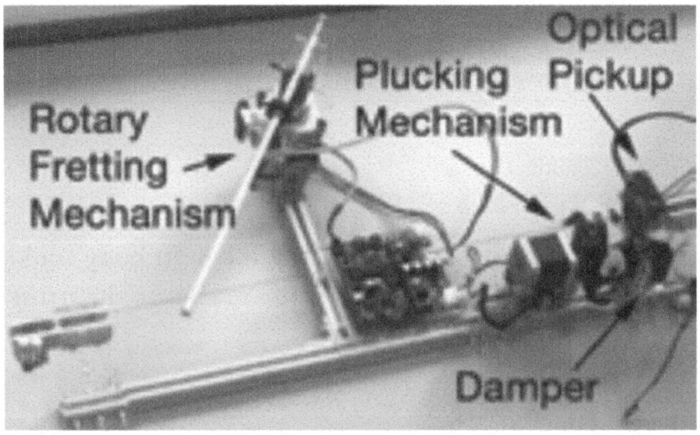

FIGURE 6.2.20. **Swivel (image from Murphy, Kapur et al., 2013).**

A slide on a string produces a characteristic sound, and the technique allows for microtonal intonation, portamento, and vibrato. These strengths become detractors if not limited appropriately. If there isn't a way to lift the slide off the string (or adequately mute the string when the slide is moving), then all musical utterances are joined in a perpetual portamento, which can become tiresome. It is also imperative that the slide is positioned precisely to produce accurate intonation.

VIBRATION

Vibrating transducers coupled to an instrument's radiating surfaces can generate sonic effects. The Overtone Fiddle (Overholt et al., 2011) features a voice-coil transducer embedded into the instrument's top and bottom resonating surfaces (Figure 6.2.21). The transducers are driven by audio signals that change the spectral character of the tones produced by the human-playable string instrument.

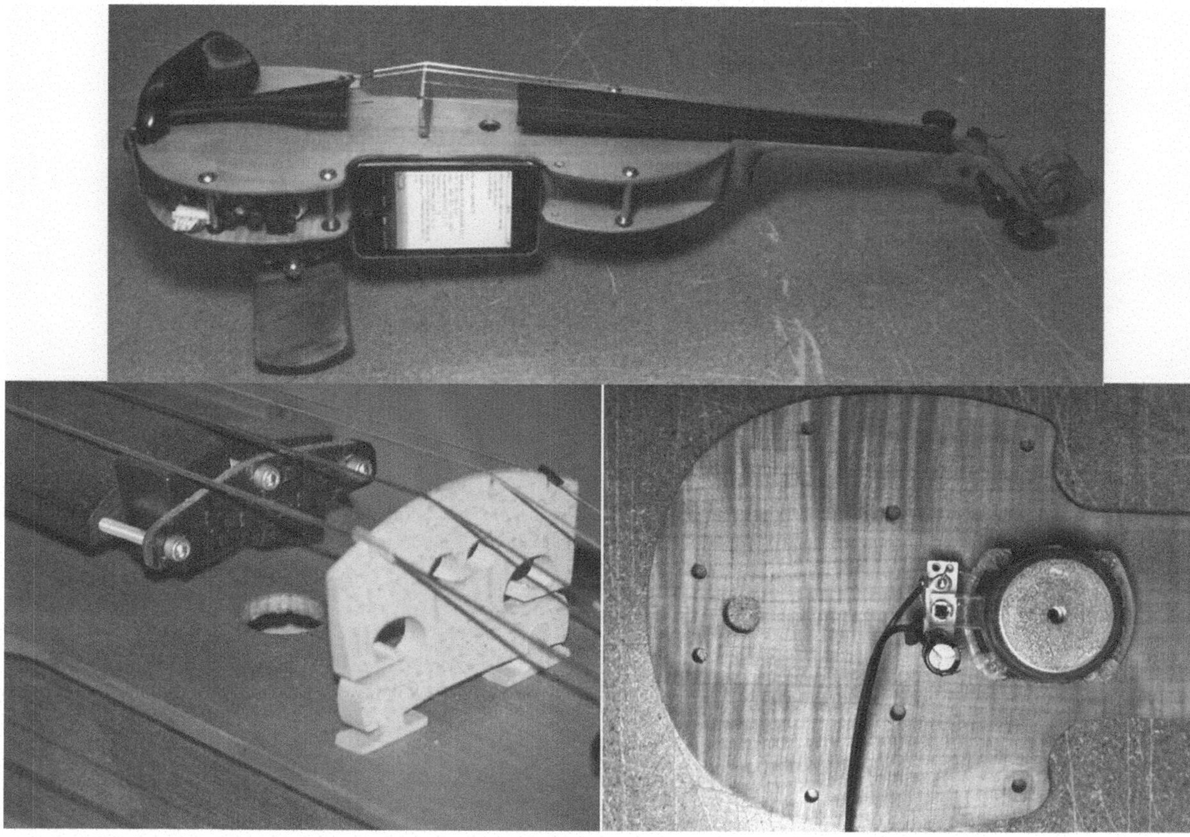

FIGURE 6.2.21. The overtone fiddle (image from Overholt, 2011).

Instead of producing a single action, such as a pluck or a strike, sustained actuation can drive a string. The Tremolo-Harp (Kemper, 2020) utilizes 11,000 rpm vibration motors suspended next to 12 chromatically tuned strings to create sustained tremolo effects. Pulse-width modulated control signals adjust the voltage delivered to the motors, expanding the instrument's dynamic range. The actuation system produces sustained, dynamically evolving sonic textures, distinguished from the discrete outputs of picking-based approaches. Kemper notes how picking-based systems compel composers to explore virtuosic possibilities, while the Tremolo-Harp encourages engagement with harmony, accompaniment, and developing contours. It is a point to keep in mind: **a tool's affordances influence what we make with it**.

INDIRECT EXCITATION: AIR, ELECTROMAGNETISM, AND SYMPATHETIC VIBRATION

The methods covered so far—plucking, hammering, and sliding—involve direct contact between an object and a string, but this is not the only way to get a string moving. Strings can also be excited by air, electromagnetism, and sympathetic vibration.

A phenomenon known as **vortex shedding** can cause a string to seemingly play itself. When air flows across a cylinder, periodic vortices (spirals) are **shed** from one side of the cylinder to the other, a phenomenon known as the **von Kármán vortex street effect** (see Figure 6.2.22).

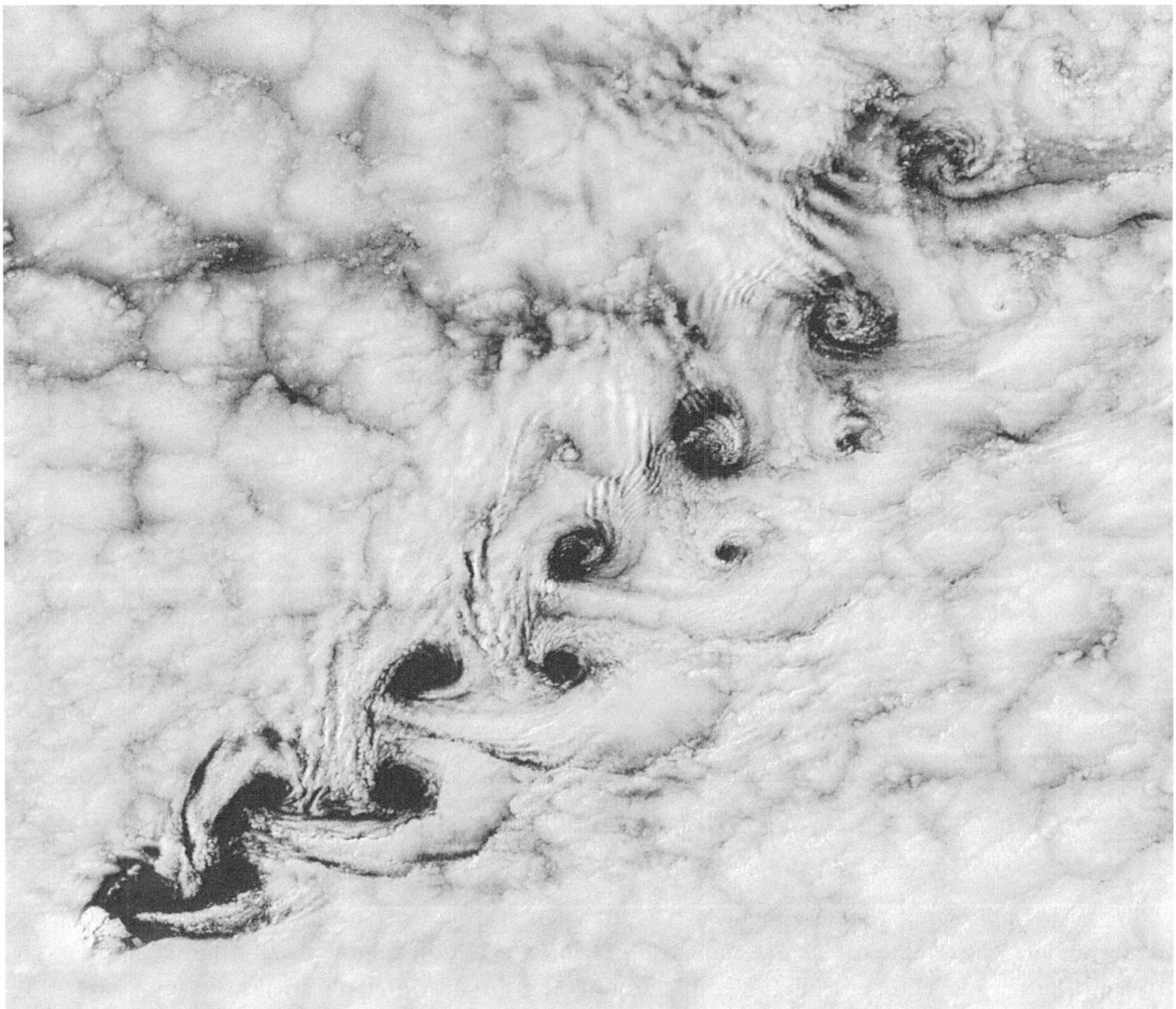

FIGURE 6.2.22. **Vortex shedding seen in the clouds as winds pass an island in the Ocean.**

This process causes areas of low pressure on the downstream side of the cylinder. Given the right physical characteristics of the cylinder (mass, tension between fixed points), the object moves toward the low-pressure areas. If the vortex shedding frequency matches the natural frequency of the object, the object will resonate. The string is excited without (visible) contact! This is how an Aeolian harp, an instrument that has existed for thousands of years, works. One of the most compelling sonic aspects of the Aeolian harp is that the interaction between the wind and the strings results in the excitation of the upper harmonics, creating timbres and harmonies unlike those that would result from plucking or bowing the string.

If the wind can already play the strings, what do we need machines for? One could manipulate the wind's velocity to create timbral effects. One of my students experimented with this idea by placing an Aeolian harp in a wind tunnel, altering the airflow's velocity, and analyzing the resulting sound spectra. Another possibility is to vary string tensions mechatronically to emphasize and attenuate parts of the frequency spectrum. Wind-driven strings could also be incorporated into a mechatronic instrument featuring plucked or bowed strings, expanding the palette of sonic options.

Another method of exciting strings without direct physical contact involves **electromagnetic** actuation. Electromagnetic actuators consist of a coil of wire wrapped around a ferromagnetic core, such as iron. Running electrical current through the wire creates a uniform magnetic field that applies force to nearby ferromagnetic objects (such as a guitar string). If the object is not ferromagnetic, a permanent magnet can be affixed to the object to produce the effect. The force applied is proportional to the electrical current and the number of turns in the wire and inversely proportional to the square of the distance between the electromagnet and the ferromagnetic object. Specific details about the design of these actuators and their accompanying circuits can be found in McPherson (2012).

Electromagnetic actuators can produce sustained sounds reminiscent of bowing but can also be used for creative exploration. Their attack can be softer and more gradual than bowing, imparting an otherworldly sonic effect. Electromagnetic actuating techniques have been incorporated into acoustic pianos (Berdahl et al., 2005; Bloland, 2007; A. McPherson, 2010; A. P. McPherson & Kim, 2010, see Figure 6.2.23), Rhodes piano (Shear & Wright, 2011), the Feedback Resonance Guitar by Edgar Berdahl (Overholt et al., 2011); robotic string instruments such as The Logos Foundation's <Aeio> and EMMI's AMI (Rogers et al., 2015); and the handheld *eBow* (Heet, 1978), commonly used to play electric guitars.

FIGURE 6.2.23. **The electromagnetically prepared piano (image from Per Bloland).**

Instead of actuating each string individually, mechanisms can be used to drive the soundboard of a string instrument. Thompson & Berdahl (2021) coupled an electrically driven shaker (*tactile transducer*) and a sensor to a piano soundboard. After a note is played, a software program analyzes the audio, "freezes" a spectral snapshot, and sends the output to the shaker, creating a feedback loop that enables variable sustain of the input sound.

Strings can also be excited through **sympathetic resonance**. All objects have a **natural frequency** (or set of frequencies) at which they tend to vibrate when excited. The vibrations of one object can excite the natural frequencies of another, given they are (close to) equal. The result is increased amplitude at these frequencies (compared to other frequencies that don't produce this tendency), known as **resonance**. Consider two strings: A (tuned to 440 Hz) and B (tuned to 880 Hz), both of which are initially at rest. If string A is excited, its primary mode of vibration is 440 Hz (the fundamental frequency), but it also vibrates at 880 Hz (the second mode of vibration), 1320 Hz (the third mode), and so forth. Given sufficient amplitude, the vibrational energy produced by string A will excite string B because the two share the *natural frequency* of 880 Hz. String B will then begin to vibrate "by itself." This phenomenon is known as **sympathetic resonance** or **sympathetic vibration**. Musically, *sympathetic vibration* can enrich an instrument's harmonic and timbral character. Instruments that contain sympathetic strings include the Indian *sitar*, the European *viola d'amore*, the *hurdy-gurdy*, the South Asian *sarangi*, and the Norwegian *Hardanger fiddle* (Figure 6.2.24).

FIGURE 6.2.24. Left: Sarangi (image By Sarangi close-up.jpg: gurmatsangeetorgderivative work: Hekerui—Sarangi close-up.jpg, CC BY 3.0, https://commons.wikimedia.org/w/index.php?curid=7052228). right: Hardanger Fiddle (By Frode Inge Helland, CC BY-SA 3.0, https://commons.wikimedia.org/w/index.php?curid=628067).

COMBINATIONS

The excitation mechanisms described above can be integrated into the same instrument. EMMI's AMI includes a picking mechanism and an electromagnetic driver to excite the string. Transducers can drive a piano's soundboard, sympathetically exciting the strings, which are captured with a microphone and fed back into the system, for example, in Chris Warren's *Feedback Piano* (Make: television, 2009). *Cyther*'s dampers are an alternative excitation system and not just a means to silence a string already in motion. Instruments need not be limited to one approach. Integrating excitation methods on the same robotic instrument is compelling because they can be precisely coordinated via computer control.

Pitch

Alongside excitation, pitch production is a fundamental consideration when making a musical string machine. Different pitches can be achieved using multiple strings, varying the string's vibrating length, adjusting the string's tension, or a combination of these options. The following sections explore these categories and the instruments that exemplify them.

MULTIPLE STRINGS

One way to produce a variety of pitches is to use multiple strings, each tuned to a different pitch (e.g., a harp or piano). The Music, Perception, and Robotics Lab's *Cyther* is an example of this approach, utilizing 10 different strings, each with a dedicated electromechanical actuator and damper (Figure 6.2.25). Multiple strings can produce chords, sweeping arpeggios, and rapid alternations between pitches regardless of pitch interval size.

FIGURE 6.2.25. **MPR Lab's Cyther.**

Space and cost are primary considerations in this approach. There may not be sufficient space or funds to implement separate stepper-motor-driven pickwheel mechanisms, along with their associated mounting hardware and motor drivers, for each string. One octave's worth of strings is rarely enough to accommodate a suitable range of harmonies and voicings. Two octaves are better, but that requires 24 strings (if adhering to the chromatic scale), 24 exciting actuators, and 24 damping actuators, which is a lot. The dimensions of the structure expand with these sonic additions, which have implications on cost and transportability. Another design consideration is whether the instrument will be strum-able or not. If a strumming mechanism is to be used, then the strings' spacing and orientation (e.g., parallel) will affect the strumming gestures that can be produced.

The configuration of actuators affects how the instrument can be played. *Cyther* features a row of hammering solenoids positioned under the strings, allowing unrestricted access to the top of the instrument for humans, machines, or both. Stepper motor pickwheels and stopping solenoids mounted on the top of an instrument would restrict such access and, thus, potential interactions.

CHANGING VIBRATING LENGTH

Changing the effective vibrating length of a string (and thus the pitch produced) is accomplished with a **tangent**, which is an object (e.g., a finger, metal or plastic rod, etc.) that contacts or "stops" the string with enough force to effectively create a new fixed end. This stopping can occur against a board (violin), a fret (guitar), or against nothing if the string is properly suspended (EMMI's PAM; *Swivel* by Jim Murphy). The advantages of varying the vibrating length to produce different pitches are that many pitch possibilities exist on a single string, requiring far fewer components (e.g., strings, actuators, tuning machines, pickups) than the multi-string approach. Disadvantages include a proportional relationship between pitch interval size and physical distance. This means that bigger pitch intervals are physically further away, requiring either more travel time for a single movable tangent or a greater number of tangents placed at disparate locations. A single string can produce different pitches but not chords. Tangents in string machines may be *fixed, adjustable, movable*, or some combination of these options.

Fixed

Tangents can be fixed at specific locations according to a tuning, so there is a 1:1 relationship between the number of tangents and possible pitches (not including the open string). The automated string instruments of Ragtime West, Vladimir Demin's *Robot Guitar*, and EMMI's *AMI* are examples that feature fixed tangents. In terms of mechanism design for fixed tangents, linear actuators (e.g., solenoids) are often used to stop the string with the armature or piston of the actuator (or a cap on top of it) acting as the tangent. More elaborate tangent mechanisms are also possible, expanding visual effects, as in EMMI's PAM.

Adjustable

Tangents can also be adjustable, allowing the instrument to accommodate different tunings. The tangents of EMMI's PAM can be manually moved to various locations along the instrument's strings, though this must be done before and not during a performance (thus, the tangents are *adjustable* during configuration and *fixed* while playing). The tangents are secured by a bolt that slides within the "t-slot" of an aluminum bar. The bolt passes through an acrylic plate to which the actuator is also attached and is secured in position with a wing nut. The tangent can be adjusted vertically and horizontally, given the tracks in the actuator plate. Such adjustments become necessary when setting the action (the distance between the tangent and the string) of the instrument. Logos Foundation's <Synchrochord> monochord takes a different approach by making the frets of the instrument adjustable to accommodate different tunings (Raes, 2012).

Movable

Movable pitch-producing systems can be based on linear tracks, rotating arms, or linkages (see Figure 6.2.26). The first method features a string-stopping mechanism, typically a single tangent or bridge, which slides along a linear track. A common design features a belt drive (whose low noise is a musical advantage) and a carriage that contains the tangent. The carriage is connected to the belt, which moves linearly as the motor rotates. The motor and idler pulley are attached to opposite sides of the track to provide adequate belt tension and allow a full range of movement. Controlling the motor's position enables the carriage to be moved to specific points along the string, producing different pitches. *GuitarBot*, *MechBass*, *BassBot*, and *Hathaani* feature this design.

Swivel exemplifies an alternative *movable* method where a fretter arm is attached to a servomotor. As the motor rotates, the arm travels along the string's plane, modulating its effective vibrating length (see Figure 6.2.26). A third possibility incorporates linkages in the form of a slider-crank mechanism (see *The Mechanical* section of the Companion Website ▶), which converts rotary to linear motion. The servomotor is coupled to the crank arm, which is connected to the slider arm at a pivot. At the end of a slider is a string-stopping mechanism that features an actuator such as a solenoid or servomotor. When the crank arm rotates, the slider moves back and forth along a track parallel to the length of the string. This design is seen in Animusic's *Resonant Chamber* (Animusic LLC, n.d.), *StrumBot*, *ServoSlide* (Placencia et al., 2020b), and *Protochord* (Placencia et al., 2020a).

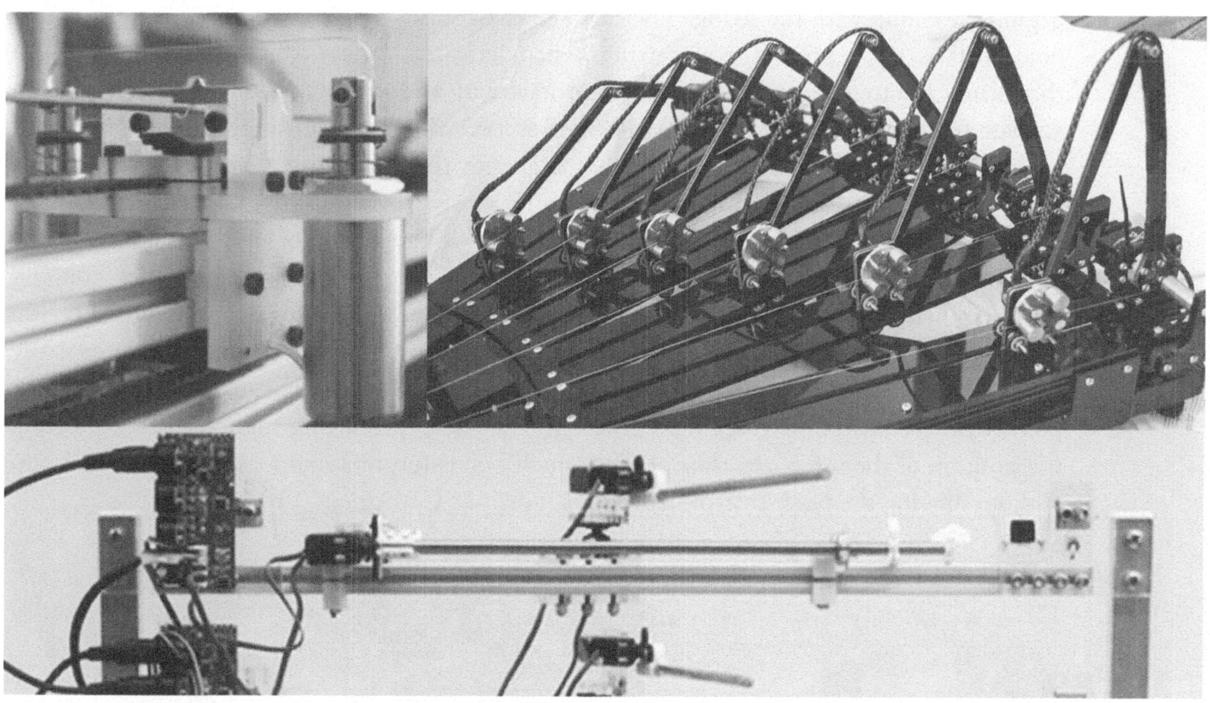

FIGURE 6.2.26. **Pitch-producing systems. Top left: linear track of mechbass (McVay et al., 2011); Top right: slider-crank of strumbot (Richard Vindriis, 2015); Bottom: rotary arm of swivel (J. Murphy, n.d.).**

The speed and accuracy of pitch production vary depending on the instrument's size, the mechanical design of the stopper/fretter system, the distance the fretter motor needs to travel, the resolution of the motor positions, the type of motor implemented, the motor control schema, and friction. Table 6.2.2 summarizes some of these differences.

TABLE 6.2.2. **Pitch-producing mechanism performance: Time required to span an octave, pitch accuracy, and possible positions across a string's length.**

Instrument	Octave (ms)	Pitch accuracy (±cents)	Positions
Swivel	82	6	625
StrumBot	144	8	
Protochord	227	4	
MechBass	360	5	593[1]

Source: *MechBass* data from McVay et al. (2011) and Murphy et al. (2015); Swivel data from Murphy et al. (2015); *StrumBot* data from Vindriis & Carnegie (2016); Protochord data from Placencia et al. (2020a).
[1] In full-stepping mode: ~9,500 positions are possible with microstepping (J. Murphy et al., 2015).

GuitarBot's travel length is ~80% of the string's length, which translates to a pitch range of over two octaves. The moving bridge can move from "end-to-end" in under 250 ms (Singer et al., 2003).

To produce pitches, *movable* systems have either a bridge that is in constant contact with the string (e.g., *GuitarBot*, *BassBot*) or a carriage that holds an independently controllable fretting clamp. In the carriage system, the carriage can move with the clamp retracted (not contacting the string). When it reaches its specified destination, the motors are activated, engaging the clamp with the string. The carriage of Baginsky's *Aglaopheme* (N. Baginsky, n.d.) holds a slide that is pressed to the strings when activated. *Swivel* contains a solenoid clamp that brings the fretting rod in and out of contact with the string. *MechBass* uses solenoids to activate the fretting clamp, while *Hathaani* uses a servo-driven rack and pinion system with a silicone-coated synthetic "fingertip" that adapts to the curvature of a violin's fingerboard (Figure 6.2.27, left). An ingenious design features a rotating clamp that has two (or two sets, in some cases) protruding cylindrical lugs that contact the string (or not) depending on the rotation of the servo (*StrumBot*, *Protochord*). One of the lugs is hard (e.g., brass) for stopping the string, while the other is covered with a soft material (e.g., felt) for damping it (see Figure 6.2.27, right). A single set of lugs (*Protochord*) allows sliding, and thus pitch glides, more easily than two sets of lugs (*StrumBot*) (Placencia et al., 2020a). Rotating the servo past the contact point with the string increases the tension on the latter, producing higher pitches. Varying the position of the servo can thus produce pitch bends (up to about a semitone) and vibrato without moving the pitch-shifting arm (R. Vindriis & Carnegie, 2016).

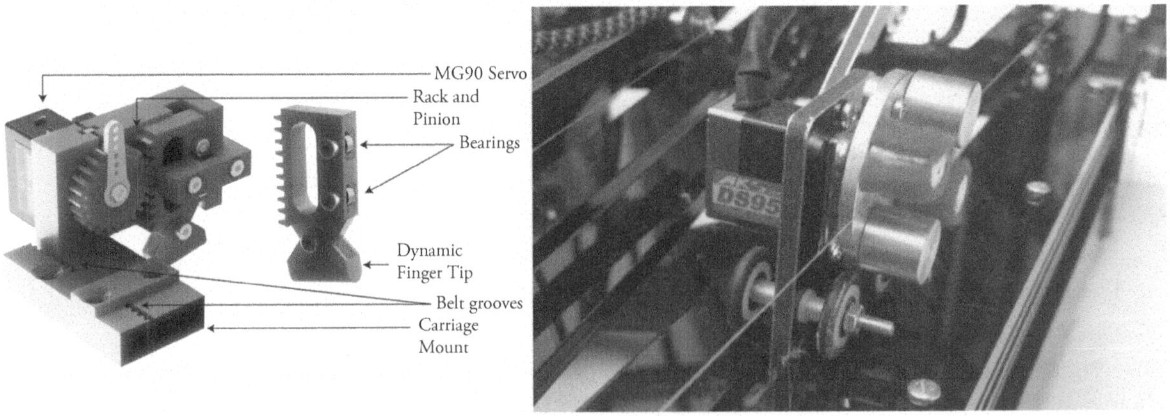

FIGURE 6.2.27. **Fretting mechanisms. Left: Hathaani (image from Sankaranarayanan & Weinberg, 2021, CC by 4.0); right: strumbot (image from R. Vindriis & Carnegie, 2016).**

These designs benefit from the simplicity of a single moving element (rather than an array of linear actuators) but avoid perpetual portamenti by decoupling the contact point from the string. The solenoid clamps also enable a variety of fretting techniques, including hammer-ons and vibrato.

Another possibility involves groups of tangent mechanisms configured in a block that moves linearly along a track, as in Z-Machines' *Mach.* This approach avoids having to place individual solenoids at each pitch location while still providing the performance benefits of a fixed-tangent design. Multiple tangent-played strings are a way to increase the pitch range of an instrument and produce chords. The balance between the number of strings and the number of tangents depends on the musical goals of the designer.

Pros and Cons

There are advantages and disadvantages to each of these approaches. A *movable* system can produce pitches along a continuum, enabling a variety of tuning systems (including microtonal ones) and portamenti. The motions of movable systems are typically larger than those of fixed-tangent ones, enhancing the visibility of the machine. Movable systems can require fewer parts and cost less than arrays of fixed tangents. A disadvantage of movable systems is that the pitch interval's size affects the time it takes to produce it: larger pitch intervals span larger linear distances, which require more time to travel. The fact that different pitch intervals require different amounts of time needs to be accounted for in either the control software or the composition. The size and direction (ascending or descending) of the pitch interval can also affect the accuracy with which it is produced: aspects such as inertia, friction, and hysteresis are in play and must be accounted for. Musically, if the movable system is incapable of disengaging from the string, perpetual portamenti will result, which can become tiresome.

With multiple fixed tangents, the speed at which a pitch interval is produced is not affected by the interval size: *PAM* can produce an octave as fast as it can produce a semitone because each actuator is located at the required pitch position and at a uniform distance from the string. This is an enormous advantage. The disadvantage of the fixed approach is that movements are often small and sometimes obscured, leaving something to be desired visually. Having one tangent per note also means more parts that require more time, effort, and money to design and assemble. Often, such systems become bulky, obscuring other aspects of the instrument, such as the neck and strings. Musically, a fixed system by itself is not capable of portamenti, vibrato, or other nuanced manipulations of pitch.

TENSION

Given a static vibrating length, the pitch produced by a string can be changed by varying its tension. This method is less common in human-played string instruments than changing vibrating length, but some examples have explored tension-based possibilities. The washtub bass and its variants are present in cultures throughout the world. The instrument features a string connected to a resonator (such as a steel washtub) and a stick. Moving the stick alters the string's tension, thereby affecting the pitch produced. The Vietnamese đàn bầu consists of a string affixed to a bamboo tube and a wooden rod perpendicularly coupled to the tube. A half coconut shell is also attached to the rod, serving as a resonator. After the string is plucked, moving the rod changes the tension on the string, varying the pitch (Media Example 6.2.7 ▶). The pedal steel guitar uses a mechanical system of pedals and levers that, when pressed, vary the tension on the attached strings (Figure 6.2.28).

FIGURE 6.2.28. Left: A washtub bass, top right: Vietnamese đàn bầu (image By Jacek Karczmarczyk—http://pl.globalquiz.org/ilustracja-pytania/wietnamski-instrument-dan-bau-1/, CC BY 3.0, https://commons.wikimedia.org/w/index.php?curid=28399766). Bottom right: undercarriage of a pedal steel (image By Patlaff—Own work, CC BY-SA 3.0, https://commons.wikimedia.org/w/index.php?curid=22760064).

Electromechanical systems have been used in both human-playable and autonomous instruments that are tuned by adjusting string tension. In the human-playable instrument category, many examples are *calibration* systems that prepare an instrument, such as a guitar, for human performance. Notable commercial solutions include from by AxCent Tuning Systems (http://axcenttuning.com), formed in 1987 as Transperformance LLC, and Tronical (https://www.tronicaltune.com), founded in 2005 in Hamburg, Germany.

Other efforts have prioritized changing the tension of strings *while* playing, creating a category of *dynamic* tuning systems. Some of these function autonomously, enabling the machine to play music without human intervention, including The Logos Foundation's *<Synchrochord>* (Raes, 2012) and *STARI* (Trail et al., 2013). Other dynamic tension-control instruments are *co-operative*, requiring interaction between human and machine to produce sound. When playing one of *The Milwaukee Servoelectric Guitar Project*'s instruments (Media Example 6.2.8 ▶), a human selects pitches by using their left hand to engage with a button interface on the top of the instrument and plucks strings with their right hand. Pitch selection is flexible and mappable: a button can correspond to a specific pitch, chord, or probability distribution of potential pitches. The abstraction of this part of the musical choice-making process, combined with the lack of obligation to physically change the pitches (the system takes care of that), transforms the interaction between the human player and the sound-making device. The novel design approach embodied in this group of instruments yields a unique type of musical expression characterized by continuous pitch glides that connect each note.

The Music, Perception, and Robotics Lab's *Cyther v3* (Barton et al., 2017) is also a *cooperative* instrument. The instrument is designed to tune its strings autonomously by adjusting string tension, but the strings are unencumbered on top, allowing human and machine to interact with each other through a shared physical medium. The MPR Lab's *Whamola* is designed with a similar philosophy by integrating a harmonic drive motor that controls the movement of the handle, which changes the tension on the string and thus modulates the pitch. The handle can also be moved by a human, facilitating *cooperation*.

Design of Tension-Modulating Systems

The fundamental components of tension-based tuning systems include transducing a string's vibrations into an electrical signal, converting the analog electrical signal into the digital domain, sensing the fundamental frequency of the signal, calculating motor positions based on a tuning system, and issuing motor control commands to activate a mechanism(s) attached to a string. The requirements of *dynamic* and *calibration* instruments differ in areas such as tuning speed and pitch range (i.e., a dynamic system requires lower latency to produce music in real time), which affects motor and mechanism selection.

Mechanisms

Changing the tension on a string requires the string to be displaced from its equilibrium position. This displacement can be achieved through an *end-movable* design, involving fixing one end of the string and varying the position of the other through *linear* (e.g., attaching the ball end to a linear actuator that pulls the string straight back) or *rotational* movement (e.g., winding a string around a post rotated by a motor). Alternatively, the approach could be *end-fixed*, where the string is anchored at both ends and force is applied *perpendicularly* along its length (e.g., by a motor-driven leadscrew). The following describes mechanisms that use these approaches, along with their advantages and disadvantages.

A popular *end-movable* approach features a worm gear-based guitar tuning machine coupled to a motor. As the motor turns, it drives the tuning machine's gears, rotating the tuning machine cylinder around which the string is wrapped. The string tension and pitch produced increase or decrease depending on the direction and speed of rotation. The advantages of this approach are (1) the gear reduction of the tuning machine decreases input torque requirements, (2) the worm gear cannot be backdriven, holding the string in place mechanically when the motor stops, and (3) tuning machines are ubiquitously available and relatively inexpensive. Disadvantages of worm gears include (1) lower efficiency due to gear friction, (2) backlash, which reduces positional precision and accuracy, (3) slower tuning speeds because of gear reduction, and (4) tuning inaccuracy resulting from friction between the string's windings as the cylinder rotates.

The tuning machine-motor system has been incorporated into calibration and fully autonomous systems. *STARI* (Trail et al., 2013) features a stepper motor coupled to a standard guitar tuning machine connected to the string. Custom 3D-printed enclosures securely connect the stepper motor to the tuning machine. *Cyther v3* uses standard tuning machines with the button (or tuning head/grip piece) removed, allowing a motor shaft to be directly coupled to the worm drive cylinder. The Logos Foundation's <*Synchrochord*> uses a conceptually similar design (but one that doesn't use a standard guitar tuning machine) that consists of a motor coupled to a worm gear that rotates a spindle around which a string is wound (Figure 6.2.29).

FIGURE 6.2.29. Left: motor-driven tuning mechanisms of Cyther v3. Right: tuning motor and gears in <Synchrochord> (Raes, 2012).

Among the *calibration* examples, the Tronical system integrates the components of the motor-machine head approach in the form of specialized tuning machines called "Powerheads," which comprise servomotors and gears (*How does TronicalTune work?*, n.d.; see Figure 6.2.30). Powerheads are secured to a guitar's headstock via a plate that connects them to a microprocessor. The design is compact but also more complex and costly. The tuning system in the Gibson Robot Guitar was primarily developed by Tronical (Borland, n.d.).

FIGURE 6.2.30. Tronical "Powerhead" (image from https://www.tronicaltune.com/how-it-works/).

An alternative *end-movable* method is exemplified by the Transperformance / AxCent tuning mechanism (N. Skinn et al., 2008; N. C. Skinn & Freeland, 1990). The core of the system (see Figure 6.2.31) consists of a string that passes over rollers and is seated in a tune arm that rotates about an upper pivot. The lower part of the tune arm has a threaded pivot pin that engages the shaft of a stepper motor. As the motor turns, the tune arm pivots, modifying the string's tension and changing the pitch it produces. A link member connects to the upper part of the tune arm and a spring, which counterbalances the forces exerted by the tensioned strings.

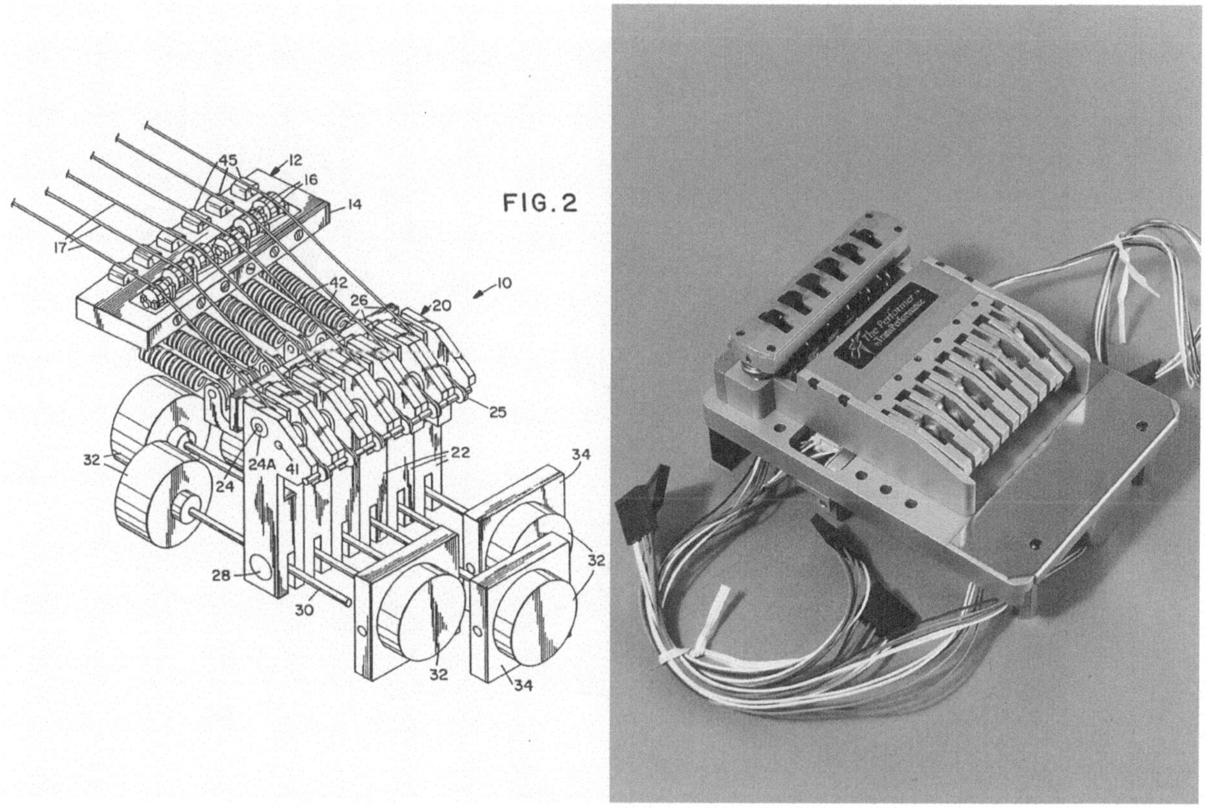

FIGURE 6.2.31. **Axcent / Transperformance tuning system (left: image from N. C. Skinn & Freeland, 1990; right: image from N. C. Skinn.)**

Tuning can be achieved through linear displacement, as seen in the Milwaukee Servoelectric Guitar's "Neko-Echo." An overview of the system is seen in Figure 6.2.32.

The ball end of a guitar string is connected to a spring on one end; the other end is mated to an aluminum bar coupled to a V-shaped bracket with an axle near its vertex. The other end of the aluminum bar is fastened to a circular hub secured to the axle of a stepper motor. As the motor turns, the V-bracket rotates, displacing the spring and the connected string, which changes the latter's tension. A bridge establishes the effective vibrating length of the string and its position relative to the body of the instrument.

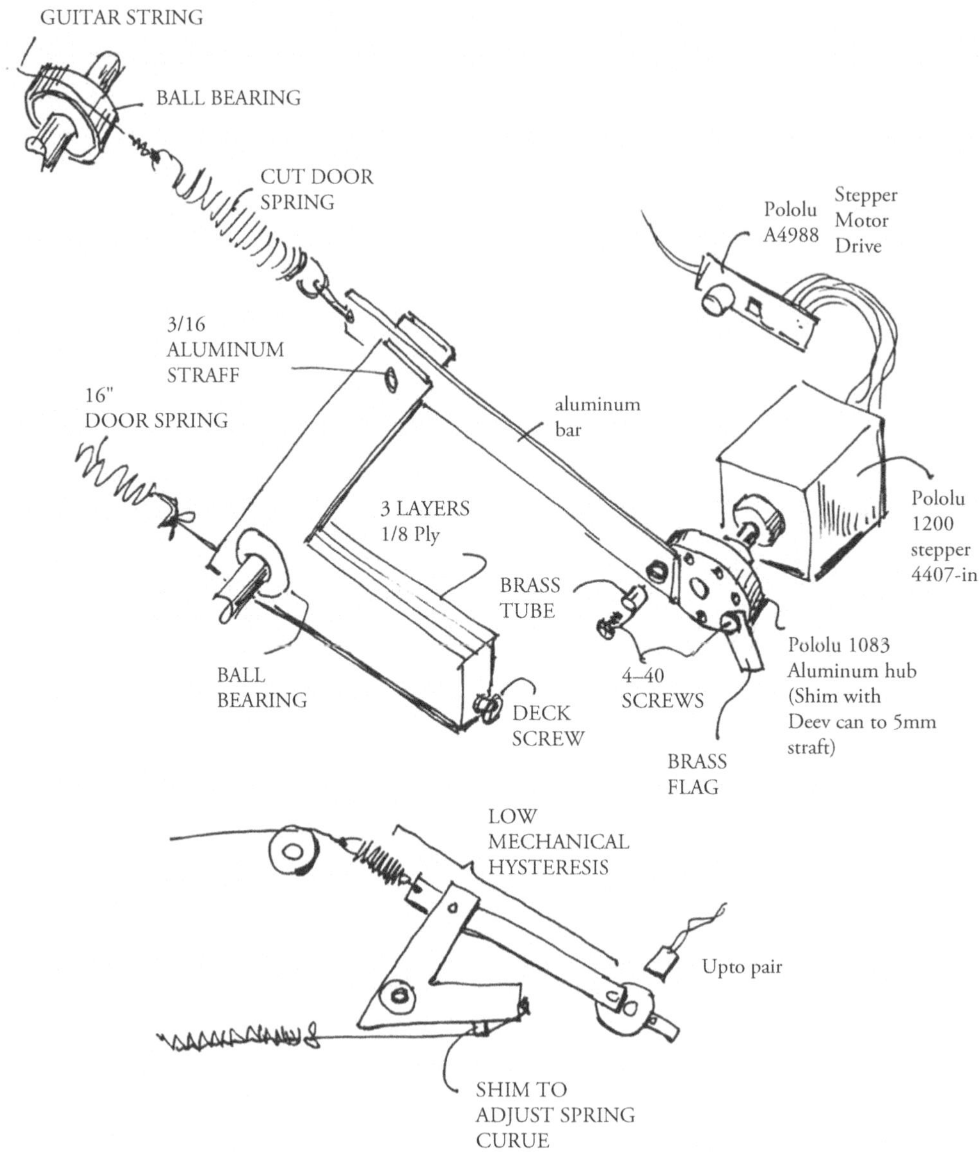

FIGURE 6.2.32. **System overview of the neko-echo by the Milwaukee servoelectric guitar project (image from http://servoelectricguitar.com/construction.htm).**

Determining Pitch

Being able to vary string tension is the first step; the next is developing a system to sense the string's fundamental frequency so adjustments can be made (if needed). There are varied approaches to this problem. The first involves determining the relationship between motor positions and string frequencies, which can be accomplished experimentally (e.g., moving a motor to a position and measuring the string frequency at that position) or calculated mathematically. Correspondences between motor positions and string frequencies can be stored in an array specified in the microcontroller code. When a particular pitch is desired, the frequency of that pitch can be used to find the corresponding motor position in the array. The difference between the current position and the goal position is determined, a control signal is output, and the motor is driven until the goal position is reached. Murphy et al. (2014) present a variation on this idea where the initial mapping stage is performed at various open string tunings, producing string characterization curves. Frequencies and their corresponding motor positions are found by interpolating between those curves and assigning virtual frets.

There are complications to consider when mapping motor positions to string frequencies. Motor positions may have corresponded to specific pitches during testing or calibration, but in usage, forces and movements cause the physical system to change, uncoupling prior associations and producing errors. Sensing string frequency and tension can help mitigate these errors.

MEASURING STRING FREQUENCY

The first step in measuring string frequency is to transduce a string's vibrations into an analogous analog electrical signal. The methods to do so include electromagnetic (*STARI*, *PAM*), optical (*Cyther*, *MechBass*), and piezoelectric (Electronic assisting violin tuner, Wang & Ortega-Sanchez, 2012) pickups and microphones. While these components can be used to amplify a signal for musical performance, detecting fundamental frequency is a different goal with a different process. For instruments with multiple strings, isolating each one helps determine individual frequencies. This can be accomplished with optical or single-string electromagnetic pickups. The fundamental frequency of the resultant electric signal is then measured, which can be achieved in various ways.

In the Transperformance / AxCent system (N. Skinn et al., 2008), a sensor transduces a string's vibrations and transmits the signal through a series of lowpass and bandpass filters. The signal is converted from the analog to the digital domain by a comparator and turned into a square wave whose frequency is determined. The frequency value is then compared to a desired reference value stored in the microcontroller's memory. If there is a difference between these values, a control signal is generated to drive the motor/gear assembly of the tuning mechanisms, compensating for the tuning error.

In *<Synchrocord>* (Raes, 2012), the signal transduced from the string's vibrations is amplified and fed into a NAND gate with Schmitt-trigger inputs (74HCT132), which transforms it into a sharply defined square wave output connected to the microcontroller. An interrupt handler on the microcontroller measures the frequency of the resultant square wave. Simultaneously, the pickup signal is fed into a tachometer (LM2907), which outputs a voltage proportional to the input frequency, and is transmitted to an analog input on the microcontroller. When the values from the frequency counter and tachometer are consistent (the tachometer is reportedly the more accurate of the two), motor control commands are implemented, the motor turns the gears, and the string's tension is adjusted.

A popular method for fundamental frequency detection utilizes the **Fast Fourier Transform (FFT)** algorithm, which converts a signal from the time domain into the frequency domain, thereby decomposing a complex waveform into its individual sinusoidal components. FFT analysis is a powerful tool that gives a detailed picture of a sound's spectral components. *STARI* (Trail et al., 2013) uses the FFT-based object **fiddle~** in the graphical programming environment Pure Data (PD). A disadvantage is that the FFT is computationally expensive, making the approach less suitable for real-time musical applications. Through mathematical refinements, the **Goertzel algorithm** (or filter) analyzes a specific frequency (rather than the entire spectrum), making the process more efficient. The approach is utilized in the *Electronic Assisting Violin Tuner* (Wang & Ortega-Sanchez, 2012).

Periodic waveforms cross zero amplitude (known as zero-crossing [ZC]) twice per waveform cycle, which the **ZC method** utilizes to calculate fundamental frequency. The frequency of the signal can be estimated by counting the number of ZCs per a given unit of time by following these steps:

1. specify a time window
2. count the ZCs within that window
3. divide by two (waveforms cross zero amplitude twice)
4. divide by the length of the observation window (in seconds) to derive the frequency in Hertz (Hz).

The ZC method is simple, fast, and computationally efficient relative to demanding options such as Fourier transforms. Disadvantages include that it performs best with pure sinusoidal signals; complex waveforms, noise, and non-periodic signals require appropriate filtering. The ZC method was employed in *ServoSlide* (Long et al., 2017), the AxCent system (as discussed in a personal conversation with Neil Skinn), and *Cyther v3* (Barton et al., 2017).

Autocorrelation analyzes a signal's self-similarity, which can determine its period and frequency. The **Hz** system (Guzman, 2024a, 2024b) uses ZC to mark a signal's potential edges and then employs autocorrelation to determine periodicity. The result is a fast and efficient process that shows good accuracy in testing in the MPR Lab. The **YIN algorithm** is also based on the autocorrelation method (De Cheveigné & Kawahara, 2002; Gylling & Svensson, 2017).

Once the current frequency is known, it is compared with the goal frequency to determine the difference. This value dictates the direction and speed at which the motor must turn to increase or decrease tension on the string to produce the appropriate pitch. A PID control scheme is a popular choice to govern this motor movement. In some cases, a frequency error threshold is established in the algorithm to specify the acceptable degree of accuracy.

MEASURING STRING TENSION

The Milwaukee Servoelectric Guitar Project utilizes string tension to measure the pitch produced by an instrument. The researchers explored a frequency sensing approach but concluded that the "capture time was too slow both because of the inherent time window required for accurate frequency detection and the lags of the filtration needed to remove guitar string harmonics" (*Frequency Sensing*, n.d.). Instead, they connected a potentiometer to a pair of helical springs. As the tension on the string changes, the spring deflects (travels), which is measured by the potentiometer. A limitation of this approach is that the logarithmic relationship between pitch and frequency does not map efficiently to the range of a linear potentiometer. In practice, this means that there is more resolution than needed in the high part of the range and not enough resolution in the low part of the range. To address this issue, two springs are used, and the extension of one of them is limited to roughly half of its range. This increases the spring constant, which provides more low-frequency and less high-frequency resolution.

The pitch produced by a string can also be determined by measuring string tension with strain gauge load cells. A strain gauge is a sensor whose resistance varies according to the force applied to it. It consists of a metallic foil that is secured in a flexible enclosure. As the foil is deformed, it becomes narrower and longer in one direction or broader and shorter in the other, increasing or decreasing its electrical resistance, respectively. The change in resistance is typically measured using a Wheatstone Bridge circuit, which comprises a network of four resistors that form two parallel voltage dividers. The strain gauges are mounted on a beam that deforms as force is applied (Figure 6.2.33).

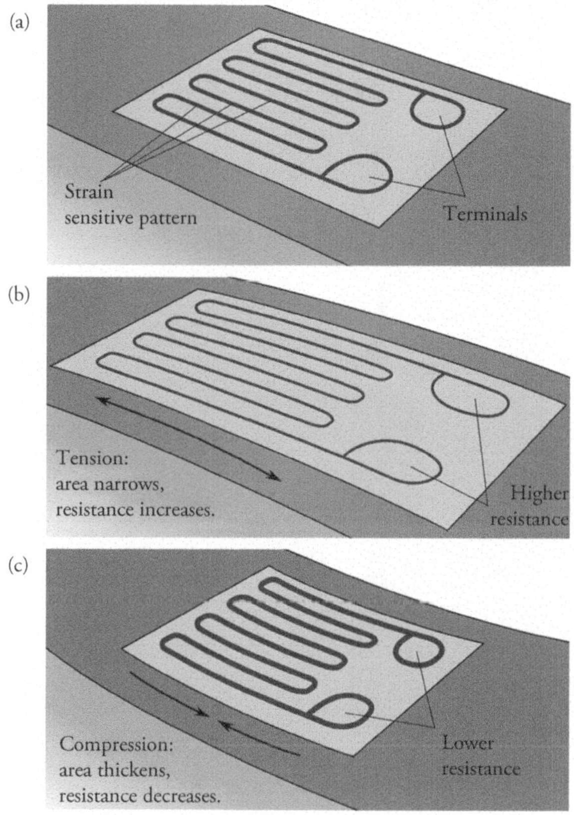

FIGURE 6.2.33. **A strain gauge response to force.**

The advantages of strain gauges include their precision, lack of moving parts, affordability, and widespread availability. Disadvantages include that they are non-linear, sensitive to temperature, and require careful calibration.

The MPR Lab experimented with strain gauge load cells to measure string tension in the instruments *Cyther v3* and *Whamola*. In these settings, the load cell was used to anchor the ball end of the string while the other end of the string was secured to a motor-driven tuning machine (Figure 6.2.34). Some experiments revealed that the load cells were capable of precise measurements. In other contexts, load cell readings were less accurate. In these cases, it was hypothesized that structural aspects were a factor in accuracy, where the tension of a string deformed not only the load cell but also the body of the instrument to which the load cell was attached (which did not have a strain gauge attached to it). This problem is potentially compounded when multiple strings are present: the tension changes on strings B and C affect the deformation of the body of the instrument and, thus, the load cell to which string A is attached, even if the tuning of string A is not intentionally changed. One solution is to strengthen the body of the instrument or to exert counterbalancing forces that oppose the string tension. Another idea is to isolate the bodies anchoring each string, although this would affect the instrument's resonance. A third path would be to place strain gauges on the body of the instrument to better understand its deformation under string tension, which could be reconciled with the string-anchor load cell values for a particular string tension.

FIGURE 6.2.34. **Load cell on MPR Lab's Whamola.**

COMBINATIONS

Combinations of the methods described help achieve musical requirements and broaden the palette of musical possibilities. For example, a guitar affords string length changes to produce different pitches, but it also consists of multiple strings to increase the pitch range, allow chords to be played, and widen timbral possibilities (as the material properties of strings differ, so do the timbres they produce). The mechatronic string instruments *GuitarBot* and *MechBass* utilize this approach. *Mach* of Z-Machines combines *fixed* and *movable* components, featuring a block of solenoids that can travel up and down the neck. This allows for rapid sequential notes within a reasonable pitch range, while also accommodating shifts in register. The speedy pitch production of fixed tangents can be combined with a movable bridge that is capable of microtonal tuning and portamenti, as in the design of EMMI's AMI (Rogers et al., 2015). A similarly capable system could be realized by swapping a movable bridge with a tension-varying device. There is no obligation to limit your design to a single approach. Different methods have different affordances and constraints; mix and match according to your musical vision.

Damping

To achieve monophonic lines, discrete durations, and musical articulations, string vibrations must be attenuated. A guitarist does this manually with the palm of his hand. Damping can also be achieved mechanically, as evidenced by the intricate and ingenious action of a piano or the jacks of a harpsichord (Figure 6.2.35).

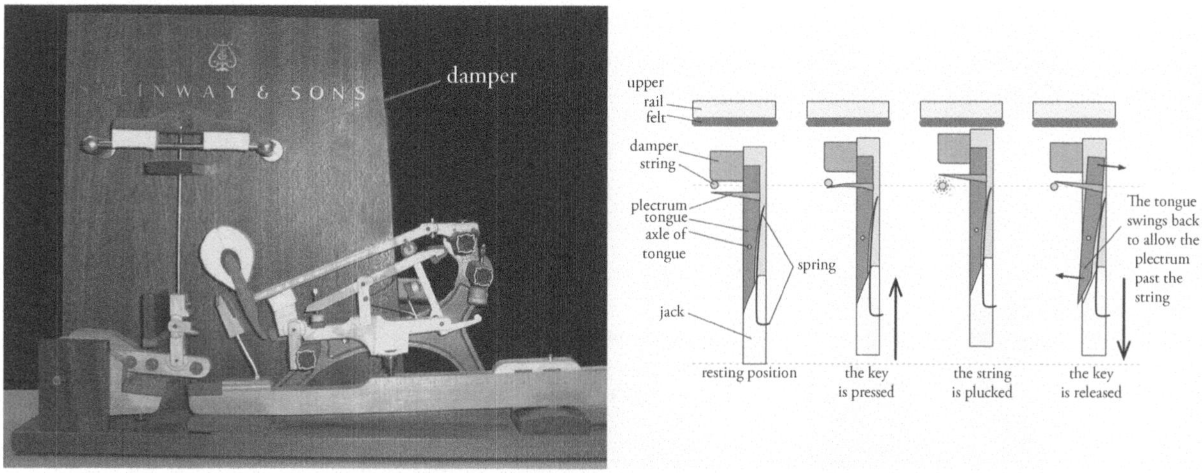

FIGURE 6.2.35. **Left: Piano action (image from By Leonce49 a.k.a Hans Weingartz—Transferred from de.wikipedia to Commons by Lukas9950 using CommonsHelper. (Original text: selbst fotografiert), CC BY-SA 3.0 de, https://commons.wikimedia.org/w/index.php?curid=8909582). Right: The action of a harpsichord jack (image by Original by Njohan, English captions and other edits by Jeff Dahl—Own work, CC BY-SA 4.0, https://commons.wikimedia.org/w/index.php?curid=3140667).**

When a piano key is pressed, the damper lifts off the string and the hammer strikes. When the key is released, the mechanism returns the damper to the string to attenuate its vibrations. The incorporation of electromechanical actuators offers new possibilities in damping. Rather than a binary scenario where an interface allows either a sustained, unmediated sound or silence, distinct excitation and damping mechanisms can be combined to produce a range of articulations and timbral effects. Controlling each actuator independently also simplifies the mechanism design. In a piano, the damper's position is coordinated with the key's position through numerous parts, including the spoon, damper lever, and damper tray. These mechanical parts can be omitted if another actuator performs the damping function.

Dampers are common in mechatronic string instruments. Damping systems are often solenoid-based. The damper of EMMI and the MPR Lab's PAM consists of a custom-made acrylic arm covered with foam, which connects perpendicularly to the armature of a pull solenoid. When the solenoid is activated, the string is damped. The damper can produce a variety of musical effects. The most obvious scenario is activating the damper at a note's offset. More nuanced usage involves modulating damper activation relative to stopping and picking actions on a scale of 1–50 ms, producing articulations that are uniquely machinic. The damper can also be held on the string during stopping and picking actions, presenting a third possibility, also discussed by Singer et al. (2003). The MPR Lab's *Cyther* features independent excitation and damping actuators: push solenoids positioned beneath the strings. Beyond the typical damping function, the dampers on *Cyther* can excite the strings, creating a softer, muted timbre that I have used in several musical works (PAM is also capable of this kind of excitation). This kind of specificity and precision is made possible by computer-controlled electromechanical instruments and is part of their artistic potential. Many of the instruments previously mentioned contain dampers using different methods: LEMUR's *GuitarBot* uses a single solenoid, and Trimpin's *Jackbox* uses two damping solenoids per string. The benefits of the solenoid approach are that it is simple and effective at delivering the essence of a damping function: to attenuate vibration. This strength is also a limitation, as the solenoid is only capable of binary states: on or off.

Damping with motors capable of incremental positions expands sonic possibilities. *MechBass* and *Swivel* use servos to attenuate string vibrations (McVay et al., 2011; Murphy, Kapur et al., 2013, see Figure 6.2.36).

FIGURE 6.2.36. **Damper of MechBass (image from McVay et al., 2011).**

As previously mentioned, *StrumBot* and *Protochord* each use a servo-driven rotating carriage that holds a stopping lug and a damping lug (*StrumBot* has two sets of lugs, while *Protochord* has one). The material covering the damping lug(s) affects how the string is damped: *StrumBot* uses felt, while *Protochord*'s lugs are covered with silicone.

The advantage of using a variable-position motor (such as a servo) versus a binary actuator (such as a solenoid) is that incremental adjustments change the relationship between the string and the damping material, resulting in varied timbres and amplitude envelopes. For example, by adjusting the clamping pressure of the dampers, *StrumBot* can modulate sustain time from 0.2 s to 1.5 s (R. Vindriis & Carnegie, 2016).

In all cases, damping must occur without changing pitch. Avoiding the latter requires applying enough force to dissipate the string's vibrational energy but not so much that it will affect the string's vibrating length and tension.

Structure

Joining sonic object(s) and excitation, pitch-producing, and damping mechanisms involves a structure that satisfies visibility requirements and holds parts together in a stable way. In Chapter 4, the following topics were introduced to help envision structural possibilities:

Summary of Form and Structure Design

- properties and relations
- case / body / frame
- mounting
- transportability

Of the *properties* and *relations* described in Chapter 4, the extent to which a structure is *conspicuous* or *invisible* contributes significantly to a string machine's physical *form*. *Emulative* examples, and particularly the anthropomorphic ones, require structures that make human-like form and motions conspicuous. Machines built around existing instruments offer more varied possibilities. In some cases, actuation components dominate the structure (e.g., Goldstein's Ro-Bow; Figure 6.1.6), while in others, they are minimally visible (e.g., the automated instruments of Ragtime West, Figure 6.1.5). The approach of Ragtime West (Caulkins, 2017) is interesting because minimizing the actuating mechanisms means that a *conspicuous* structure (e.g., a cabinet) is required to *invisibly* house the complex network of necessary components. The conspicuous and the invisible are defined by each other.

The extent to which a structure will be *conspicuous* or *invisible* also depends on the number, scale, and positioning of actuating mechanisms, as well as the way they move. The greater the number of actuators and the smaller their movements, the easier it is to obscure the visibility of the instrument and sonic object(s). Actuator positioning is also essential: the movements of push solenoids oriented perpendicular to the strings will be barely detectable if looking from above. As previously mentioned, some pitch-producing methods generate larger and more visible movements than others (e.g., a single moving bridge vs. an array of fixed tangents); thus, *function* and *form* are related in string machines via the method by which pitches are produced. The same could be said of excitation: a four-bar bowing mechanism generates more visible motion than a push solenoid positioned next to a string.

Integrated electrically amplified machines (*GuitarBot*, PAM, *Swivel*, *MechBass*) reduce the need for the large acoustic resonators that define form and structure in many acoustic instruments. In these examples, structures holding the strings and actuating mechanisms are sleek, pragmatically satisfying their functional requirements. They appear as configurations of skinny rectangles (strings occupy a fair amount of space in one spatial dimension but little in the others). Without relying on the acoustic properties of wood, these structures can be made from materials such as metal. The result is a visual aesthetic that references past instrumental practices, but also industrial, technological, and minimalist qualities that characterize modern expression. Physical form is emphasized even more dramatically when structure is an artistic priority, as in Andy Cavatorta's *Gravity Harp* (Figure 6.1.8). An approach that prioritizes *integration* thus allows *form* and *function* to be balanced, with structure becoming a meaningful aspect of a work's creative expression.

Where a string instrument falls on the *single / complex ⇔ distributed / simple* dimension has implications for its structure. The core of many string machines is a (relatively) *simple* module that consists of excitation, pitch-producing, and damping mechanisms that interact with a string. Modules are duplicated and combined to make larger structures. *GuitarBot*, *MechBass*, and *Swivel* follow this paradigm, featuring multiple nearly identical modules that are physically connected and synchronized to create a unified instrument. These individual modules are networked to allow for coordination of multiple, physically *distributed* actuators, and electric transduction and amplification enable sound to be projected without each string driving a shared acoustic resonator (as is the case with most human-playable string instruments). Duplicating a simple string module results in a larger, distributed system.

I am inspired by this idea when envisioning new instruments. Years ago, I imagined a *distributed / simple* instrument containing a single string whose vibrations are transduced into an electrical signal that drives tactile transducers coupled to resonant bodies of different sizes and shapes (tin cans, acoustic instruments, glass vases, walls, windows, etc.), which are distributed throughout a room (David Tudor's *Rainforest* was a source of inspiration). The sonic object and its resonator(s) are physically detached, allowing for an array of timbres that depend on the resonating objects, their spatial position, and the acoustic properties of the room. The myriad timbral iterations connect back to a single source of sound, bringing coherence to the system as a whole.

The structure of *distributed / simple* instruments is determined by the extent to which modules are separated from one another. For *GuitarBot, MechBass,* and *Swivel,* the modules are relatively proximate, so the structural design was a matter of coupling modules together. In the conceptual multi-resonator string instrument described in the previous paragraph, the distance between resonators precludes their connection via physical linkages or beams. Instead, structure becomes a question of relative spatial configuration and the way that individual elements are networked.

Conversely, a *single / complex* system could consist of one mechanism with protean capabilities. Imagine a multi-DOF actuator that could play or damp multiple strings in different locations. This would require that the strings are closer together (compared to a *simple / distributed* system) and even joined, which would have implications for the instrument's structure. This, in turn, could affect the body's resonance and require it to withstand the combined tension of all the strings. A more complex actuator likely requires a bigger and more robust structure to secure it, given its size and range of motion. This may throw the instrument's visual proportions out of balance: imagine a string secured by a minimal, sleek resonator that also features a bulky actuator protruding from its base. These aspects can be brought into balance through structural modifications, for example, by making the resonator / string anchor more complementary in terms of size, materials, and shape.

The extent to which actuating mechanisms are *specific* or *general* will affect structure. While actuating systems designed to play particular instruments are necessarily *specific*, their structures may have some degree of *generality* by being *adjustable*. For example, a guitar-playing machine may have a clamping system that accommodates different body sizes and shapes. String-stopping actuators may also be adjustable, allowing a machine to play string instruments with different fret spacing (e.g., guitar, bass). *Adjustability* and *generality*, in this case, are a matter of both actuator and structure design to allow configurability. This is a case where simultaneously designing the structure and actuating mechanisms would be beneficial, rather than doing one before the other.

References

Animusic LLC (Director). (n.d.). *"Resonant Chamber"—Animusic.com* [Video recording]. Retrieved August 13, 2021, from https://www.youtube.com/watch?v=toXNVbvFXyk

Baginsky, N. (n.d.). *The Three Sirens—A robotic rock band.* Retrieved June 5, 2020, from http://www.the-three-sirens.info/binfo.html

Baginsky, N. A. (n.d.). *aglaopheme—Slide guitar robot.* Retrieved January 25, 2022, from http://www.baginsky.de/agl/agl_index.html

Barton, S., Prihar, E., & Carvalho, P. (2017). Cyther: A human-playable, self-tuning robotic zither. *Proceedings of the International Conference on New Interfaces for Musical Expression*, 319–324. https://doi.org/10.5281/zenodo.1176266

Berdahl, E., Backer, S., & Smith III, J. O. (2005). If I had a hammer: Design and theory of an electromagnetically-prepared piano. *ICMC*. International Computer Music Conference, 1–4.

Bloland, P. (2007). The electromagnetically-prepared piano and its compositional implications. *ICMC*. International Computer Music Conference, 125–128.

Borland, J. (n.d.). *Gibson's self-tuning guitar*. MIT Technology Review. Retrieved January 25, 2017, from https://www.technologyreview.com/s/408771/gibsons-self-tuning-guitar/

Caulkins, K. (2017, October 30). *Ragtime West Catalog*. http://www.ragtimewest.com/

Crandall, R. (2013, March 23). Ultimate music collectible: Automatic violin-piano. *Music Machines*. https://www.rickcrandall.net/ultimate-music-collectible-automatic-violin-piano/

Daito Manabe (Director). (2013, September 4). *Squarepusher × Z-MACHINES* [Video recording]. Rhizomatiks. https://www.youtube.com/watch?v=VkUq4sO4LQM Daito Manabe (Director). (2013, September 4). Squarepusher × Z-MACHINES [Video recording]. Rhizomatiks. https://www.youtube.com/watch?v=VkUq4sO4LQM

De Cheveigné, A., & Kawahara, H. (2002). YIN, a fundamental frequency estimator for speech and music. *The Journal of the Acoustical Society of America*, *111*(4), 1917–1930.

Frequency Sensing. (n.d.). Milwaukee Servoelectric Guitar Project. Retrieved February 2, 2022, from http://servoelectricguitar.com/construction4.htm

Gill, H. (2021). *Stepper motor or servo motor: Which should it be?* [White Paper]. Kollmorgen. https://www.kollmorgen.com/en-us/service-and-support/knowledge-center/white-papers/stepper-motor-or-servo-motor-which-should-it-be/

Guzman, J. de. (2024a, September 18). Pitch perfect: Enhanced pitch detection techniques (Part 1). *Cycfi Research*. https://www.cycfi.com/2024/09/pitch-perfect-enhanced-pitch-detection-techniques-part-1/

Guzman, J. de. (2024b, October 2). Pitch perfect: Enhanced pitch detection techniques (Part 2). *Cycfi Research*. https://www.cycfi.com/2024/10/pitch-perfect-enhanced-pitch-detection-techniques-part-2/

Gylling, M., & Svensson, R. (2017). *Robotic electric guitar tuner* [Bachelor's, KTH—Royal Institute of Technology]. https://www.diva-portal.org/smash/record.jsf?pid=diva2%3A1200615&dswid=-4499

Heet, G. S. (1978). *String instrument vibration initiator and sustainer*. Google Patents.

How does TronicalTune work? (n.d.). Tronical—Autotunes your guitar in seconds. Retrieved January 6, 2022, from https://www.tronicaltune.com/knowledgebase/how-does-tronicaltune-work/

Kamatani, T., Sato, Y., & Fujino, M. (2022). *Ghost Play—A violin-playing robot using electromagnetic linear actuators. Proceedings of the International Conference on New Interfaces for Musical Expression*. https://doi.org/10.21428/92fbeb44.754a50b5

Kapur, A., Trimpin, E. S., Suleman, A., & Tzanetakis, G. (2007). A comparison of solenoid-based strategies for robotic drumming. *ICMC, Copenhagen, Denmark*.

Kemper, S. (2020). *Tremolo-Harp: A vibration-motor actuated robotic string instrument*. New Interfaces for Musical Expression, Birmingham, UK. https://www.nime.org/proceedings/2020/nime2020_paper57.pdf

Li, C.-H. G., Lin, M.-C., Bautista, B. A. P., & To, B. E. (2019). A low-noise guitar robot featuring a new class of silent actuators. *IEEE/ASME Transactions on Mechatronics*, *24*(4), 1577–1585.

Long, J., Skelly, N., Murphy, J., Carnegie, D., & Kapur, A. (2017). Improving robotic chordophones with closed-loop embedded audio feedback systems. *Proceedings of the International Computer Music Association Conference 2017*. https://quod.lib.umich.edu/i/icmc/bbp2372.2017.055/1/--improving-robotic-chordophones-with-closed-loop-embedded?rgn=full+text;view=image;q1=robot

Make: television. (2009, February 27). How-to—Build a feedback piano. *Make: DIY Projects and Ideas for Makers*. https://makezine.com/article/craft/music/how-to-build-a-feedback-piano/

McPherson, A. (2010). The magnetic resonator piano: Electronic augmentation of an acoustic grand piano. *Journal of New Music Research*, *39*(3), 189–202.

McPherson, A. (2012). *Techniques and circuits for electromagnetic instrument actuation*, 6. https://www.nime.org/proceedings/2012/nime2012_117.pdf

McPherson, A. P., & Kim, Y. E. (2010). Augmenting the acoustic piano with electromagnetic string actuation and continuous key position sensing. *NIME*, 217–222.

McVay, J., Carnegie, D. A., Murphy, J. W., & Kapur, A. (2011). MechBass: A systems overview of a new four-stringed robotic bass guitar. *Proceedings of the 19th Electronics New Zealand Conference (ENZCon), Dunedin, New Zealand*, 145–150.

More about StringThing. (n.d.). Retrieved July 22, 2021, from http://martinriches.de/stringmore.html

Moy, N., Ngo, A., Robinson, J., & Tan, T. (2021). *Strumuké: Self Strumming ukulele. Musical Machines*. https://wp.wpi.edu/musicalmachines/2021/10/11/strumuke-self-strumming-ukulele/

Murphy, J. (n.d.). *Swivel*. Jim Murphy. Retrieved February 16, 2022, from http://jim-murphy.com/

Murphy, J., Kapur, A., & Carnegie, D. (2013). Swivel: Analysis and system overview of a new robotic guitar. *Proceedings of the 2013 International Computer Music Conference*. International Computer Music Conference, 445–448.

Murphy, J., McVay, J., Kapur, A., & Carnegie, D. (2013). Designing and building expressive robotic guitars. *Proceedings of the 2013 Conference on New Interfaces for Musical Expression (NIME), Daejeon, Korea*. http://www.nime.org/2013/program/papers/day4/paper12/36/36_Paper.pdf

Murphy, J., McVay, J., Mathews, P., Carnegie, D. A., & Kapur, A. (2015). Expressive robotic guitars: Developments in musical robotics for chordophones. *Computer Music Journal*. http://www.mitpressjournals.org/doi/abs/10.1162/COMJ_a_00285

Murphy, J. W., Mathews, P., Kapur, A., & Carnegie, D. A. (2014). Robot: Tune yourself! Automatic tuning for musical robotics. *In Proceedings of the International Conference on New Interfaces for Musical Expression*, 565–568. http://www.nime.org/proceedings/2014/nime2014_345.pdf

Overholt, D. (2011). The overtone fiddle: An actuated acoustic instrument. *New Interfaces for Musical Expression*, 4–7.

Overholt, D., Berdahl, E., & Hamilton, R. (2011). Advancements in actuated musical instruments. *Organised Sound*, *16*(2), 154–165. https://doi.org/10.1017/S1355771811000100

Placencia, J. P. Y., Murphy, J., & Carnegie, D. (2019). Exploring dynamic variations for expressive mechatronic chordophones. In M. Queiroz & A. X. Sedó (Eds.), *Proceedings of the International Conference on New Interfaces for Musical Expression* (pp. 413–418). UFRGS. https://doi.org/10.5281/zenodo.3673017

Placencia, J. P. Y., Murphy, J., & Carnegie, D. (2020a). Designing an expressive pitch shifting mechanism for mechatronic chordophones. In R. Michon & F. Schroeder (Eds.), *Proceedings of the International Conference on New Interfaces for Musical Expression* (pp. 309–314). Birmingham City University. https://doi.org/10.5281/zenodo.4813375

Placencia, J. P. Y., Murphy, J., & Carnegie, D. (2020b). Survey of hardware and software Design approaches for mechatronic chordophones. *Computer Music Journal*, *43*(1), 38–58. https://doi.org/10.1162/comj_a_00499

Raes, G.-W. (2012). *Synchrochord*. The Logos Foundation. https://www.logosfoundation.org/instrum_gwr/synchrochord.html

Richard Vindriis (Director). (2011, October 31). *BassBot: A self playing robotic bass guitar* [Video recording]. https://www.youtube.com/watch?v=P0qR0Ksiwyo

Richard Vindriis (Director). (2015, November 18). *StrumBot Presentation* [Video recording]. https://www.youtube.com/watch?v=lbwNu_Vg5ic

Rogers, T., Kemper, S., & Barton, S. (2015). MARIE: Monochord-aerophone robotic instrument ensemble. *Proceedings of the International Conference on New Interfaces for Musical Expression*, 408–411. https://pdfs.semanticscholar.org/de7a/ff6fdb9ac3e2cc0dfa5fd269368780944f27.pdf

Sankaranarayanan, R., & Weinberg, G. (2021, April 29). Design of Hathaani—A robotic violinist for Carnatic music. *International Conference on New Interfaces for Musical Expression*. NIME 2021. https://doi.org/10.21428/92fbeb44.0ad83109

Seth Goldstein (Director). (n.d.). *RO-BOW—A kinetic sculpture that plays a violin by Seth Goldstein* [Video recording]. Retrieved July 21, 2021, from https://www.youtube.com/watch?v=EPTUM2_bxnQ&t=1s

Shear, G., & Wright, M. (2011). The electromagnetically sustained Rhodes piano. *NIME Proceedings, Oslo, Norway*. New Interfaces for Musical Expression, Oslo, Norway.

Shibuya, K. (2011). Violin playing robot and kansei. In J. Solis & K. Ng (Eds.), *Musical robots and interactive multimodal systems* (pp. 179–193). Springer Berlin Heidelberg. http://link.springer.com.ezproxy.wpi.edu/chapter/10.1007/978-3-642-22291-7_11

Singer, E., Larke, K., & Bianciardi, D. (2003). LEMUR GuitarBot: MIDI robotic string instrument. *Proceedings of the International Conference on New Interfaces for Musical Expression*, 188–191.

Skinn, N. C., & Freeland, S. J. (1990). *Automatic musical instrument tuning system*. https://www.google.com/patents/US4909126

Skinn, N., Strazzabosco, F., Kith, S., & Johnson, B. S. (2008). *Apparatus and method for self-tuning stringed musical instruments with an accompanying vibrato mechanism* (United States Patent US7446248B2). https://patents.google.com/patent/US7446248/en?oq=axcent+tuning+systems

Tait, A. (2023). Firmware design for a mechatronic chordophone. *Wellington Faculty of Engineering Symposium*. https://ojs.victoria.ac.nz/wfes/article/view/8381

Thompson, W., & Berdahl, E. (2021, June 1). An infinitely sustaining piano achieved through a soundboard-mounted shaker. *International Conference on New Interfaces for Musical Expression*. NIME 2021. https://doi.org/10.21428/92fbeb44.2c4879f5

Toyota unveils personal-transport, violin-playing robots. (2007, December 6). Toyota Motor Corporation Official Global Website. https://global.toyota/en/detail/275807

Trail, S., Tzanetakis, G., Jenkins, L., Cheng, M., MacConnell, D., & Driessen, P. (2013). *STARI: A self tuning auto-monochord robotic instrument*. 405–409. http://ieeexplore.ieee.org/abstract/document/6625511/

Vindriis, R., & Carnegie, D. (2016). StrumBot—An overview of a strumming guitar robot. *Proceedings of the International Conference on New Interfaces for Musical Expression*, 146–151. https://doi.org/10.5281/zenodo.1176135

Vindriis, R. G., Carnegie, D. A., & Kapur, A. (2011). A comparison of pick-based strategies for robotic bass playing. *Proceedings of the 18th Electronics New Zealand Conference*, *21*, 22.

Wang, Z. J., & Ortega-Sanchez, C. (2012). Electronic assisting violin tuner. *TENCON 2012-2012 IEEE Region 10 Conference*, 1–6. http://ieeexplore.ieee.org/abstract/document/6412214/

Wouter van Veldhoven (Director). (2014). *Automated composition for violin bow, zither, toy piano and tape recorders* [Video recording]. https://www.youtube.com/watch?v=iCI9xCfTE0Q&list=PLHs4qcpz6eqKfcxIniDxlAKpZnIGGVU7W&index=5

CHAPTER 6

STRINGS

Part 3: Making

Realization

BOM, FABRICATION, AND COST

The BOM and fabrication considerations associated with making a machinic string instrument are similar to those of other kinds of projects, so revisit the general descriptions of those topics in Chapter 4: The Process of Making a Musical Machine. For string instruments, a primary consideration in the fabrication process is string tension. The cumulative tension exerted by strings on the structure that secures them can be significant, particularly if they are thick and there are many of them, which can affect the instrument's intonation. This tension can be balanced by the quality and thickness of materials (e.g., thick steel will flex less than thin acrylic), structural reinforcements, a counterforce device (such as a spring), or a combination of these elements. Constant retuning is annoying, so make your structure resilient. Prototyping in scrap wood is often a good way to get a sense of how the instrument will lay out. In terms of assembly, adjustability in string-stopping and picking mechanisms is usually beneficial, as small position changes can meaningfully affect the sound produced. Regarding cost, many machines use solenoids to change the effective vibrating length of a string. In most cases, the force required to stop the string is relatively minimal, so smaller, less powerful actuators can often do the job, which can help save money.

Control and Feedback

String machines, like their percussion counterparts, can be controlled by programs stored on microcontrollers or networked to other computers that send instructions via communication protocols such as MIDI or Open Sound Control—OSC. For those requiring a primer (or who would benefit from a refresher), the Companion Website ▶ contains sections devoted to *Sensing, Communications, and Control*, which contain technical descriptions and projects that can help you collect and interpret information, as well as connect and control system components.

Specific to string instruments, feedback and control are essential aspects of both excitation and pitch-producing actions. Regarding excitation, it is desirable to know the specific location of components (such as a pick) to ensure proper timing and dynamic levels of notes produced. Consider a stepper motor-driven pickwheel. The number of steps required for each picking action can be calculated precisely. The problem is that sometimes steps are skipped, which causes the pick to come to rest in unintended locations, leading to notes played early, late, or not at all. To avoid such inaccuracies, a motor with an encoder can provide feedback on the pickwheel's location, allowing for position adjustments as needed. Similarly, if the picking mechanism has height-displacement capabilities, knowing the precise vertical position of the pick will allow the implementation of a control system that ensures proper note onsets and dynamics.

At "higher" levels of pitch-producing actions, questions of control include who or what determines when and how the action occurs. Sometimes, humans originate such commands. In a human-playable instrument equipped with a *calibration* system, this action may originate from a player selecting a tuning, for example, via a button on the instrument. This paradigm is evident in dynamic cooperative systems such as the original Milwaukee Servoelectric Guitar, in which a tuning action is initiated by manually selecting a pitch preset via a keyboard or keypad (*Theory*, n.d.). In *autonomous* systems, pitch-producing actions emanate from computers that contain either predetermined musical works encoded in software, algorithms that generate notes in real time, or some combination of the two. In *cooperative* systems, control of a string instrument is shared between human and machine. Often, this involves dividing string-exciting and pitch-producing duties, which must be synchronized (see the *Prior Art* section of Chapter 6, Part 1, for examples). Possibilities that synthesize these approaches, where human interface elements interact with computer algorithms, represent fertile ground for future efforts to explore.

At "lower" levels, feedback and control affect how systems move to reach positions that produce desired pitches, which usually involves varying string length or tension. Most pitch-altering machines follow a control paradigm in which the system senses a proxy for the string's pitch (e.g., its vibrational frequency or tension) and compares it to the desired pitch value. Discrepancies between the two result in commands that drive a motor-mechanism to vary the length of or tension on a string (e.g., via a moving bridge or rotating a tuning machine) to produce the desired pitch. As explained in Project 16 (Companion Website ▶), a control loop feedback system such as PID is typically required to reach a desired position as rapidly and accurately as possible.

It is also possible to generate pitch-altering motor commands via electrical circuits. The control system of the Milwaukee Servoelectric Guitar is shown in Figure 6.3.1. Each pitch is associated with a string tension stored on a microcontroller. These tension signals are used to generate command voltages to independent closed-loop servo systems that interact with each string. Each command signal is summed with a feedback signal from a potentiometer (which is inverted), creating an "error voltage" that is fed into a high-gain amplifier. The signal from the amplifier drives the motor that controls the tension on the string. As the motor moves, the potentiometer value changes until the error voltage becomes 0, at which point the motor stops. Oscillations that result from over- or under-rotating are minimized by setting an appropriate amplifier gain (*Theory*, n.d.).

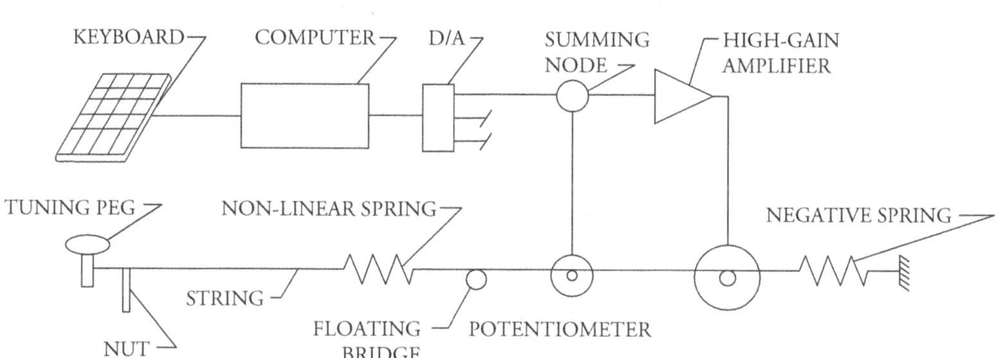

FIGURE 6.3.1. **Milwaukee servoelectric guitar control system (*Theory*, n.d.).**

Evaluation

The methods used to evaluate string machines depend on the nature of the design in question, but two areas need to be addressed in most cases. The first is pitch accuracy. If there are no frets, strings can produce minute variations in pitch, making proper intonation problematic. Even if there are frets, intonation can vary from structural problems. Test pitch accuracy across the instrument's range using different starting pitches, intervals, and directions (up and down).

The second area is time, specifically speed and precision. Regarding speed, it is desirable to know how fast an instrument can play when repeating a single pitch and when playing pitch intervals of various sizes. For string instruments, this typically involves two distinct systems responsible for excitation and pitch production. Test each one independently; test them together. They will likely have different capabilities (e.g., fretting tangents can move faster than the picking mechanism), which should be considered when specifying the control system. Regarding precision, determine the accuracy of temporal intervals of different sizes, as well as the system's latency. For string instruments, this will vary depending on the design. A system with one tangent per pitch is going to produce intervals more quickly than one that incorporates a sliding carriage or tunes via string tension changes. These aspects of temporal performance can be evaluated using audio recordings. Programs such as Audacity allow you to record audio and precisely measure the time intervals between events at the millisecond level.

The other aspects of your machine that need to be evaluated should align with the requirements you specified. Dynamic range will vary depending on whether the instrument is acoustic or electric. Follow the guidelines given in Chapter 4 to ensure that the methods you use are appropriate, rigorous, and repeatable.

Example

The ideas from this chapter and the framework introduced in Chapter 4 will guide the ideation, design, and realization of a string-playing machine.

IDENTITY, PURPOSE, AND USAGE

This string instrument will allow composers to explore possibilities of rhythm, pitch, melody, and articulation within a single voice; thus, it will be a monochord. It should be able to play virtuosic lead and solo lines. It is not intended to be associated with a particular musical style, so it should be capable of timbral and dynamic variability. It will primarily inhabit studio settings, providing a tool for composition and recording, interfacing with popular DAWs via MIDI communication. It should have some visual presence for creative inspiration, but it should not emulate human or animal forms; instead, it should convey a machine aesthetic.

REQUIREMENTS

From this initial vision, we can establish musical, sonic, visual, and interaction requirements. The instrument will be equipped with steel strings that can be electrically amplified to produce glistening clean sounds, textured distorted sounds, and everything in-between. It should be able to produce plucked and hammered timbres, with sustain or damping that can be controlled. It should be able to play rapid passages with IOIs of <100 ms with temporal precision within 30 ms. The time required to produce all pitch intervals should be uniform, and the pitch range should be at least two octaves, chromatically tuned. The instrument will take a soprano role; thus, the bottom of the range should start at the pitch C5. Following the perceptual literature, pitch accuracy should be within ±12 cents. As the instrument is envisioned as playing lead lines, it will be monophonic. Some variation in dynamic range is desired, particularly to produce different kinds of articulations, such as accents. A wide dynamic range isn't as necessary for projection purposes, given the output will be electrically amplified. Visually, the instrument will be minimalistic, avoiding extraneous structure wherever possible. It will be made of metal and plastic, embracing its machinic nature. The range of motion can be small, but it should be made apparent to the greatest extent possible. The machine should be able to interpret and autonomously act on musical instructions in the form of MIDI messages sent from a host computer, enabling interactions with a wide variety of musical software environments.

PROPERTIES AND RELATIONS

Instead of being a player of an instrument, the instrument's sonic object and actuating system will be *integrated*. As mentioned in the requirements, the machine should function autonomously and thus will not be human-playable. The sonic output will be amplified electrically, so the *form* of the instrument does not need to accommodate a large acoustic resonator. Instead, the design will be minimal, generally providing only what is *functionally* necessary while at the same time *conspicuously* showing the means of action. The instrument is not intended to *emulate* an existing string instrument. Given the monochord requirement, the actuators will be compactly configured, oriented around a *single* string. The preceding implies that the mechatronic system will be *specific* to the sonic object, as opposed to something that could fit a range of existing instruments. The chromatic tuning requirements mentioned above can be realized by *fixed* stopping mechanisms. Some degree of adjustability may be needed to fulfill the dynamic range requirements.

DESIGN

We begin the design with the sonic object: the string. As we aspire to a minimalist aesthetic, we need a structural piece that will secure the string and withstand the tension exerted by the tuned string without significant warping. We need a material that is strong, easy to work with, and embraces a machinic visual identity: an aluminum bar is a good choice. On one end of the bar, a perpendicular metal plate will secure the ball end of the string. The other end of the string will be tensioned by a standard guitar tuning machine. A second perpendicular plate, located near the tuning machine, will provide symmetry and balance. The hole securing the ball end of the string will be aligned with the hole in the post of the tuning machine so that a separate bridge is not necessary (see Figure 6.3.2).

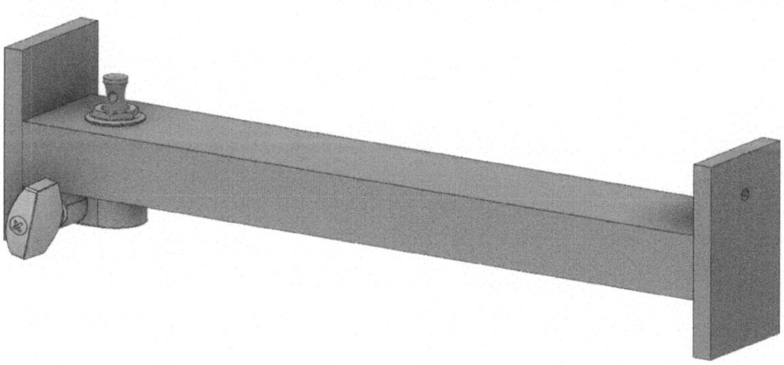

FIGURE 6.3.2. **String machine frame and tensioning components.**

Since this will be an electric instrument, a large acoustic resonator is not required, but the frame will have its own resonances and idiosyncratic timbral character. As there is only one string, we will use a single-string electromagnetic pickup.

The next part of the design involves determining how the string will be excited. It was established in the requirements that plucked timbres were desired. As we learned in the *Plucking* section of Chapter 6.2, this can be achieved using rotational, oscillating, or reciprocating systems. Here, we will integrate a reciprocating system that uses a harpsichord jack mechanism. The advantages of this approach are that the actuation is simpler than other methods, as the motion is linear; the harpsichord mechanism can both pluck and damp, thus obviating the need for a second damping system; and the actuator can be located underneath the instrument, avoiding a bulky structure on the instrument's surface. Harpsichord jacks are rare in musical string machines, so this choice will shape the device's unique visual and sonic character. Making your own jack is a possibility, but there are nuances in configuring the spring return. Alternatively, jacks are commercially available, such as from harpsichord repair companies.

When selecting an actuator, the jack must move up and down rapidly within a limited range, making a solenoid a suitable choice. The harpsichord jack will attach directly to the armature of the solenoid, as seen in Figure 6.3.3.

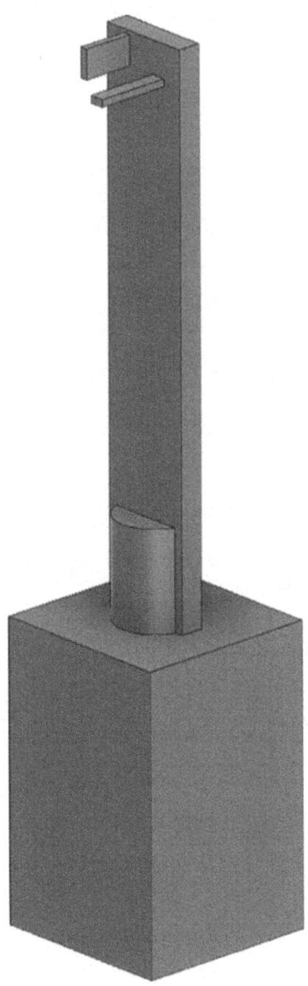

FIGURE 6.3.3. **Plucking mechanism using a solenoid and harpsichord jack.**

When selecting a specific type of solenoid, we must consider its power consumption. Will activating the solenoid result in a plucking or damping action? The former is attractive because we want precise control of the timing, force, and velocity of a pluck, which can be achieved through electrical means. In addition, the plucker will be at rest more than it will be activated, so the rest state should be driven mechanically (e.g., a spring), while the active state can be driven electrically. We will, therefore, use a push solenoid. Make sure to consider this question! I have seen students neglect it, only realizing they would prefer the reverse configuration after the machine was built.

Lesson: Figure out what actuator state (on / off) is best suited to a musical action (e.g., plucking, damping), keeping in mind that it is generally best to minimize power consumption.

The force and velocity required to pluck the string must also be determined. This could be accomplished experimentally. Create a slot in a string test rig that allows the jack to slide in and out, and attach the jack to a force gauge to measure the amount of force required to pluck the string. This picture is not complete as we lack an understanding of actuator velocity requirements, but it is a starting point. Alternatively (or in addition), you can experiment with solenoids in your possession to provide a general sense of what specifications might work. Here, we will assume that the conclusion of these experiments indicates a force of 7 N is needed to pluck the string. Looking at the products of Ledex, a respected solenoid manufacturer, we see that the STA tubular solenoid 102M–STA 26 × 30 is 25.91 mm wide (there are no adjacent actuators, so we have the width of the neck to work with) and can produce 8 N of force, given a 25% duty cycle and nominal stroke (how far the armature is extended). The force-stroke curves in a solenoid datasheet indicate that the amount of force exerted depends on power, duty cycle, and stroke (the smaller the stroke, the greater the force exerted; see Figure 6.3.4). In this case, there will be some distance between the string and the resting position of the plectrum, which should be considered when examining the force/stroke curves.

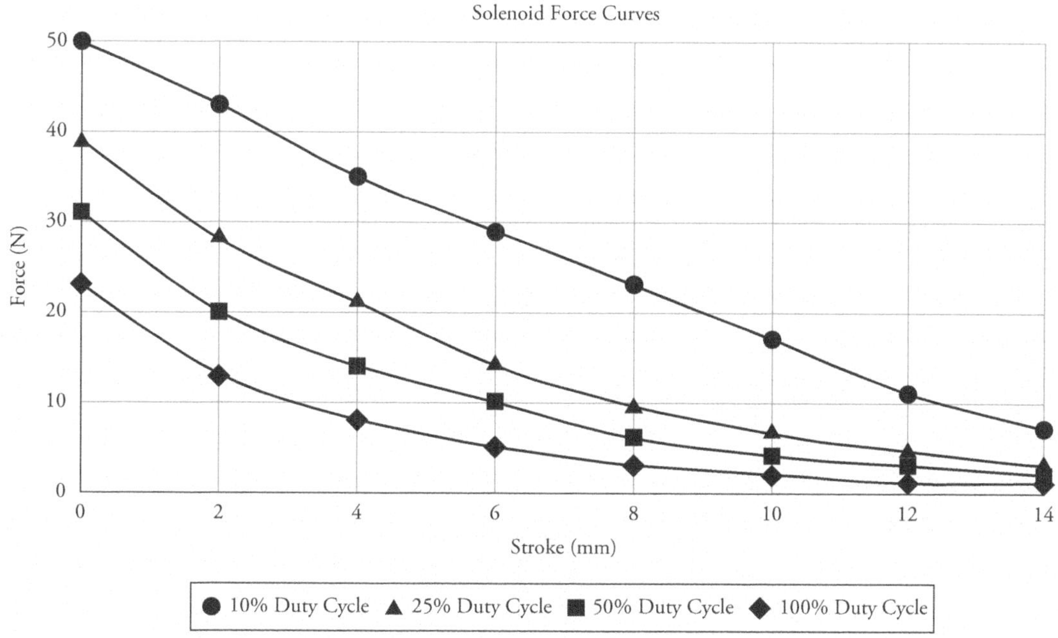

FIGURE 6.3.4. **Solenoid force-stroke curves (based on the performance of ledex solenoid).**

The armature can rotate freely within the body of the solenoid, so a method is needed to secure the jack's position so that it reliably contacts the string. Here, we will accomplish that with a slot in a *guide plate,* separated from the instrument neck by standoffs. The stopping plate will be made of clear acrylic to avoid visually obscuring the actuators. One of the musical requirements of the excitation system was dynamic range. In this case, we can vary the force and velocity of the solenoid armature, and thus the harpsichord jack, through PWM. We can now add the plucking mechanism to the design, as seen in Figure 6.3.5:

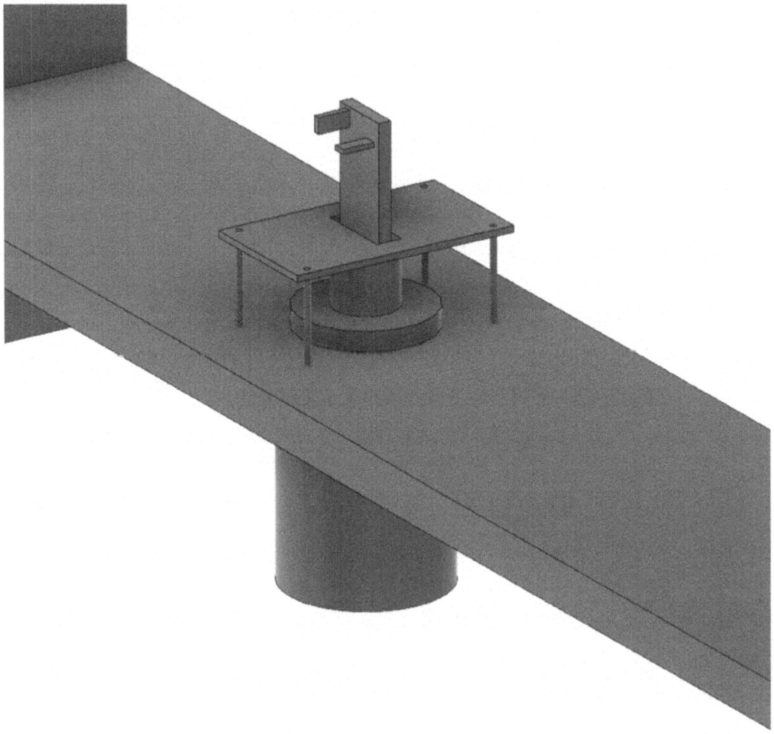

FIGURE 6.3.5. **Plucking mechanism with a guide.**

The next significant choice involves how pitches will be produced. The musical requirements state that pitches should be produced rapidly at periods that do not vary depending on the pitch interval. The latter precludes both movable / track and tension-based systems. A fixed-tangent configuration can achieve the stated requirements, so it is selected here. The pitch range of the instrument is to be two octaves, starting at C5, chromatically tuned. Visually, the design aspires to be minimalistic, but at the same time, motion should be apparent to make the causality of sound production clear. One option would be to have perpendicular tangents attached to the armatures of solenoids, as in EMMI's *PAM.* A sleeker design consistent with the plucker features solenoids positioned underneath the neck, with a "C"-shaped effector attached to the top of each armature (an upside-down "U" shape is also a possibility). Each effector's rotational movement would be constrained by a slot in the *guide plate,* as in the case of the plucking jack. The underside of the effector contacts the underside of the *guide plate* when the solenoid is inactive, driven by the spring return, ensuring that the armature does not travel past its desired position. One side of the C-effector would be open, allowing it to be removed from the string if needed. The string-stopping mechanism is pictured in Figure 6.3.6.

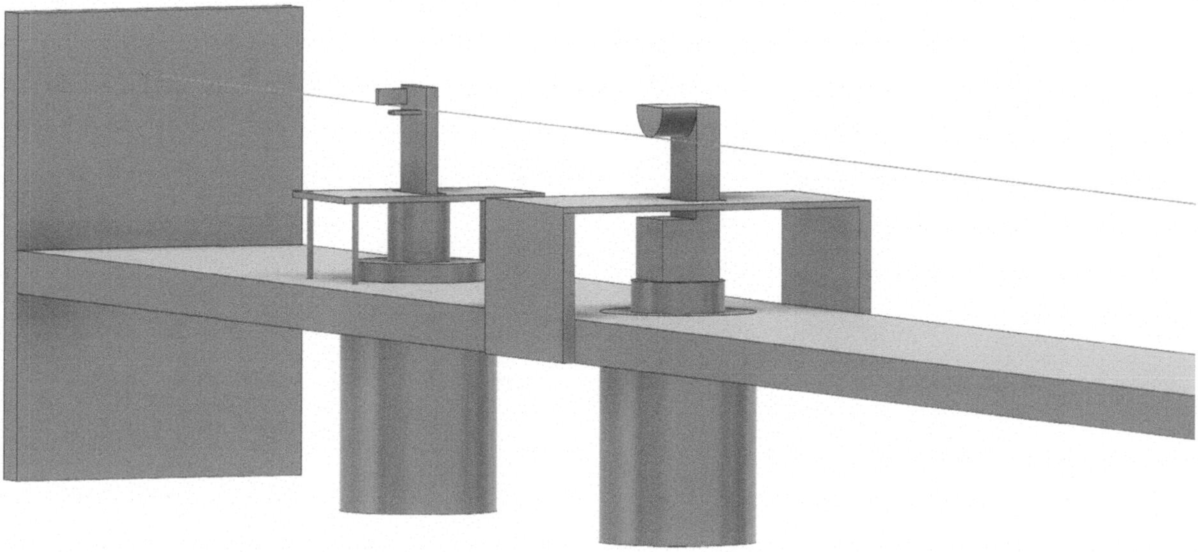

FIGURE 6.3.6. **String-stopping mechanism with a guide plate.**

Note the two different designs for the guide plates: the plucker guide plate uses standoffs, while the string-stopping mechanism guide plate is attached to the frame with brackets. Either design could work; which one you choose (or they could be combined) depends on visibility requirements, aesthetic preferences, availability of materials, and so on. Guide plates could be foregone completely if the armatures of the solenoids are manufactured to limit rotation (though these solenoids are not as common).

The string-stopping mechanisms need to be spaced across the neck to produce the two-octave chromatic scale desired. Since we are using 12-TET, the position of the stopping mechanisms can be calculated using a factor of 0.94387431, as described in the Instrument Design section of the Companion Website ▶. When determining the scale length, it is essential to consider that the spacings must accommodate the dimensions of the solenoid enclosures. Twenty-four pitches and a scale length of 25″ or 635 mm (which is similar to a PRS electric guitar) produce the following spacings seen in Table 6.3.1.

With stopper positions defined, the next step is to specify the solenoids that will actuate the stopping mechanisms. The tightest spacings at the highest pitches are ~10 mm: solenoids that narrow are difficult to procure, so the scale must be lengthened or the solenoids must be positioned on either side of the string in alternation. Making the scale length longer will make transportation and setup more challenging (it is easier to work with standard dimensions around which cases, stands, tables, etc., are also designed), so we will alternate solenoids here. This design produces the smallest spacings of 18.88 mm, which becomes the maximum width specification for our solenoids (in reality, we require less than this to allow room to work within). The remaining specification to be determined is the holding force, as the solenoid must be powerful enough to stop the string cleanly. We can determine this value experimentally using a force gauge (a scale or weight would also work) on an existing instrument or test rig that has similar string qualities (e.g., vibrating length, gauge, tension) to those of the design. Assume this experiment reveals that 2 N are required to produce clear and clean (e.g., free of buzzing or unwanted harmonics) pitches across the range of the string. Stopping forces may vary along the string, so test at the extremes (e.g., near the nut and at the highest pitch) and a few intermediate positions. You can search for a solenoid with these specifications

TABLE 6.3.1. **Stopper positions for two octaves of a chromatic scale.**

Stopper	From previous (mm)	From nut (mm)	Stopper	From previous (mm)	From nut (mm)
1	35.64	35.64	**13**	17.82	335.32
2	33.64	69.28	**14**	16.82	352.14
3	31.75	101.03	**15**	15.88	368.02
4	29.97	131.00	**16**	14.98	383.00
5	28.29	159.29	**17**	14.14	397.14
6	26.7	185.99	**18**	13.35	410.49
7	25.2	211.19	**19**	12.6	423.09
8	23.79	234.98	**20**	11.89	434.99
9	22.45	257.43	**21**	11.23	446.21
10	21.19	278.62	**22**	10.6	456.81
11	20	298.62	**23**	10	466.81
12	18.88	317.50	**24**	9.44	476.25

on vendor sites (see the Companion Website for links ⯈), but I prefer manufacturers such as Ledex, which has established a reputation for producing high-quality actuators. Looking through their catalog (https://www.johnsonelectric.com/en/product-technology/motion/solenoids) reveals that the function and form of the STA Tubular Linear DC series, specifically, the 51M–STA 13 × 27, which has a diameter of 13.2 mm and a holding force of 2 N at 100% duty cycle (4 W), meets our requirements.

With the stopping mechanisms specified, they can be added to the design. This can be done manually by measuring and marking the center holes using the calculations in Table 6.3.1 and then drilling each hole with the corresponding diameter bit. Alternatively, the layout can be specified precisely in CAD software (Figure 6.3.7), which can then be used to laser-cut or machine the desired material.

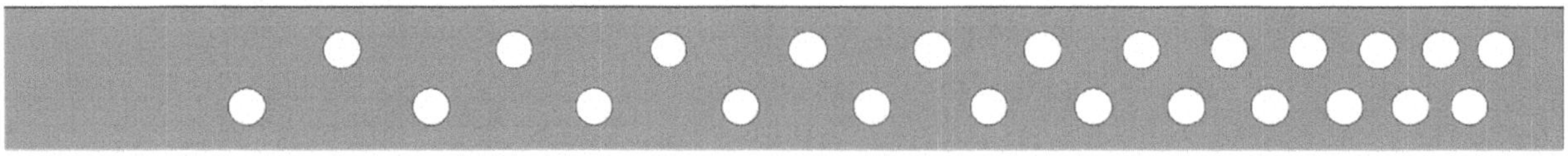

FIGURE 6.3.7. **Stopper layout.**

Electronics

The actuators in this design are solenoids, allowing us to utilize the circuit and code described in Project 1 on the Companion Website ⯈. The primary difference here is that we have 25 actuators to think about rather than just one, which has implications for the design of the electrical system.

The first is the power supply, which must accommodate all the actuators that will be simultaneously energized. If this was a zither or a piano, all 25 solenoids could be active, which means the power supply would need to be able to deliver the current drawn by all solenoids in total. Here, the instrument is a monochord, so only one of the stopping solenoids will be on at a time. Plucking and string stopping often coincide, so the power supply must be able to accommodate the total current drawn by both. Assume we test the solenoids, rated for 12 V, which together draw 750 mA; therefore, a 1 A power supply should be sufficient. Importantly, the code must be written to prevent multiple string-stopping solenoids from being activated simultaneously. If this occurs, the actuators will draw more current than the power supply can deliver, potentially leading to damage.

With components selected, the next step is to design the circuit. Determine how the MOSFETs will be configured. One option is to build multiple versions of the circuit described in Project 1. This would work, but it is not particularly economical in terms of space and connections. Alternatively, we could use an integrated circuit (IC) that contains multiple MOSFETs in one package, as described in Project 7. With that choice made, make a schematic showing how the various components will connect. This can be done with pencil and paper; software tools such as KiCad are also useful.

With the schematic finished, the circuit can be built. Connect the power and ground wires from each solenoid to the circuit board (you will need to measure and cut different lengths of wire), and the circuit board to the Arduino (output pins of the Arudino -> gate pins of the MOSFETs, common ground) as in Project 1. Assembling the circuit on a breadboard first is a good idea to ensure everything works properly. A breadboard can get you up and running, but connections can come loose, which can be annoying and time-consuming to troubleshoot. Eventually, you will want to move toward something more permanent. One option is to solder the components in place on a protoboard (some mimic the structure of breadboards, which makes the transfer between the two easier). An even better option is to have a PCB manufactured, which will save a lot of soldering time and effort (and potentially board space). Many PCB manufacturers can realize your board and ship it to your door; they are just a Gerber file upload away! (See the Companion Website ▶ for vendors.) Wire the electromagnetic pickup to a standard ¼″ jack, which can be connected to an external amplifier.

Structure

While we aspire to a minimalist design, the visual form of the current design is somewhat utilitarian. To address this, we can introduce more complex shapes and curved contours into the design, such as the one seen in Figure 6.3.8.

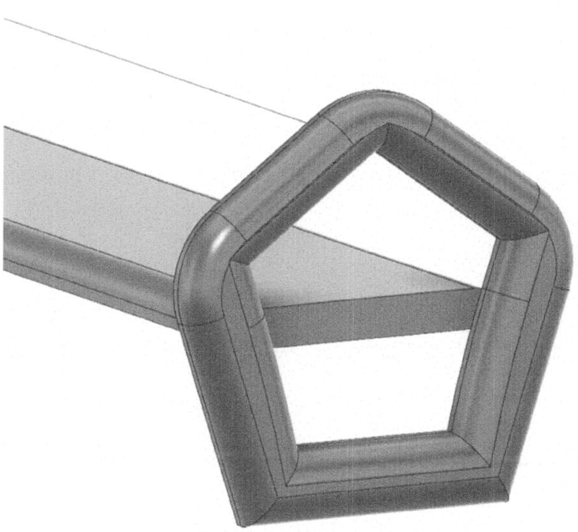

FIGURE 6.3.8. **Shaped and contoured endplate.**

The Arduino and circuit board can either be mounted on the underside of the frame or housed in a separate enclosure such as the one described in Chapter 5.3. If mounting components under the frame, ensure that live connections cannot be inadvertently touched, which can be dangerous. Seat the electromagnetic pickup in the frame directly underneath the string. Cut holes in the back of the instrument to secure the three jacks (USB, ¼″, and 2.1 mm barrel for power). The instrument will be self-contained; no other case, frame, or mounting accessories are needed. Because the dimensions are similar to an electric guitar's and the electronics are secured within the aluminum neck, the instrument should be relatively easy to transport. We may even be able to find a case for it!

FABRICATION AND COST

The body of the instrument is primarily aluminum, which can be cut either at a machine shop, using a third-party service (see the Companion Website ▶ for vendors), or by hand. The "C-effectors" that stop the strings and the endcap that holds the plucking jack can be 3D-printed and press-fit onto the solenoid armatures. The guide plate(s) can be made from laser-cut acrylic and attached to the neck using standoffs or brackets, as previously described. All components in this design can be secured with screws. Connect the host computer to the Arduino via a USB cable, a standard ¼″ audio jack carries the output of the electromagnetic pickup to an amplifier, and a 2.1 mm barrel jack joins the power supply to the circuit board.

CONTROL

To control the machine, we can incorporate code from Projects 1 and 2 on the Companion Website ▶. The solenoids can be turned on and off with the digitalWrite function, and we will use variables, arrays, and functions to handle stopping and plucking commands. Given that one of the requirements was translating MIDI messages into musical actions, we will also use code from Project 9 on the Companion Website ▶. The pitch range of the instrument is C5–C7, so the 24 solenoids will be mapped to MIDI notes 71–96 (72 is the open string), per the MIDI standard. We must ensure that only one stopping solenoid can be played at a time and that MIDI notes outside the specified range have no effect. Stopping a note is a binary operation (either on or off), so MIDI velocities that are >0 will activate the solenoid (on) and 0 will turn it off. Plucking can be controlled in several ways. One option is to have a toggle, which, when enabled, plucks every note. Another possibility is to dedicate a MIDI note (outside of the pitch range of the instrument) to plucking, such as MIDI note 0. Both options can be incorporated in an interface, but the latter is essential as it enables precise, independent control of the plucking mechanism. The plucker will be sensitive to velocity, so we will adapt code from Project 10 on the Companion Website ▶ in this control scheme.

Music: Strings

The abilities of machinic and human string players are represented in the Venn diagram that has been referenced throughout the book:

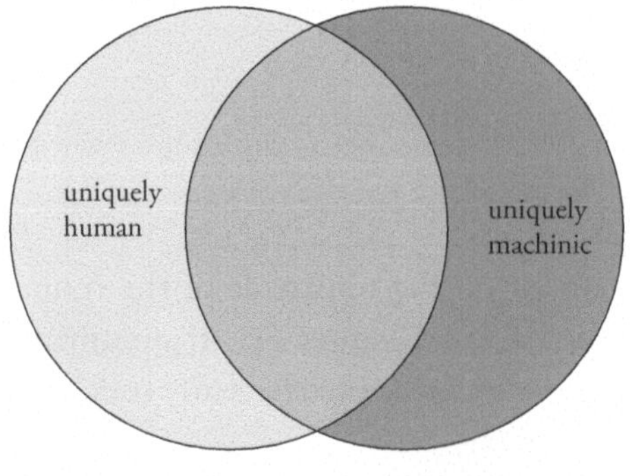

What is shared and what is unique depends on the machine's design. Aspects unique to human performers include the fine, nuanced movements that characterize string stopping and activation. Human string players stop a string using hammer-ons, pull-offs, variable pressure, vibrato, slides, and bends. They excite strings in equally diverse ways, using a panoply of objects (picks, fingers, fingernails, bows, ebows) at varied positions, angles, and forces. The temporal synchronization between stopping and picking produces a range of musical articulations. Together, the wide range of possibilities (partially) explains how human guitarists can be so different beyond the equipment they use or the sequence of pitches they play.

String machines also have unique musical capabilities depending on the particulars of their design. Fixed solenoid arrays perform pitch patterns and speeds unattainable by human hands, transforming the discrete into the textural. For example, fixed-solenoid string machines can trill so fast that the output no longer appears as the alternation of two notes. Instead, a unified sonic stream emerges where manipulating the actuation rate or actuation proportions (i.e., duty cycle) produces timbral effects rather than rhythmic ones. Similar effects can be achieved by playing extended pitch sequences with only the stopping mechanisms, resulting in legato textures. Multi-string machines can voice expansive harmonies and chord forms that are not bound by the number of digits on the human hand. Damping systems are critical in these designs, or else sustain can become unwieldy. As with percussion machines, the temporal precision of machines enables complex rhythms and new types of articulation. My compositional interest has been stoked by the musical effects of varying the time between stopping, damping, and picking. Human performers also vary these aspects, but machines do so precisely and repeatably, resulting in unique sonic statements. Human players spend years learning how to synchronize their stopping and excitation (e.g., picking, bowing) hands. Machines can produce such synchronization, but they are capable of offsets between the two that are unnatural for most human players.

In the middle is an expanse of territory that is available to both humans and machines, populated by reasonable speeds and familiar chord voicings. It is on this "common ground" that unique characteristics become apparent. In my composition, *Rise of a City* (Media Example 6.1.3 on the Companion Website ⓑ), a human guitarist and the string machine *PAM* share much of the same musical material. Juxtaposing the two reveals similarities, differences, and idiosyncrasies in fuller light. Reflexively, these compositional efforts help inform and guide the design process of subsequent instruments. After hearing the two together, you gain insight into formative questions: How similar should human- and machine-played instruments sound? Where do unique characteristics shine? What deficiencies are revealed? There is no single answer to any of these questions; rather, one gains a greater perspective on them by engaging in the dialectic between design and realization in both technical and artistic domains.

Summary: Strings

In this chapter, the template for design and realization developed in Chapter 4 was applied to the world of string instruments. The history of string-playing automata reaches back centuries. To navigate this rich lineage, ideas from the *Properties and Relations* section of Chapter 4 were employed, with a particular focus on the relations between *players* and *instruments*. This approach yielded four primary categories of string machines: *independent* players that often emulate humans, systems *attached* to extant string instruments, novel configurations that *integrate* sonic objects and actuators (which include kinetic musical sculptures), and *cooperative* instruments where human and machine interact through a shared physical medium. Establishing requirements begins with musical performance, and within that category, timbre is a primary consideration. Temporal requirements follow psychological thresholds, while pitch requirements should address range, set, tuning, and modulation. The requirements for dynamic range depend on the context within which the instrument is used. A range of articulations is possible on a string instrument: *legato*, *staccato*, and *accent* are among the most prevalent. Visibility and musical performance are often at odds in the case of string machines, as speed and accuracy are most easily realized through small movements. Establishing visibility requirements involves reconciling the two. The machine's visibility influences how it interacts with others. Internally, interaction results from sensing and feedback systems that integrate information about the string's vibration and tension, as well as the positions of the motors that stop and excite it.

Requirements provide a guide for designing a musical machine. Starting with the sonic object, the basics of string vibration were covered. The next step is to choose whether to acquire an instrument or build one from scratch. Building one from scratch is a reasonable possibility, which may involve establishing string anchors, bridges, nuts, tuning pegs / pins / machines, frets, resonators / radiators, and electric transduction and amplification. Considering the affordances and constraints of the instrument in terms of what is *idiomatic* for the player can help clarify direction when addressing whether to acquire or build. Strings can be excited directly through plucking, strumming, bowing, hammering, scraping, and sliding, or indirectly through air currents, sympathetic vibration, and electromagnetic manipulation. Methods of excitation expand the palette of timbral and articulatory possibilities. Plucking is perhaps the most common in musical machines, enacted through rotational, oscillating, and reciprocating mechanisms. Dynamics can be varied through electrical (e.g., PWM) or mechanical (e.g., height-adjusting mechanisms) means. Noise is often produced by these mechanisms, which can be attenuated by modifying existing actuators or designing new ones.

Different pitches can be produced on a string instrument by using multiple strings, changing the effective vibrating length of the string (typically with an actuated *tangent*), changing the tension of the string, or some combination of these options. Length-modifying tangents may be *fixed*, *adjustable*, or *movable*. *Movable* pitch-producing systems utilize linear tracks, rotating arms, and / or linkages. Tension-adjusting mechanisms are constituted by motor-driven gears or linkages controlled by algorithms that incorporate string frequency and / or tension feedback. These systems have been used to calibrate instruments for performance but have also been featured in *dynamic* (which change string tension while playing) and *cooperative* (which involve human–machine interaction) systems.

Damping is another primary component of string instruments used to control the duration of notes. Damping in string machines can be either binary (on/off) or variable, depending on the actuator and mechanism used. Varying a damper's position and force can produce a range of timbral and dynamic effects.

The structure of string machines can be envisioned by applying the *properties* and *relations* described in Chapter 4. Of these, the extent to which a structure is *conspicuous* or *invisible* is a significant contributor to a string machine's physical *form*. The *single / complex* ⇔ *distributed / simple* dimension is also meaningful, as many existing string machines are the concatenation of simpler modules. *Adjustability* and *generality* are also important structural considerations when thinking about the types of string instruments the machine might interface with.

The chapter concludes with an example of a string machine designed to play monophonic, rapid passages and wide intervallic leaps within two octaves of a soprano register. Notes are produced through stopping and plucking and attenuated via damping. The output is amplified electrically, and the instrument's visual aesthetics tend toward minimalism. The result of these requirements was a design that features 24 fixed pitch-producing tangents, a plucking / damping mechanism based on a harpsichord jack, an electromagnetic transducer, and a sleek design that minimizes superfluous structure while making the movements of the actuators apparent.

Reference

Theory. (n.d.). Milwaukee Servoelectric Guitar Project. Retrieved January 11, 2022, from http://servoelectricguitar.com/theory.htm

CHAPTER 7

AEROPHONES

Part 1: Context and Ideation

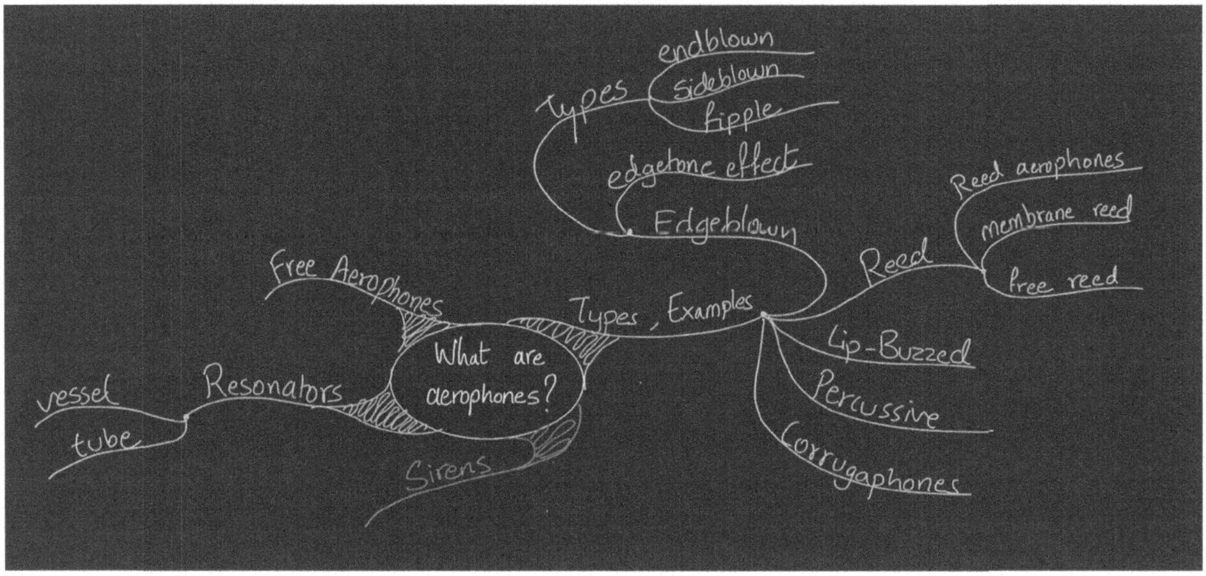

FIGURE 7.1.1. **Mindmap by Yash Garje.**

Aerophone Basics

Like all instruments, **aerophones** produce sound by exciting a body of air. Unlike **chordophones**, **membranophones**, or **idiophones**, the primary sound in an aerophone is generated by a volume of air rather than a string, membrane, or solid object. In most aerophones, the vibrating body of air is enclosed in a chamber. The essential components of these instruments are an **energy source** that generates a flow of air that is periodically interrupted by an **exciter** (e.g., a reed), which energizes an adjacent body of air enclosed in a **resonator**. Pitches are produced by changing the effective vibrating length of the air column, such as through tone holes, valves, or a slide mechanism. Producing sound on an aerophone is complex in ways that producing sound on percussion and string instruments is not, requiring balancing air pressure in the mouth, the materials and configuration of the lips, the pressure of the lips on the mouthpiece or reed, and the materials and dimensions of the resonator. Within the larger category of aerophones, subgroups include **edge-blown, reed, lip-buzzed, percussive, corrugaphones, sirens,** and **free**.

Musical Machines. Scott Barton, Oxford University Press. © Oxford University Press 2026. DOI: 10.1093/oso/9780197599891.003.0007

EDGE-BLOWN AEROPHONES

Given the right conditions, a narrow stream of air traveling over an opening with an edge border produces the **edgetone effect**. Air flows to one side of the edge (e.g., top) and accumulates pressure, which in turn forces the air to the other side of the edge (e.g., bottom), where the pressure accumulates and forces the air back in the opposite direction (e.g., top); the cycle continues (Figure 7.1.2).

FIGURE 7.1.2. **Edgetone effect.**

The result is a rapid and periodic fluctuation of air pressure, which, when fed into a resonator, can produce rich, musical sound. Resonators assume a variety of shapes and sizes, including cylindrical, conical, or globular, and are commonly stopped at one end (e.g., flutes). The pitch, volume, and timbre of the resultant sound are a function of the size, materials, and shape of the resonator as well as the air beam, which can be characterized in terms of length, thickness, angle, and velocity (Solis & Takanishi, 2011). Edge-blown aerophones can be **end-blown** (e.g., bottle, jug, shakuhachi)**, transverse** or **side-blown** (e.g., flute), or can use a **fipple** (e.g., recorder, ocarina), which is an object that focuses the air stream over the edge (see Figure 7.1.3).

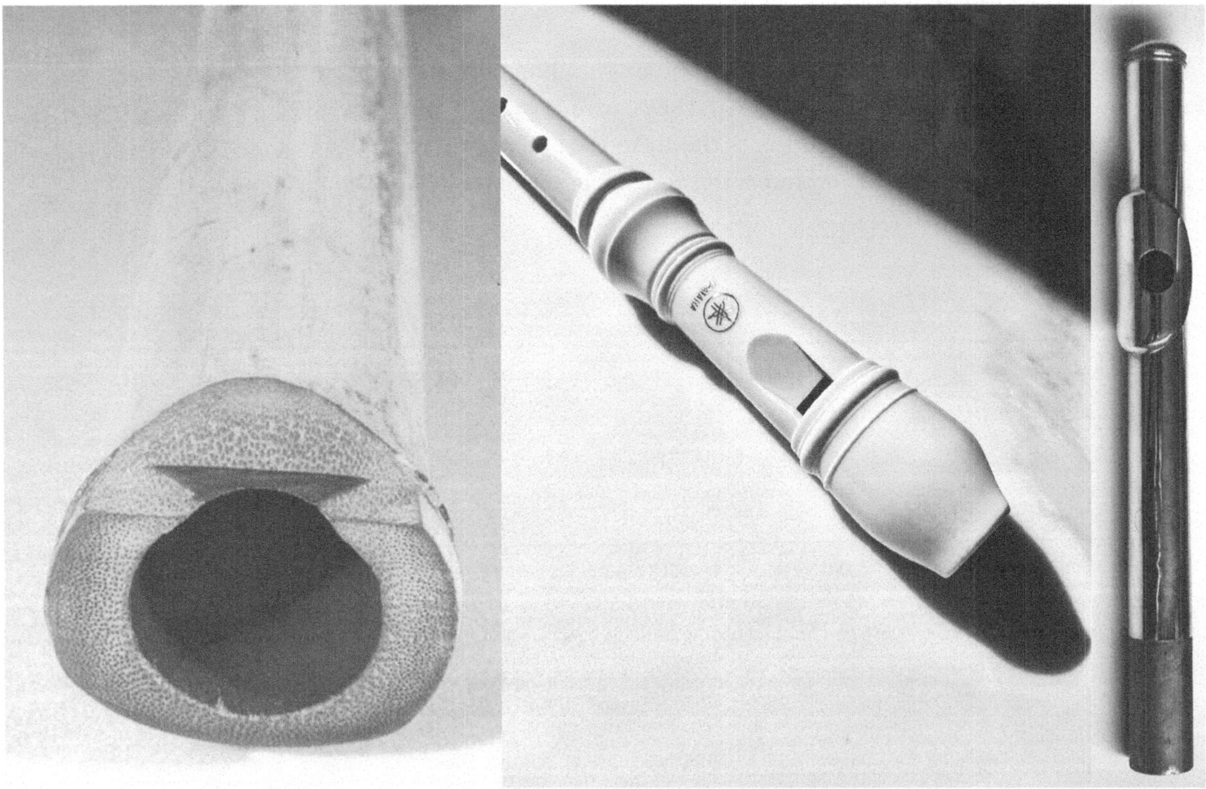

FIGURE 7.1.3. Left to right: end-blown Shakuhachi, fipple of a recorder, transverse flute.

REED AEROPHONES

Reed aerophones consist of a **lamella** (a thin layer or plate of material such as wood) that is set into vibration by a stream of input air. A player produces the airflow, resulting in a pressure difference between the mouthpiece and the oral cavity, which sets the reed into vibration. As the reed vibrates against another object (a mouthpiece for single-reed instruments, another reed for double-reed instruments), it acts as a gate, opening and closing a passage for the air to flow through, resulting in a series of pulses (Figure 7.1.4).

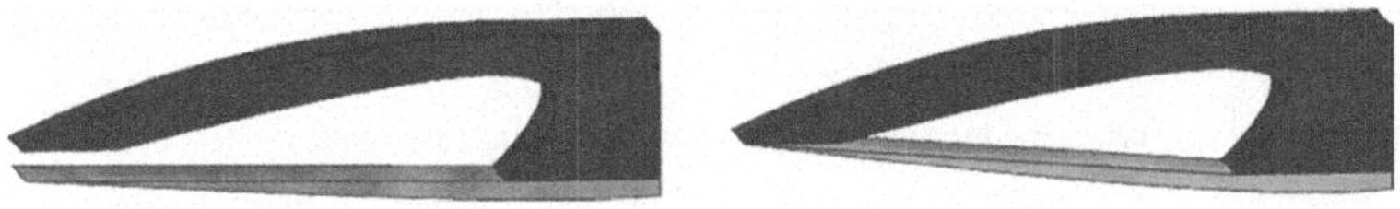

FIGURE 7.1.4. Reed opening and closing against a mouthpiece.

These pulsations excite the air column enclosed in the instrument vessel (e.g., the bore of a clarinet). Reciprocally, the resonant frequencies of the air column drive the reed to vibrate, resulting in a sustained tone that is colored by the resonant properties of the system. The sound produced by a reed instrument depends on airflow, mouth pressure, lip force, lip position, tonguing, and reed hardness (Almeida et al., 2004, 2007, 2010, 2013) as well as the internal shape of the mouthpiece and the **lay**, which is the aperture that the reed vibrates against (Hopkin, 1996). Single-reed instruments include the clarinet and saxophone, while double-reed instruments include the oboe, bassoon, English horn, and shawm.

Different pitches and dynamic levels require different amounts of oral pressure, depending on pitch register. For example, the range of usable oral pressures in saxophone playing is larger in middle pitch registers as compared with the high and low extremes (Lin et al., 2016). Intraoral pressure is modulated by the airflow into the cavity directed by the air supply. Increasing the airflow into the mouth increases intraoral pressure, effecting expanded reed motion and producing proportionally louder sounds (Figure 7.1.5).

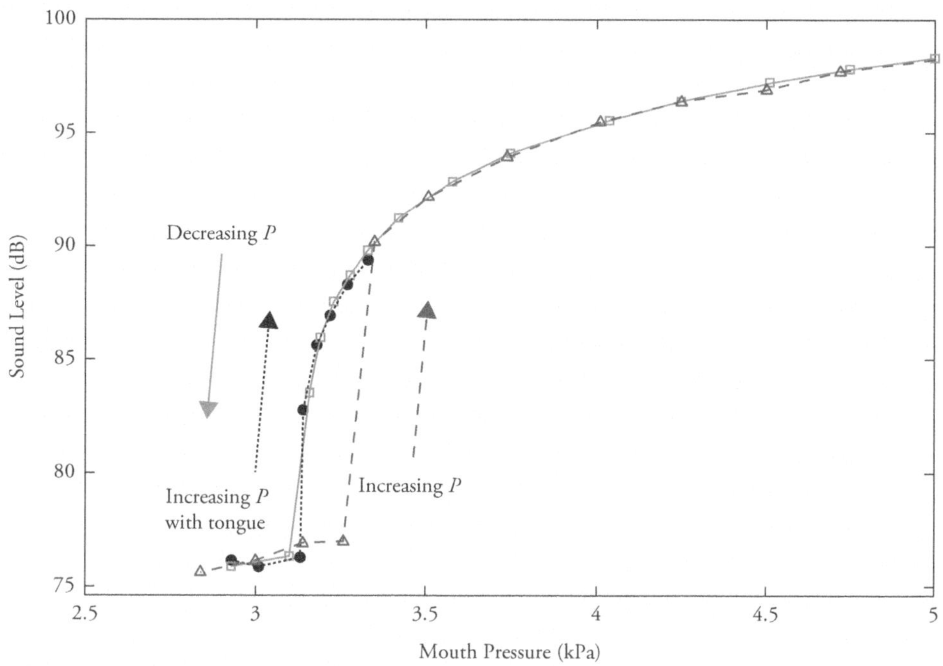

FIGURE 7.1.5. **Sound level of notes produced by a machinic clarinet at different intraoral pressures with a constant lip force of 1N (W. Li et al., 2014).**

The red line (squares) shows decreasing pressure, the blue line (triangles) increasing pressure, and the dotted black line (circles) indicates increasing pressure using an artificial tongue to initiate notes. The latter is significant because it demonstrates that quieter notes require tonguing, not just increased pressure, to initiate oscillation (W. Li et al., 2014).

A limited range of lip pressures on a reed can produce a note: too little and the lip will not act as a pivot for vibration; too much will inhibit the movement of the reed (W. Li et al., 2014; Lin et al., 2016; Lin, Kawai, Nishio, et al., 2019). These lip forces work in coordination with the pressure generated in the intraoral cavity by the air supply (e.g., pump). The combinations of mouth pressure and lip force that produce a note (or not) can be seen in Figure 7.1.6.

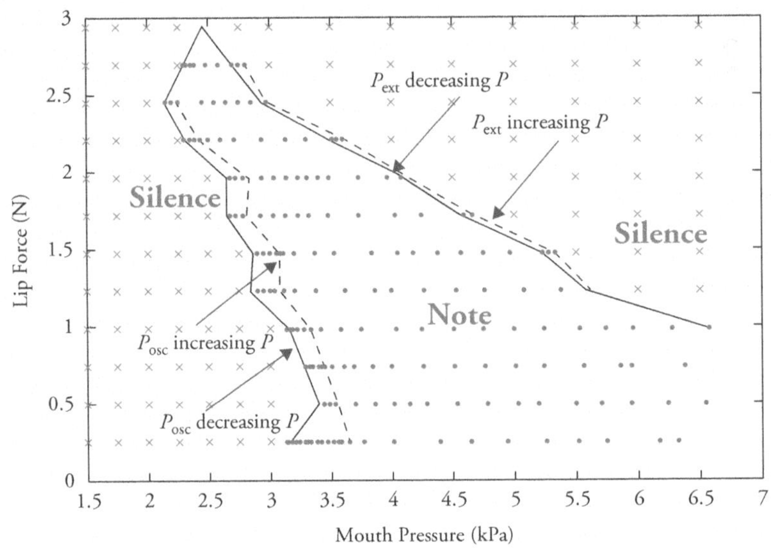

FIGURE 7.1.6. **Blowing pressure and lip force needed to produce sound in a reed instrument where P_{osc} and P_{ext} are the thresholds of reed vibration (W. Li et al., 2014).**

Note the hysteresis regions at the boundaries, meaning the thresholds for producing a note are slightly different when pressure is increasing vs. decreasing. Musically, this is particularly relevant when producing dynamic gestures, such as *crescendos* and *diminuendos*.

These factors work together: increased intraoral pressure produces lower pitches, which must be compensated by providing more force to the reed (W. Li et al., 2014). The material composition of the force-applying lip also plays a role in sound production. The malleability of the lip helps to attenuate high frequencies (i.e., "squeaks") that are produced when a harder object, such as the teeth, applies force directly to the reed (W. Li et al., 2014). The range of pitches, articulations, and dynamic gestures demanded in musical performance requires a complex modulation of these parameters.

Tonguing is a human performance technique in which the tongue is used to stop or damp the oscillations of a reed. It is fundamental in creating a clear tone onset, distinguishing adjacent notes, and producing soft dynamic levels. Human players use the front and back of the tongue and mouth in various combinations to perform techniques such as *double-tonguing*. Producing articulations on a reed instrument such as *staccato* or *portato* depends on contact time between the tongue and reed—30 ms for *portato*, less for staccato (Chatziioannou et al., 2019)—in combination with precise control over airflow (Lin, Kawai, Matsuki, et al., 2019). Details about the physical factors involved in the production of transients on reed instruments have been the focus of numerous research studies (W. Li et al., 2016; Almeida et al., 2017).

While significant attention has been devoted to the development and study of reed aerophones such as clarinets, less common instrument categories also have musical potential. **Membrane reeds** (Hopkin, 1996, 2021) are another class of reed aerophone that involves stretching a membrane (e.g., a balloon) over a tube and then positioning the opening to the membrane (e.g., the part of the balloon that one typically blows into) below the rim of the tube. When the player blows into the membrane, air forces its way over the rim of the tube in a series of pulses, given the membrane's elasticity. The pulses then excite the air column in the tube. The quality of the sound produced can be surprisingly compelling, almost bassoon-like, when an adequately sized PVC tube is used (Media Example 7.1.1 ▶). These instruments are cheap and easy to make, but intonation is challenging to control.

A **free reed** differs from the instruments previously discussed in that it does not need to be coupled to an air column to produce sound. Instead, one end of the reed is fixed in a frame, allowing it to oscillate and create pulses in an input air stream. Typically, free reeds come in banks where each reed is tuned to a specific pitch, such as in harmonicas, accordions, and harmoniums (Figure 7.1.7).

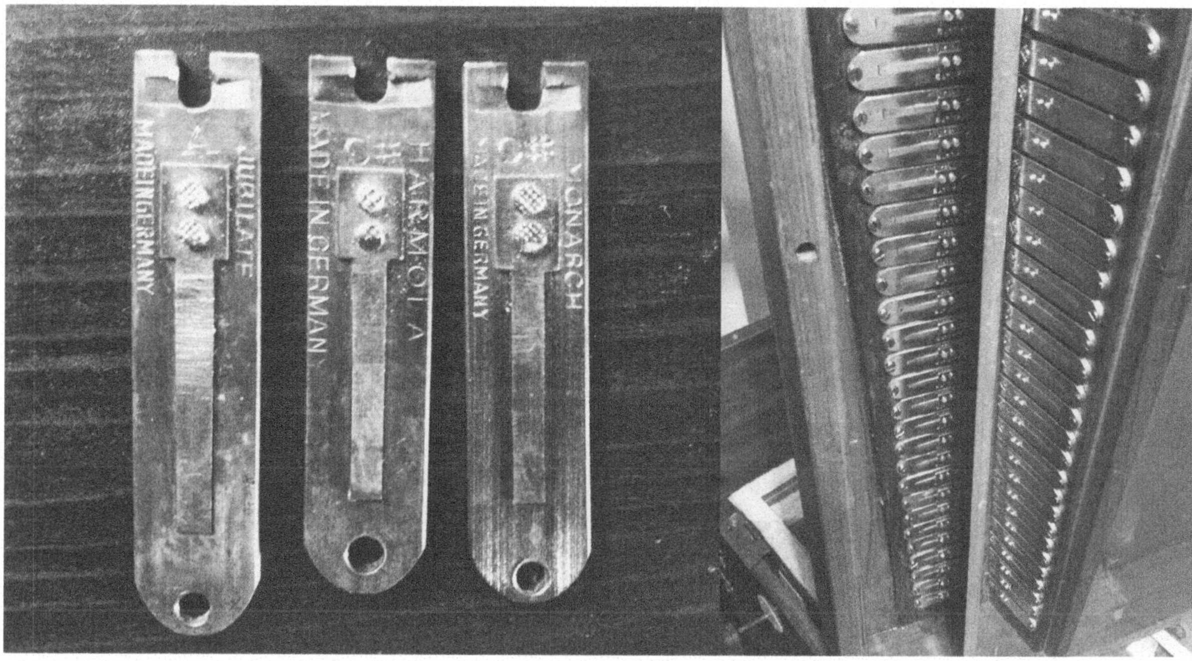

FIGURE 7.1.7. **Left: close-up of harmonium reeds (image By Bubble71—Own work, CC BY-SA 3.0, https://commons.wikimedia.org/w/index.php?curid=29734548). Right: bank of harmonium reeds (IMAGE By David Estes—Own work, CC BY-SA 3.0, https://commons.wikimedia.org/w/index.php?curid=16772314).**

LIP-BUZZED AEROPHONES

As the name suggests, a **lip-buzzed aerophone** (a **trumpet** in the language of Hornbostel & Sachs, 1961) generates sound when a player's lips rapidly move together and apart, acting as a gate (similar to a reed) to create vibrations in an air stream. This buzzing creates pulses of air that excite the air column in the adjacent tube, whose resonances reinforce the lips' vibration. Adjusting lip tension and airflow enables the player to sound pitches from a harmonic series. Changing the length of the tubing through valves, slides, or keys allows intermediary pitches to be produced, enabling the realization of full musical scales. Examples in this category include the cornet, euphonium, flugelhorn, French horn, trumpet, horn, sousaphone, trombone, and tuba.

A key factor in lip-buzzed sound production is the relationship between the player's **embouchure**—which comprises the lips, facial muscles (there are at least eight groups), tongue, and teeth—and the pressure in the system. Players use different embouchure shapes (e.g., "smiling," "puckered") and vary the amount of pressure that the lips apply to the mouthpiece. When *mouth overpressure*, the difference between the pressure in the mouth and atmospheric pressure, exceeds a threshold, lip oscillation occurs. The range of frequencies and dynamics of the lip opening (area, width, height, and shape) depends on the player's physiology and the pitch produced (see Campbell et al., 2021, for a detailed review). Producing a transient requires a rapid increase in pressure (~10 Pa/s) over a short period (<100 ms), depending on factors such as slide position or valve configuration (Stevenson, 2009).

The timbre produced by brass instruments depends on the dynamic level at which notes are played. Playing loudly produces a "brassy" timbre that is characterized by higher energy in the upper harmonics. Such timbres are achieved through changes in mouthpiece pressures and embouchure configurations (Stevenson, 2009). Mouthpieces provide enhanced resonance in a relatively high part of the audible frequency range, adding brightness to the timbre (Hopkin, 1996).

PERCUSSIVE AEROPHONES

Percussive or **plosive** aerophones work by striking the end of an enclosure (such as a tube or a vessel), which excites the volume of air within it. The percussive action is performed by a **beater**, which can range from the human hand to a foam paddle. The sound of percussive aerophones (when done right) is a compelling mixture of a clear transient with a rich, sustained pitch complex. Anyone who has seen the Blue Man Group perform will know this truth (Media Example 7.1.2 ▶).

CORRUGAPHONES

Corrugaphones (also known as whirly tubes or bloogle resonators) are an unusual form of aerophone. As the name suggests, they consist of corrugated tubing, characterized by evenly spaced ridges and grooves perpendicular to the length of the pipe (Figure 7.1.8). As air flows across the ridges, standing waves are created according to the spacing of the ridges and the speed of the airflow. Changing the air flow rate can excite different resonant frequencies of the tube, resulting in a harmonic series. For a whirly tube, this is accomplished by swinging the tube around in a circle at varying rates.

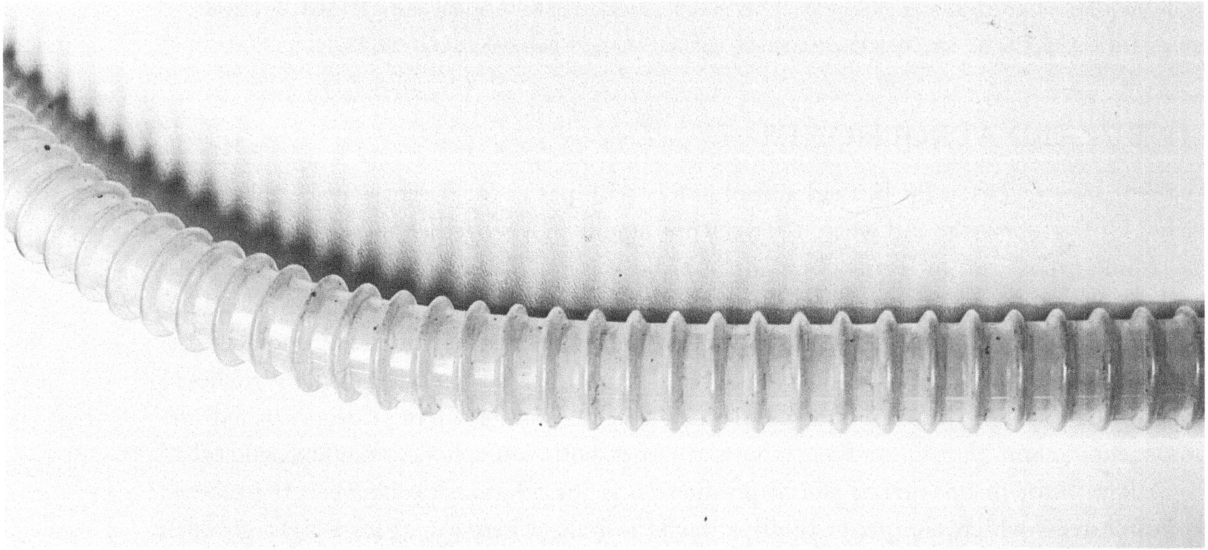

FIGURE 7.1.8. **Corrugated tubing.**

SIRENS

Sirens are not typically recognized for their musical utility, but they, too, have artistic potential. A promising subset of this group directs air through a series of evenly spaced holes, creating periodic air pulsations that can become sound at the right frequencies. This excitation method is theoretically possible with a linear system (e.g., a belt and pulley), but it is far more efficient to use a spinning disc that contains a circle(s) of holes within it. The pitch produced depends on the disc's speed and the number of holes it contains. For example, a disc containing 44 holes rotating at 10 revolutions per second will produce a tone with a frequency of 440 Hz (A4).

RESONATORS

Non-free aerophones require an enclosure within which a volume of air is excited. Two forms of enclosures are **vessels** and **tubes**. Globular vessels (often referred to as **Helmholtz Resonators**) typically produce a single frequency with few upper partials. The smaller the chamber, the higher the pitch. The smaller, thicker, and deeper the opening to the chamber, the lower the pitch. Tubes are capable of richer harmonic spectra and are commonly found in musical instruments such as flutes and trumpets. The pitch the vessel produces depends on the enclosed volume of air; changing the vessel's capacity changes the pitch. Generally, lower volume means higher pitch, achieved by shortening the tube. The shape of the tube affects the harmonicity of the partials produced: cylindrical and conical shapes produce a harmonic series; therefore, they are commonly used in musical aerophones.

FREE AEROPHONES

In **free** aerophones (Hornbostel & Sachs, 1961) or **outer air instruments**, the body of vibrating air is not enclosed in the instrument. An example is the bullroarer (Figure 7.1.9), which contains a blade attached to a string that is swung in a circle; the speed of rotation determines the pitch and volume of the sound.

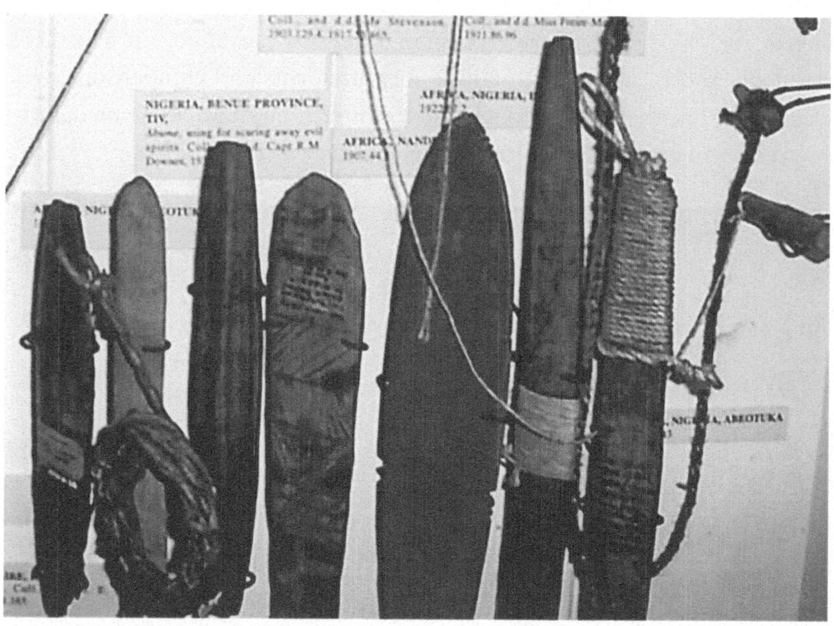

FIGURE 7.1.9. **African bullroarers (image By Throwawayhack at the English-language Wikipedia, CC BY-SA 3.0, https://commons.wikimedia.org/w/index.php?curid=4529623).**

Identity, Purpose, and Usage

Musical and creative expression have motivated designers and builders of machinic aerophones. The *Percussive Aerophone* by the MPR Lab enables expansive chords and complex rhythms that would be difficult, if not impossible, for human performers to play. The Logos Foundation equipped their aerophones with rich parametric controls, offering composers a wide array of sonic possibilities. Revisions to the Waseda Flutist (WF-4RIV) were purported to produce "more natural" sounds, double-tonguing, smoother transitions between notes, and vibrato. The Waseda Anthropomorphic Saxophone robot, version 5 (WAS-5), was designed to enhance performance quality through improved articulation abilities (Lin, Kawai, Matsuki, et al., 2019).

Machinic aerophones are vehicles for human expression, not only through the sound they make, but also through their appearance, movement, and connection to human audiences. Some builders have made cultural connections across geography and time. The bamboo flute-playing robot (J. Li et al., 2019) was inspired by ancient Chinese woodworking and costuming practices. Kato et al. (2018) developed a robot capable of playing Thai flutes (khlui phiang aw) in traditional Thai style. Others have used the physical presence and interactive possibilities of machinic aerophones as vehicles to entertain and educate audiences. For example, the flute robot built by Huang et al. (2014), based on Waseda's flutist robot No. 4 Refined IV (WF-4RIV), was featured in a Science and Technology Museum.

Machinic aerophones provide a platform for research. Some were created to understand human physiology better, particularly the breathing system (Vaucanson's *The Flute Player*). Some have been motivated by the desire to model human motor control (Chatziioannou et al., 2019; NICTA-UNSW Robot Clarinet, 2008; Solis & Takanishi, 2011). Others provide a vehicle to study the acoustic principles involved in wind instruments (NICTA-UNSW Robot Clarinet). Engineering problems motivate some, such as the coordination of multiple DOFs (Solis & Takanishi, 2011). There are advantages of machinic platforms in scientific inquiry and musical performance (Richards, 2003; Stevenson, 2009). Artificial mouths can accommodate internal sensors that provide valuable feedback on physical processes, which can subsequently be used to adjust the system's state. Artificial mouths can invariantly produce air and sustain an embouchure for long periods. They allow precise control over parameters, such as the mouthpiece position or airflow rate. They are accessible and configurable in repeatable and reliable ways, making measurements and comparisons over time easier and more accurate. While their limited pitch and dynamic range have been noted, artificial mouths reveal new avenues of creative exploration and scientific experimentation.

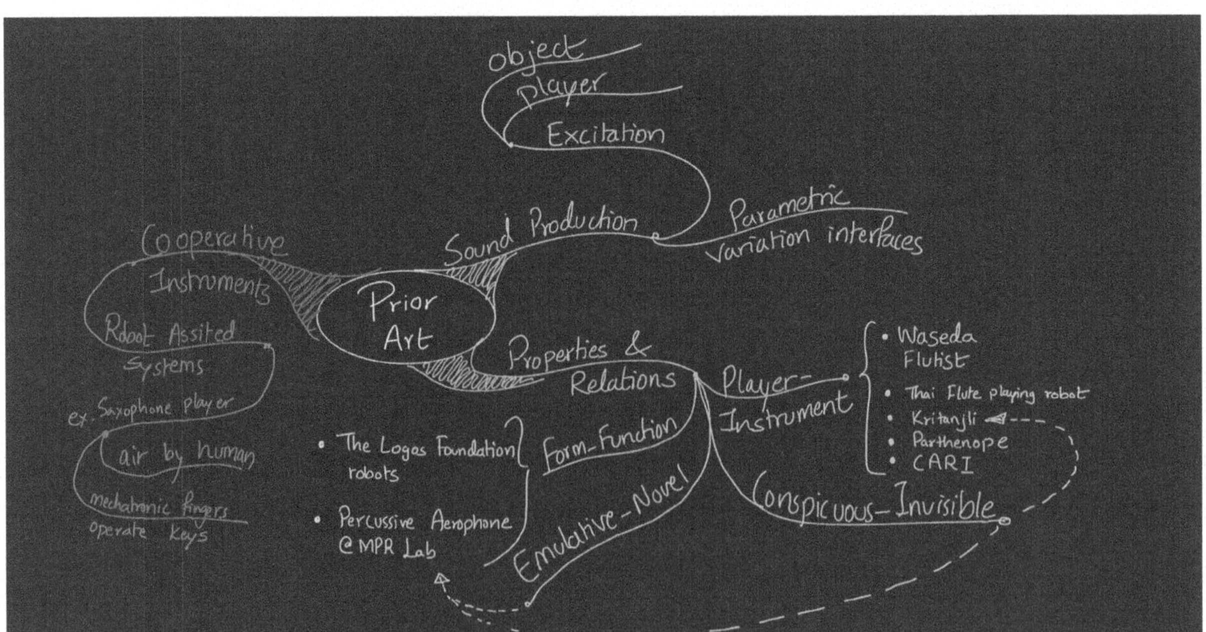

FIGURE 7.1.10. **Mindmap by Yash Garje.**

Prior Art

Human-playable aerophones are ancient instruments found in Greece, India, Egypt, and Europe, dating back roughly 40,000 years. Mechanical aerophones also have a long history, dating back to the steam-powered automatic flute player built by the Persian Banū Mūsā brothers, as featured in *The Book of Ingenious Devices* (850), and later *The Flute Player* (1737), built by French artist and inventor Jacques de Vaucanson. Mechatronic and robotic aerophones have been the subject of more recent efforts, though not on the same scale as string and percussion instruments (the complexity of the sound production process is part of the reason, as we will learn later in the chapter). As with percussion and strings, there are different ways to organize past examples to direct analysis and design. The *Properties and Relations*, as described in Chapter 4, provide structure to help interpret the myriad examples within this category.

SOUND PRODUCTION

When considering how aerophones produce sound, we examine the properties of the sonic object, as well as the methods used to excite the air column within it. Doing so reveals **excitation** and **pitch production** as two categories that are useful in parsing existing examples and guiding future designs.

Excitation

Methods of aerophonic excitation are varied and consequential. Categories introduced by Hornbostel & Sachs (1961) ("H & S"), including *edge-blown, reed, trumpet, rotating,* and *plosive* aerophones, capture these differences and can be used to group examples. To that list, we add *membrane* and *electroacoustic drivers,* as well as *siren discs,* which have been featured in more recent instruments. Keep in mind that instrument classification systems, such as H&S, focus on the *sonic object* and not the *player*, which is assumed to have a standard set of characteristics. In the world of musical machines, we can design both the *object* and the *player*, so we need to separate these two when considering how sound is produced. Table 7.1.1 presents examples of aerophonic musical machines categorized by the most common excitation methods (edge-blown, reed, and lip-buzzed) and instrument classifications.

TABLE 7.1.1. **Aerophonic musical machines by excitation and instrument type.**

Instrument	Excitation	Musical machines
Flute	Edge-blown	*Automatic Flute Player* (850, Banū Mūsā), *The Flute Player* (1737, Jacques de Vaucanson), *Flute-Playing Machine* (1979–1982, Martin Riches), *Waseda Flutist* (Takanishi & Maeda, 1998; Solis et al., 2006, 2007; Solis & Takanishi, 2011), *<Flut>* (2020, Logos Foundation), and other examples (Huang et al., 2014; Kato et al., 2018; J. Li et al., 2019; Zhao et al., 2020).
Slide flute	Edge-blown	d'Andréa-Novel et al. (2016)
Recorder	Edge-blown	MPR Lab (2015)

(continued)

TABLE 7.1.1. **Continued**

Instrument	Excitation	Musical machines
Pipe organ	Edge-blown	*<Bomi>, <Bourdonola>, <Krum>, <Piperola>, <Ror>, <vox Humanola>, <Pos>, <Trump>, and <Qt>*—Logos Foundation (Raes, n.d.).
Clarinet	Reed	NICTA-UNSW Robot Clarinet (2008); *<Klar>* (2012, Logos)
Bassoon	Reed	*<Fa>* (2011, Logos)
Oboe	Reed	*<Ob>* (2010, Logos)
Saxophone	Reed	*Waseda Anthropomorphic Saxophonist* (2007–current); *<Autosax>, Logos, 2009; <Asa>, Logos, 2013*
Accordion	Reed	*<Ake>* (2008, Logos), *<Bako>* (2007, Logos)
Melodica	Reed	*<Melauton>* (2017, Logos)
Bagpipe	Reed	McBlare (Dannenberg et al., 2005)
Reed organ	Reed	*Kritaanjli* (Murphy et al., 2014), *<Harma>* (2005, Logos), *<HarmO>* (2010, Logos), *Harmonium* (2016, MPR Lab)
Brass instrument research[1]	Lip-buzzed[2]	S. Bromage, 2007; S. R. Bromage et al., 2003; Ehara et al., 2001; Gilbert et al., 1998; Lopes et al., 2013; Richards, 2003; Teissier et al., 2003; Vergez & Rodet, 1997
Trumpet	Lip-buzzed	*<Trumpeter>* (2021, Logos)
Tuba	Lip-buzzed	*<So>* (2003–2007, Logos)
Valve trombone	Lip-buzzed	*<Bono>* (2007–2010, Logos)
Cornet	Lip-buzzed	*<Korn>* (2008–2010, Logos)
Horn	Lip-buzzed	*<Horny>* (2013, Logos)
Flugelhorn	Lip-buzzed	*<Bug>* (2016–2017, Logos)
Saxhorn	Lip-buzzed	*<Cornalto>* (2021, Logos)

[1] Research platforms used to study the acoustics and mechanics of brass instrument-playing.

[2] "trumpets" in the classifications of Hornbostel & Sachs (1961).

Aerophones can also produce sound from a single compression and release of air, known as *plosive* or *percussive* aerophones. Examples include Logos Foundation's *<Puff>*, whose solenoid-controlled bellows produce puffs of wind into brass pipes, capable of 84 different pitches over 3.5 octaves (quartertone equal temperament) (Raes, 2007). The MPR Lab's *Percussive Aerophone* (Sundberg et al., 2018) is another example, featuring two octaves of PVC pipes that are sounded by foam discs that slap the top of the pipes (and thus excite the air columns within) when driven by pneumatic cylinders.

A more recent group of machinic aerophones has emerged with the development of *electroacoustic compression drivers*. These drivers consist of an energized coil that moves in a magnetic field proportional to the electrical current, which concurrently applies force to a coupled diaphragm (*cone* in the language of speakers), creating sonic fluctuations in air pressure within an enclosed air column (e.g., a pipe or tube). Such drivers have been used to excite aerophonic instruments instead of traditional air supply-mouth systems. The Logos Foundation has pioneered this design, featuring it in <*Hybr*>, <*HybrHi*>, and <*HybrLo*>. EMMI also used this approach in *CARI* (Rogers et al., 2015). A variation on this idea is found in Logos Foundation's <*Pi*>, <*2Pi*>, and <*3Pi*>, which use electromagnetic buzzers to excite aluminum pipes into resonance (Raes, n.d.).

Within the category of **free** aerophones are rotating aerophones such as **siren discs**. The MPR Lab's *Parthenope* (Sidler et al., 2020) employs this approach, generating a range of pitches through two motor-driven discs whose rotation rates can be precisely modulated. Each disc has four circles of holes that are aligned with four solenoid valves connected to a source of compressed air. Because the spacing of holes differs on each circle, four different pitches can be produced, given a particular disc rotation speed. By using different hole patterns and varying rotational speed, a wide variety of scales and pitches can be played.

Pitch Variation

An aerophonic instrument can produce different pitches utilizing tone holes, valves, or slides. **Tone holes** allow the effective vibrating length of the air column, and thus the pitch, to be changed. They are opened and closed using binary actuators (such as a solenoid), either directly or via an intermediary mechanism (such as a key). Examples in this category include many of the *flutes* and *reed aerophones* listed in the previous section.

Another way to change the length of an air column is to use a **valve** system or a **slide**. When a valve is opened, it allows air to flow through the tubing in the valve block, lengthening the air column and lowering the pitch. Different valves can lower the pitch by varying amounts, allowing combinations of valves to produce a full scale. The mechanics of pressing a valve is similar to pressing a key, which can be accomplished with binary actuators (such as solenoids) or rotational motors (such as servos). Examples in this category are listed in the "trumpets" section above. Instead of changing the tube length by a discrete amount, as in the case of a fully depressed valve, a slide allows for positions along a continuum, making microtonal tunings and portamenti possible. The mechatronic control of such systems usually requires a motor, a mechanism to convert rotational to linear motion, and a sensing system to determine the position of the slide. An example of this kind of instrument is the *RoboTrombo* (*Robotic Trombone*, n.d.).

Some aerophones have elaborate mechanical interfaces in the form of keyboards, such as the pipe organ and harmonium. Sound production in these instruments involves a means to generate airflow, for example, a bellows or an air compressor. The source of air pressure differential (e.g., a vacuum chest connected to a bellows) is connected to reeds or enclosed air columns through a mechanism (e.g., a key or valve), which excites vibration and produces sound. Solenoids are often chosen to actuate this open / close mechanism because of their speed, small form factor, and binary nature. As these are complicated instruments to make, existing instruments are usually outfitted with machinic actuation systems. Often, the original instruments are modified to accommodate the electromechanical systems, such as removing the key mechanisms or bellows. Examples in this category include harmoniums (*Kritaanjli*, Murphy et al., 2014; <*Harma*> and <*HarmO*>, Logos; the *robotic harmonium* of the MPR Lab), pipe organs (<*Piperola*>, <*Vox Humanola*>, <*Bomi*>, Logos), accordions (<*Ake*> and <*Bako*>, Logos), and melodicas (<*Melauton*>, Logos).

PROPERTIES AND RELATIONS

Another way to make sense of the long and varied history of machinic aerophones is to consider them in terms of *Properties and Relations,* as introduced in Chapter 4. We can categorize machinic aerophones according to the purpose for which their builders intended them and the ways in which they have been utilized. As with machinic percussion and string instruments, design choices have been driven by the relationships between *player* ⇔ *instrument*, *form* ⇔ *function*, *emulative* ⇔ *novel*, and *conspicuous* ⇔ *invisible*.

Player ⇔ *Instrument*

The machinic *players* of aerophones have largely been *emulative* and specifically anthropomorphic. The Waseda Flutist and Saxophonist are two prominent and sophisticated examples. The Waseda Flutist has been in development since 1990, starting with WF-1; its progress has been regularly reported since then (Solis et al., 2006, 2007, 2009; Solis & Takanishi, 2007, 2011; Takanishi & Maeda, 1998). The 41-DOF robot contains artificial lips (3-DOF), lungs (powered by motor-controlled bellows), tongue, neck, vocal cords (including a glottis with motor-controlled vocal folds), arms, and fingers (Media Example 7.1.3 ▶). The Waseda Anthropomorphic Saxophonist robot (WAS) has been under development since 2009, beginning with the 15 DOF WAS-1, equipped with mechanisms to control the lower lip, tongue, artificial lungs, and fingers (Solis et al., 2009). WAS-2 improved mouth and finger mechanisms, enabling more stable sound production (Solis & Takanishi, 2011, Media Example 7.1.4 ▶). The fifth iteration (WAS-5) features 31 DOF and mechanisms that replicate human organs in the production and control of airflow as well as articulation and dynamics control systems (Lin, Kawai, Matsuki, et al., 2019; Lin, Kawai, Nishio, et al., 2019). An anthropomorphic flute robot built by Zhao et al. (2020) is equipped with flexible four-DOF hands, each DOF driven by a servo motor, and passive joints that allow the robot to be configured as desired (Figure 7.1.11).

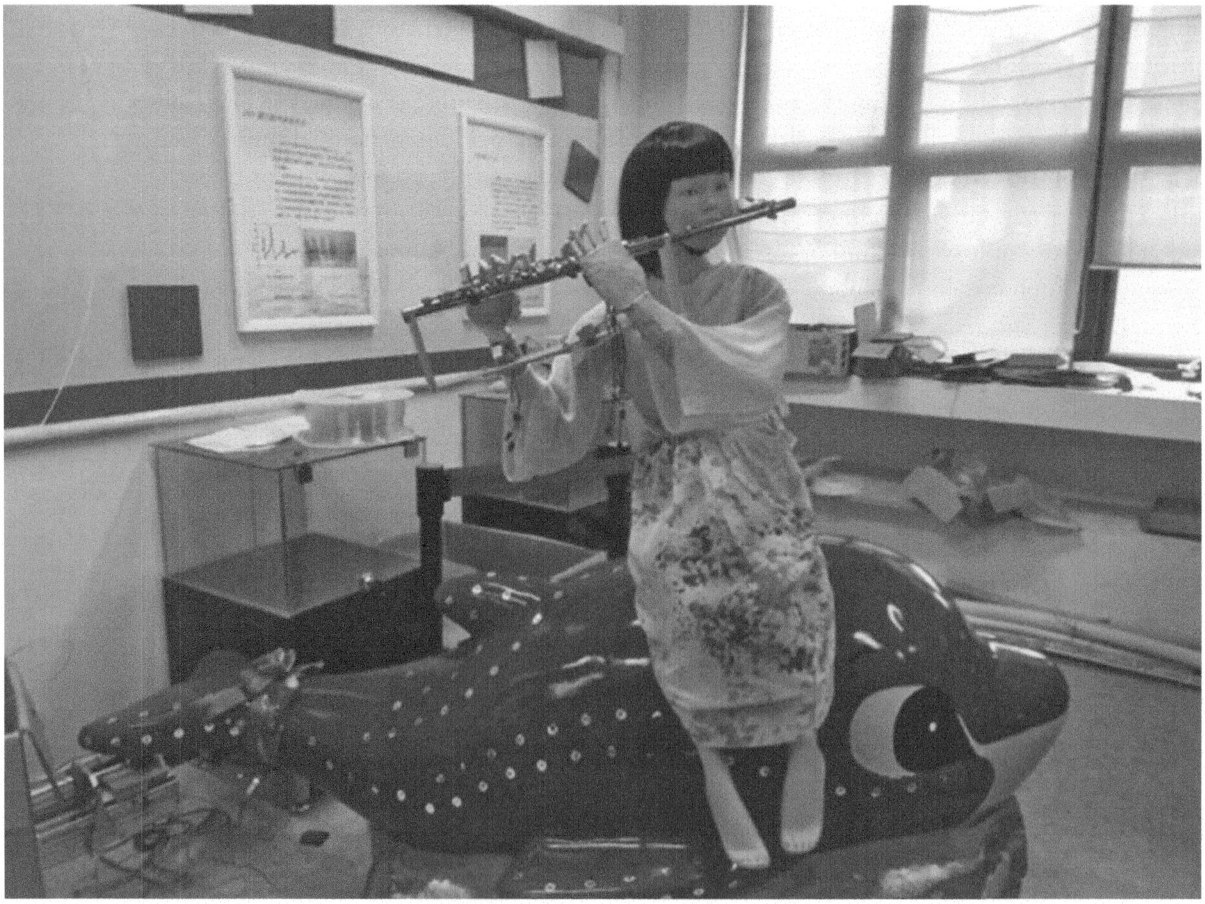

FIGURE 7.1.11. **Flute robot (Zhao et al., 2020).**

Some builders of machinic *players* envision their machines within specific performance contexts, utilizing audience interaction to guide the design of visual gestures and aesthetics. Some seek to arouse audience interest through the creation of characters that convey personalities through physical gestures and costuming. Li et al. (2019) drew inspiration from ancient Chinese wood carvings and costuming in designing a flute-playing robot that resembles a knight-errant adorned with headwear and armor on its 2-m-tall frame (Figure 7.1.12). Huang et al. (2014) developed a flute-playing robot for a science and technology museum. Given this context, the aesthetic goal was for the machine to appear "approachable" (p. 2221). The researchers developed six distinct visual designs, which were subsequently used in a survey administered to potential audience members; the results informed the final design choices. A notable feature is the screens used for eyes, which provide a means of visual expressivity that complements the music played by the machine (Figure 7.1.12).

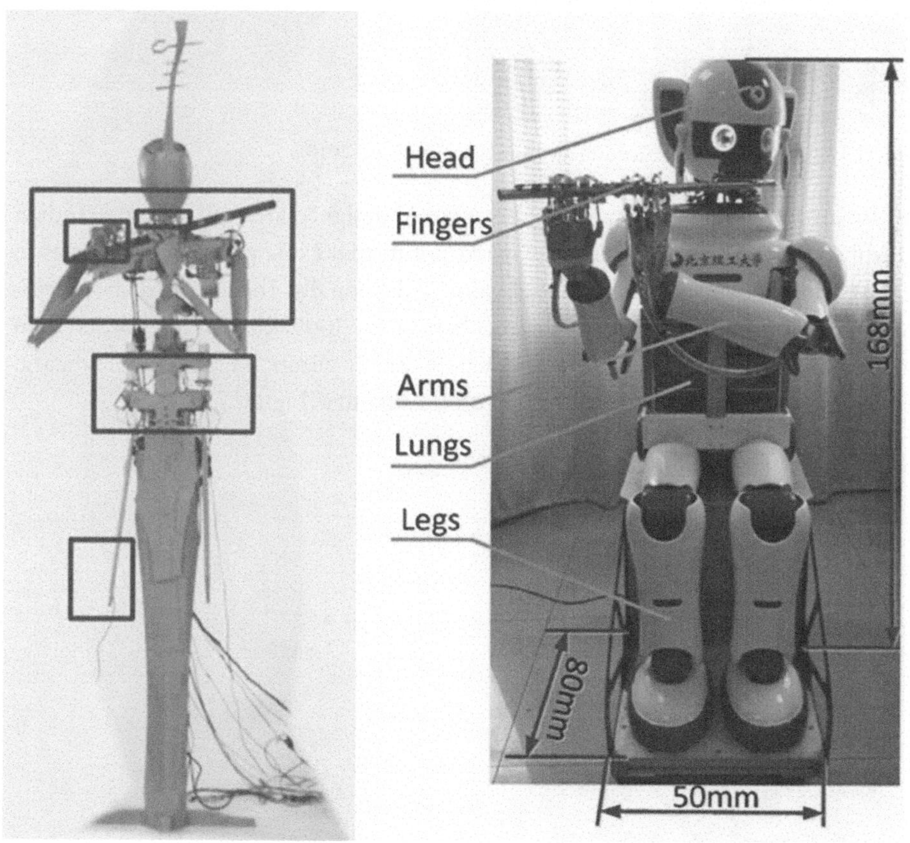

FIGURE 7.1.12. **Flute-playing robots. Left: Li et al. (2019); right: Huang et al. (2014).**

Robotic aerophones have been developed as *attachments* to acoustic instruments originally designed for human players. The NICTA-UNSW Robot Clarinet (Figure 7.1.13) was created as a platform for studying human performance and instrument acoustics and, therefore, was built around a human-playable clarinet. Because the project's focus was sound production rather than visual form, the anthropomorphism of the machine is limited to structures such as the mouth rather than tangential physiological systems, such as the head and arms. The *Thai flute-playing robot* (Kato et al., 2018) and *mechatronic slide flute* (d'Andréa-Novel et al., 2016) were similarly built around existing instruments designed for human players.

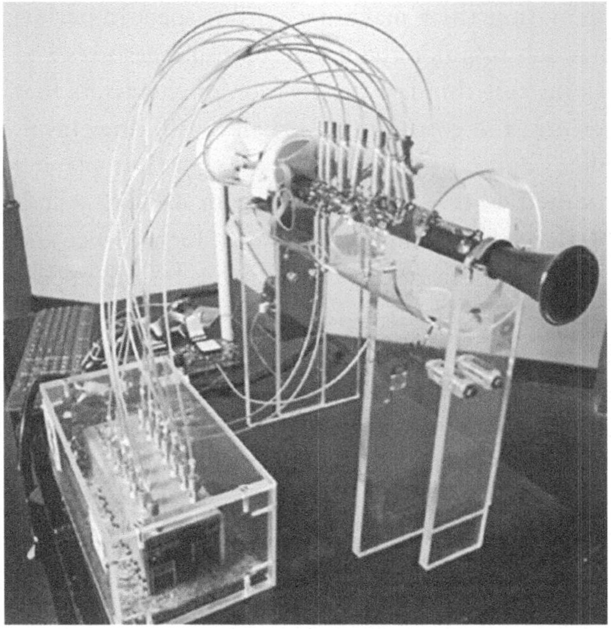

FIGURE 7.1.13. **NICTA-UNSW Robot Clarinet (image from Almeida et al., 2010).**

Kritaanjli is also an attachment, with an actuation system designed around a portable Indian harmonium (Murphy et al., 2014). The authors aspired not to make any permanent changes to the instrument, given its monetary and emotional value. Additionally, they wanted the instrument to remain playable by humans. *McBlare* (Dannenberg et al., 2005, 2011) is a robotic system that attaches to conventional bagpipes. The system includes an air pump, pressure regulator, and electromechanical relays to open and close tone holes on the chanter (Figure 7.1.14).

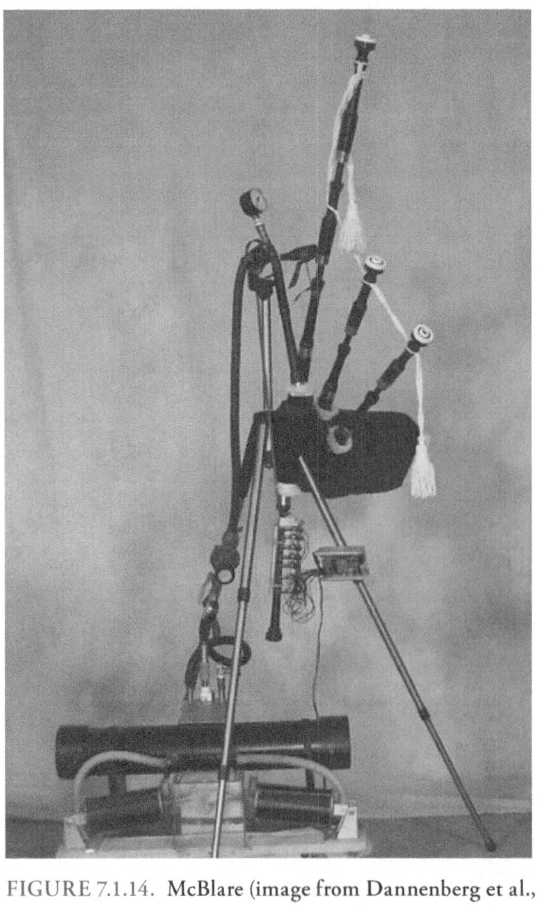

FIGURE 7.1.14. **McBlare (image from Dannenberg et al., 2011).**

The Waseda Anthropomorphic Saxophonist robot X (WAS-X) marked a shift in design philosophy from *emulative* toward *attachment*. As described in Uchiyama et al. (2023), the complexity and size of the previous iteration of the machine (WAS-5) inhibited HRI. In response, WAS-X was conceived using a *Design Thinking* approach with the instrument at the center, resulting in a minimal open-frame structure to which a craned air delivery unit and fingering mechanisms are attached.

Attachments often have a utilitarian aesthetic, with designs optimized for musical performance, cost, or ease of fabrication. Some attachments emphasize the instrument by making actuation and control systems minimally apparent (*McBlare* leans in this direction) or potentially *invisible*. Regarding advantages, builders of aerophone attachments can circumnavigate the design and construction of a complicated acoustic instrument, minimize a system's spatial footprint, and facilitate transportability. A disadvantage of most aerophone machine attachments is that their movements are often minimal, and thus, they must rely on other aspects to engage viewers visually.

Some machinic aerophones *integrate* the sonic object and actuation systems to form *novel* instruments. An example is the MPR Lab's *Parthenope* (Sidler et al., 2020), a robotic siren that produces sound by releasing compressed air through perforated spinning discs via solenoid valves. These discs are a focal point in the visual form of the instrument; they are the only component whose movement is visible, and they are a primary source of sound. In *Parthenope,* there is little distinction between the actuating system and the sonic object (Figure 7.1.15).

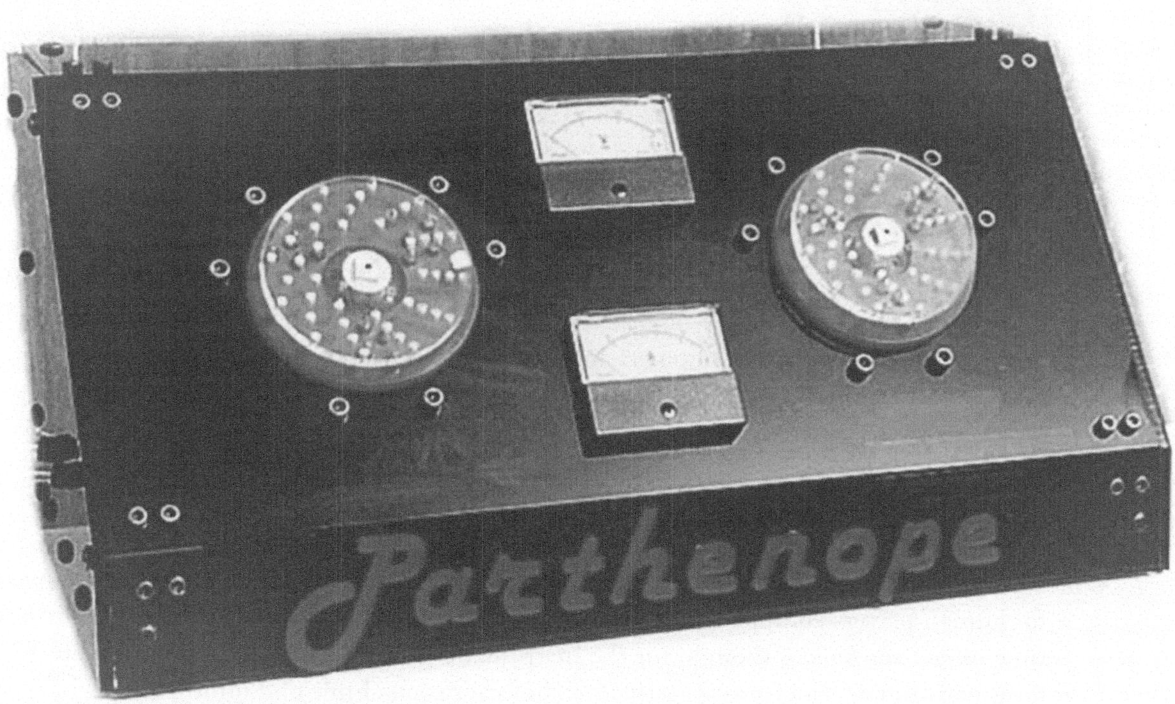

FIGURE 7.1.15. **Parthenope by WPI's MPR Lab.**

CARI (Cylindrical Aerophone Robotic Instrument) (Rogers et al., 2015) by *EMMI (Expressive Machines Musical Instruments)* is also an integrated instrument. The sonic object and the means of excitation were envisioned and developed together; here, a cylindrical air column with tone holes is excited by a compression driver. The instrument's machinic identity is evident in its acrylic and aluminum materials and its perpendicular angles, making little distinction between the excitation system, air column, and tone hole mechanisms (Figure 7.1.16).

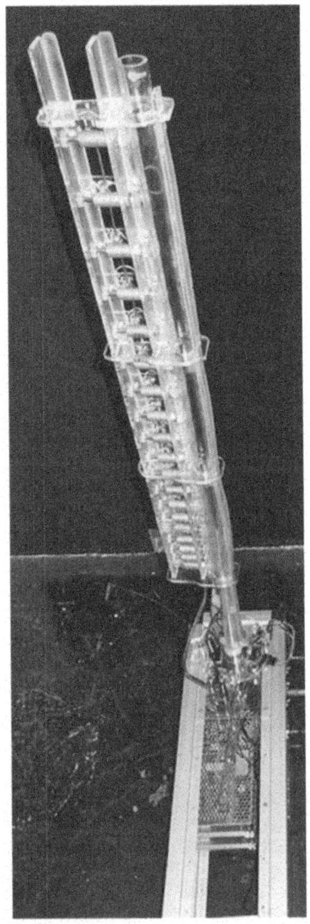

FIGURE 7.1.16. Emmi's CARI.

The novelty of *integrated* instruments is a point of interest in presentation and performance. They escape many of the associations that are connected to traditional wind instruments, which affords composers and performers expressive freedom. At the same time, *integrated* instruments take significant time and effort to develop. Due to their novelty, they typically require several iterations before reaching a musically usable point.

FORM ⇔ FUNCTION

The wind instruments made by The Logos Foundation feature sonic objects designed for human players equipped with mechatronic actuation systems. The latter are not just functional *attachments* though, for the instruments combine with machinic actuators in forms that project a definitive aesthetic. The latter is machinic and minimalistic, marked by sleek, unadorned metal and a pragmatism expressed in encompassing structures seated on large, conspicuous wheels that evoke notions of *carts*. Structural arcs extend and intersect parallel and perpendicular planes. In many cases, the electronics (circuit boards, power supplies) are conspicuously mounted, openly acknowledging machinic identity (Figure 7.1.17).

FIGURE 7.1.17. **Robotic aerophones of The Logos Foundation. From left to right: <ob>, <Horny> and <Flut> (Raes, n.d.).**

While these machines feature instruments designed for human players and, in some cases, move horizontally and vertically somewhat as humans do, they do not mimic human forms and, therefore, are more *novel* than *emulative*. The result is a visual identity that values both *function* and *form*, exemplified in much of Logos's orchestra.

As the MPR Lab's *Percussive Aerophone* was iteratively developed, we paid increasing attention to the configuration of the pipes and the holistic form of the instrument. Earlier versions contained two rows of pipes along a single plane. This arrangement was problematic because the frame was excessively conspicuous, and some pipes were more visible than others. We wanted to make the frame as invisible as possible, with the pipes appearing as if suspended in the air. At the same time, a robust structure was needed because some of the pipes were heavy. As a compromise, we used a minimal amount of black (painted) aluminum channel for the frame, which acted as a background for blue and white pipes secured by rear-facing pipe clamps. The combination emphasizes the pipes and minimizes the frame. The pipes, which vary in length, are arranged across five different platforms at varying heights, creating an undulating horizontal contour across the instrument that is complemented by the curves of the longer pipes from front to back (Figure 7.1.18).

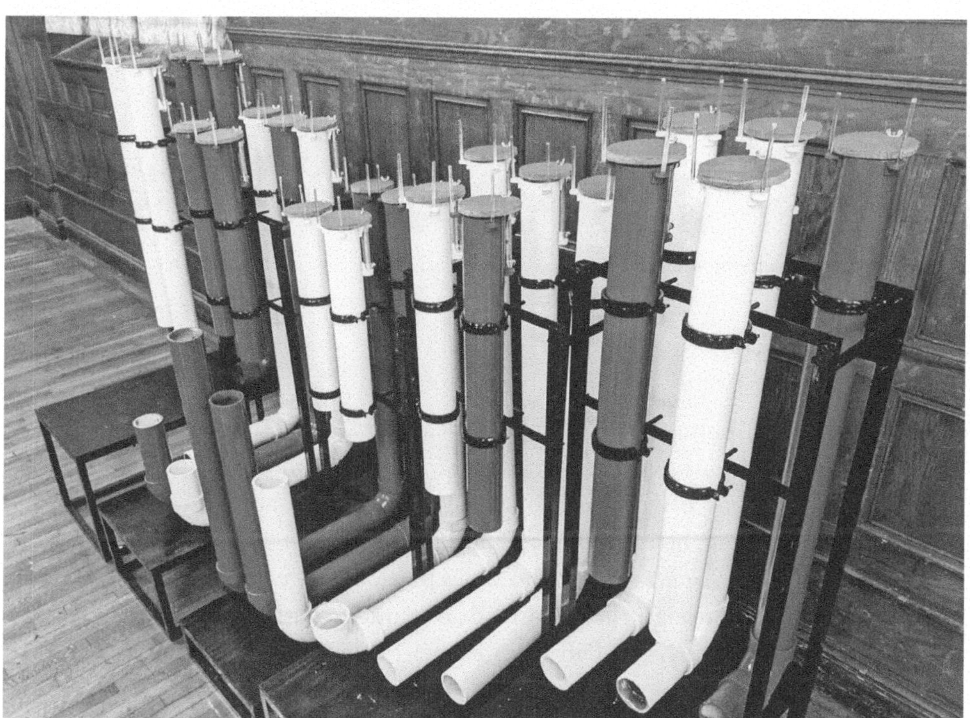

FIGURE 7.1.18. **Percussive aerophone by the MPR Lab.**

CONSPICUOUS ⇔ INVISIBLE

The philosophy behind the automated harmonium of the MPR Lab differed from that of the builders of *Kritaanjli*. The 150-year-old instrument was donated to the lab and was in a state of disrepair that required invasive action. Our desire was not to ruin the craftsmanship of the original builders. Instead, we sought to bring it new life by infusing it with technologies of the modern day. I was inspired by *xenochrony*, a compositional technique used by Frank Zappa, where a musical part from one piece is extracted and placed in a different context. Here, a parlor organ of years past is transported into a future populated by microcontrollers and LEDs. The instrument sheds its anachronistic identity and becomes a tool of modern music-making, but retains a story to tell. In keeping with this transformation, we removed human interfaces, notably the keyboard and stops, and made the machinic processes of the instrument conspicuous. The result of this was a design that externalizes the bellows pumping mechanism and reveals the key- and stop-operating actuators. One can thus see *into* the machine, which weaves old and new technologies together at its core (Figure 7.1.19).

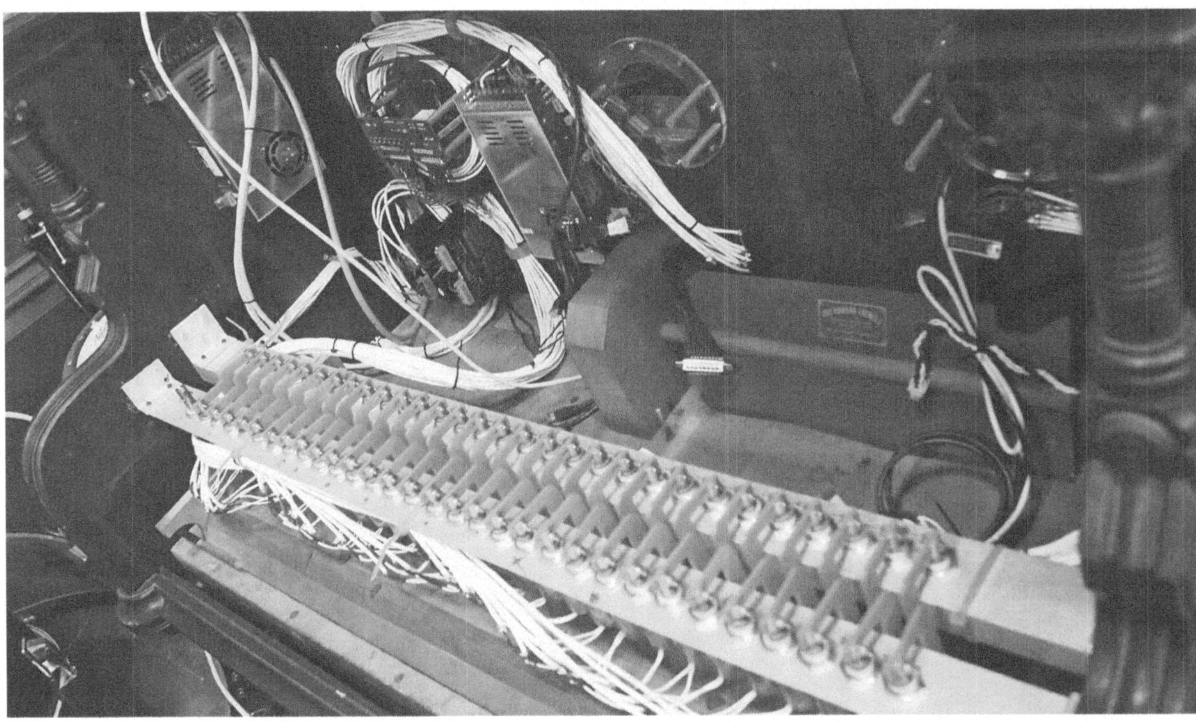

FIGURE 7.1.19. **The MPR Lab's harmonium.**

OTHER RELATIONS

Most aerophone machines fall into the *single / complex* category, given that excitation systems require a specific set of conditions to excite air columns whose length is manipulated by a set of coordinated fingers. "Fingers" often consist of relatively *simple* actuating mechanisms (e.g., a solenoid positioned over a tone hole) replicated and *distributed* over the body of the instrument. In this way, a *simple / distributed* system becomes part of a larger system that is *complex* and *single* as it is coordinated through the commands of a microcontroller. Instruments such as the *Percussive Aerophone* by the MPR Lab feature a relatively *simple* actuating system distributed across the instrument's 25 pipes. The *specific* ⇔ *general* dimension is interesting in the case of aerophones because, on the one hand, producing precise intonation, a wide dynamic range, and varied articulations means tailoring a system to the design of a specific instrument. With that said, there are many similarities between different members of an instrumental family as defined by the excitation method. That is, single-reed instruments work in a general way, as previously described, as do brass instruments. Making a design sufficiently *adjustable*, for example, the airflow's positioning relative to a mouthpiece, might, therefore, allow a machine to interface with a wider variety of acoustic instruments.

COOPERATIVE AEROPHONES

Some builders have developed *cooperative* machinic aerophones. The purpose of these systems ranges from teaching beginners to enabling individuals with physical limitations to make music. For example, Kurosawa and Suzuki (2010) developed a robot-assisted saxophone system in which a human provides the air source while mechatronic "fingers" open and close the keys. This division of labor is well-suited to the players' capabilities. Minute changes in air pressure and embouchure shapes are idiomatic to human performance but are difficult to realize mechanically. Conversely, activating an array of binary interfaces (tone holes, keys, valves) in different configurations with speed and precision is easily accomplished by machines.

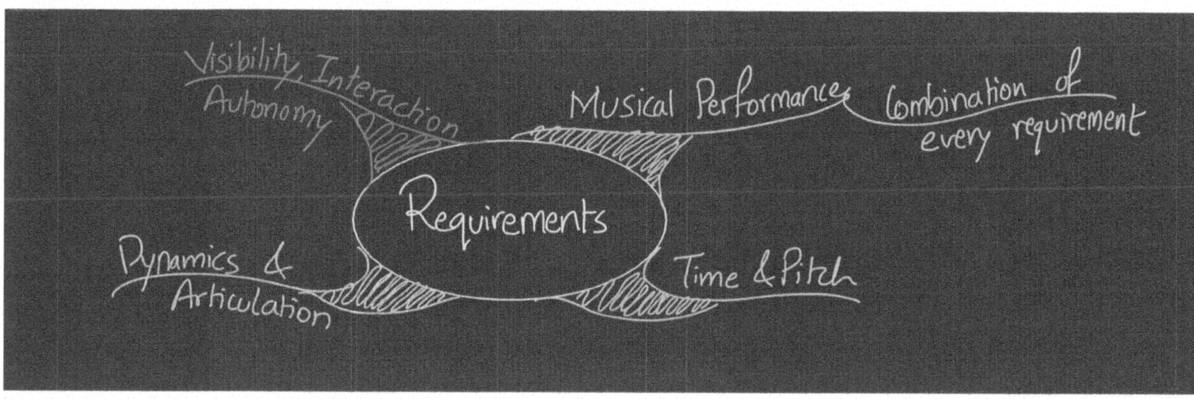

FIGURE 7.1.20. **Mindmap by Yash Garje.**

Requirements

MUSICAL PERFORMANCE

Time

As with all instruments, timing requirements for aerophones are set by our perceptual capabilities; thus, a 100-ms IOI is a speed goal to aspire to. Human performers can play faster than this rate (browsing through YouTube will provide numerous examples), but the perceptibility of individual notes becomes compromised as IOIs shorten. Fifty milliseconds (20 Hz) is another critical threshold, serving as the lower boundary of pitch perception; thus, some (e.g., Lin et al., 2019) have used this rate as a goal. Timbral and harmonic possibilities come to light at these more rapid speeds, which is a possibility to consider as you set your temporal requirements.

Pitch

The accurate production of pitches is a must on most musical instruments; aerophones are no exception. Producing accurate pitches on aerophones is challenging because intonation depends on factors such as air flow rate, fingering, and mouth configuration; thus, multi-parametric control is usually necessary. As with other instruments, the range of tolerable pitch variation is set according to human perception; a machine should be able to produce any pitch within ±12 cents of the ideal frequency, as described in Chapter 4. Variations in airflow rate can produce variations in pitch, enabling musical effects such as vibrato and portamento. The continuous nature of aerophonic sound and the precision of computer control inspire the imagination about the panoply of pitch contours that are possible. It also presents a necessity to make notes discrete so as not to be confined in perpetual portamento.

Dynamics and Articulation

The dynamic range of human-played aerophones varies widely depending on the instrument. Table 7.1.2 shows sound power measurements of orchestral woodwinds (Weinzierl et al., 2018):

TABLE 7.1.2. **Sound power levels for orchestral aerophones determined for single notes over the range of an instrument.**

Instrument	$L_{W\_ff\_max}$	$L_{W\_ff\_min}$	Range	$L_{W\_ff\_av}$	$L_{W\_pp\_av}$	Range
Flute	105	77	28	101	89	12
Oboe	101	80	21	99	84	15
Clarinet (Bb)	110	53	57	102	69	33
Bassoon	104	82	22	101	86	15
French Horn (double, F/Bb)	114	71	43	112	85	27
Trumpet	112	74	38	107	83	24
Trombone (modern, Bb/F)	113	68	45	112	82	30
Tuba	122	71	51			
Saxophone (alto)	111	82	29			

Note: From left to right, the columns indicate the instrument, the maximum value reached playing as loud as possible, the minimum value reached playing as soft as possible, and the range in between the two. The next set of columns shows the average values and their difference over the range of the instrument with pitch distributions of the symphonies of Beethoven used as weights (Weinzierl et al., 2018).

As Weinzierl et al. (2018) point out, the minimum and maximum values are of technical interest, but the average values are more musically relevant. This is because the minimum and maximum values typically occur within contrasting pitch regions that are sometimes extreme and little used in musical practice. The dynamic range for a particular pitch is far smaller. To yield more musically relevant results, the authors of the study calculated a weighted average of the measurements using a pitch distribution found in the classical repertoire (Beethoven's Symphonies). The results of these calculations, as shown in the columns on the right of Table 7.1.2, are the range to aspire to first.

Because aerophones can generate sound continuously, gestures such as crescendo and diminuendo are possible. The expanse of these gestures depends on the dynamic range of the instrument, so they should be considered when determining dynamic range requirements. Tremolo, the periodic oscillation of sound intensity, is another sonic goal. Tremolo does not transpire over a wide range, but it does require high-frequency oscillation. Taken together, changes in intensity and speed are needed to create dynamic musical gestures and must be considered when designing system components, such as the device that generates an airstream.

Articulations on aerophones involve interrupting airflow to generate an envelope consisting of an attack, sustain, and decay. A human player creates articulations by vocalizing syllables like "*tu*," "*du*," "*gu*," "*ku*," "*ee*," or "*ah*" and positioning and flicking their tongue in various parts of their mouth. These methods enable notes to be played staccato, legato, accented, marcato, portato, bright, sharp, or full. Special techniques unique to aerophones, such as double-, triple-, and flutter-tonguing, are also possible. Become familiar with the sounds of these articulations, particularly if you do not play the instrument that you purport to build. Figure out which are musical necessities and which are stretch goals. Craft your requirements accordingly.

VISIBILITY

Unlike percussion performance, where physical gesticulations can be dramatic and encompassing, the motions involved in aerophonic sound production are typically small, sometimes even invisible. The energy source, which involves blowing or pressurizing air, is usually hidden within an enclosure (e.g., a bellows or an air compressor). The movements of an *exciter*'s structural parts (e.g., the reed) are microscopic, and the resultant effects on the stream of air are invisible. The currents of the air volume in the resonator are similarly inaccessible to the naked eye. Modifying the length of an air column provides an opportunity for visible movement. The problem is that in most cases, the movements required involve closing a hole or pressing a key quickly, which means a short distance to travel.

Aerophonic performance does not need to be static; some techniques can enhance visibility. One possibility is to position cameras close to the instrument and project images of mechanisms, such as the reed or artificial fingers, on a larger scale. Airflow can be visualized, for example, by using microparticles illuminated with a sheet of laser light. Instruments that change the length of the air column through a slide, such as a trombone, offer the greatest possibility for visible gestures, given the larger scale of their movements. Machinic aerophones could also move in ways unrelated to sound production, such as gliding across a surface, bobbing and turning their heads, bending their legs, or rotating their torsos. Non-anthropomorphic designs require creativity, but one can imagine elaborate structures whose inner workings are made explicit. Examples include machinery needed to operate a bellows or branch-like structures that extend over the body of an instrument to open and close keys. Consider potential performance scenarios and the scale and types of movements that are required (or afforded) by them.

INTERACTION AND AUTONOMY

The general observations about interactivity and autonomy discussed in Chapter 4 hold when designing a machinic aerophone. Category-specific *internal* interactions involve sensing the conditions within parts of the instrument (such as mouth pressure) and adjusting the motoric system accordingly (these will be discussed in greater detail in subsequent sections). Regarding *external* interactions, machinic aerophones are capable of the same kind of autonomy as other musical machines. They can be driven by music-generating algorithms or can play a score from a DAW sequence. Because their movements are typically minute or invisible, aerophones on stage benefit from interactive relationships with humans, other machines, or the environment. Sonic nuance can be challenging to achieve on a machinic aerophone; thus, the instruments often work best in a complementary role, particularly when the collaborators are human. They can also act as input or in response to other machines. For example, the output of a string instrument could serve as the input signal for a compression driver coupled to a resonator, as seen in *AMI* and *CARI* by EMMI (Rogers et al., 2015). Finally, since aerophones fundamentally involve the excitation of volumes of air that have a reciprocal effect on the exciter, these physical interactions can be visualized to illuminate how the machine's actions result from the affordances of the natural world.

References

Almeida, A., George, D., Smith, J., & Wolfe, J. (2013). The clarinet: How blowing pressure, lip force, lip position and reed "hardness" affect pitch, sound level, and spectrum. *The Journal of the Acoustical Society of America, 134*(3), 2247–2255.

Almeida, A., Lemare, J., Sheahan, M., Judge, J., Auvray, R., Dang, K. S., John, S., Geoffroy, J., Katupitiya, J., & Santus, P. (2010). Clarinet parameter cartography: Automatic mapping of the sound produced as a function of blowing pressure and reed force. *Proceedings of the International Symposium on Music Acoustics*, 1–6.

Almeida, A., Li, W., Smith, J. R., & Wolfe, J. (2017). The mechanism producing initial transients on the clarinet. *The Journal of the Acoustical Society of America, 142*(6), 3376–3386.

Almeida, A., Vergez, C., & Caussé, R. (2004). Experimental investigations on double reed quasi-static behavior. *Proceedings of ICA, 2004*, 1229–1232.

Almeida, A., Vergez, C., & Caussé, R. (2007). Quasistatic nonlinear characteristics of double-reed instruments. *The Journal of the Acoustical Society of America, 121*(1), 536–546.

Bromage, S. (2007). *Visualisation of the lip motion of brass instrument players, and investigations of an artificial mouth as a tool for comparative studies of instruments* (Dissertation, University of Edinburgh). https://era.ed.ac.uk/bitstream/handle/1842/1966/Bromage%20S%20thesis%2007.pdf?sequence=1&isAllowed=y

Bromage, S. R., Richards, O. F., & Campbell, D. M. (2003). Reproducibility and control of the embouchure of an artificial mouth for playing brass instruments. *Proceedings of the Stockholm Musical Acoustics Conference.*

Campbell, M., Gilbert, J., & Myers, A. (2021). Buzzing lips: Sound generation in brass instruments. In *The Science of Brass Instruments* (pp. 63–100). Springer.

Chatziioannou, V., Schmutzhard, S., Pàmies-Vilà, M., & Hofmann, A. (2019). Investigating clarinet articulation using a physical model and an artificial blowing machine. *Acta Acustica United with Acustica, 105*(4), 682–694.

d'Andréa-Novel, B., Joly, C., Noël, T., & Senpauroca, J. (2016). Control of a slide flute: A mechatronic project. *Proceedings of the 3rd International Symposium on Movement and Computing*, 1–5.

Dannenberg, R. B., Brown, B., Zeglin, G., & Lupish, R. (2005). McBlare: A robotic bagpipe player. *Proceedings of the 2005 Conference on New Interfaces for Musical Expression*, 80–84. http://dl.acm.org/citation.cfm?id=1085962

Dannenberg, R. B., Brown, H. B., & Lupish, R. (2011). McBlare: A robotic bagpipe player. In J. Solis & K. Ng (Eds.), *Musical Robots and Interactive Multimodal Systems* (pp. 165–178). Springer.

Ehara, F., Nagai, K., & Mizutani, K. (2001). Relationships between vibration of artificial lips and sound frequency of a trumpet. *Proceedings of the International Symposium on Musical Acoustics, Perugia, Italy*, 513–516.

Gilbert, J., Ponthus, S., & Petiot, J.-F. (1998). Artificial buzzing lips and brass instruments: Experimental results. *The Journal of the Acoustical Society of America, 104*(3), 1627–1632. https://doi.org/10.1121/1.424375

Hopkin, B. (1996). *Musical Instrument Design*. See Sharp Press. https://wpi.primo.exlibrisgroup.com/permalink/01WPI_INST/dbokq8/alma9936814463004746

Hopkin, B. (2021). Membrane reeds: Indonesia and Nicasio. In B. Hopkin & S. Tewari (Eds.), *Sound Inventions* (pp. 178–189). Focal Press.

Hornbostel, E. M. von, & Sachs, C. (1961). Classification of musical instruments (A. Baines & Klaus P. Wachsmann, Trans.). *The Galpin Society Journal, 14*, 3–29.

Huang, G., Zhang, W., Meng, F., Tian, Y., Shi, Q., Petersen, K., Takanishi, A., & Huang, Q. (2014). Development of an anthropomorphic flute robot for the application in science and technology museum. *2014 IEEE International Conference on Robotics and Biomimetics (ROBIO 2014)*, 2220–2225.

Kato, T., Shimazaki, K., Higashijima, K., Tokunaga, R., Ayuthaya, P. T. N., & Thumwarin, P. (2018). Refinement of a Thai flute-playing robot in Thai style. *2018 International Conference on Engineering, Applied Sciences, and Technology (ICEAST)*, 1–4.

Kurosawa, Y., & Suzuki, K. (2010). Robot-assisted playing with fingering support for a saxophone. *Proceedings of the International Computer Music Conference, 2010*, 1–4. https://quod.lib.umich.edu/i/icmc/bbp2372.2010.001/1/--robot-assisted-playing-with-fingering-support-for?rgn=full+text;view=image;q1=robot

Li, J., Hu, T., Zhang, S., & Mi, H. (2019). Designing a musical robot for Chinese bamboo flute performance. *Proceedings of the Seventh International Symposium of Chinese CHI*, 117–120. https://dl.acm.org/doi/abs/10.1145/3332169.3332264

Li, W., Almeida, A., Smith, J., & Wolfe, J. (2014). Tongue, lip and breath interactions in clarinet playing: A study using a playing machine. *Proceedings of the 21st International Congress on Sound and Vibration, ICSV21, Beijing, China.*

Li, W., Almeida, A., Smith, J., & Wolfe, J. (2016). How clarinettists articulate: The effect of blowing pressure and tonguing on initial and final transients. *The Journal of the Acoustical Society of America, 139*(2), 825–838. https://doi.org/10.1121/1.4941660

Lin, J.-Y., Kawai, M., Matsuki, K., Cosentino, S., Sessa, S., & Takanishi, A. (2019). Musical articulation system on humanoid saxophonist robot. In V. Arakelian & P. Wenger (Eds.), *ROMANSY 22–Robot Design, Dynamics and Control* (pp. 392–399). Springer.

Lin, J.-Y., Kawai, M., Nishio, Y., Cosentino, S., & Takanishi, A. (2019). Development of performance system with musical dynamics expression on humanoid saxophonist robot. *IEEE Robotics and Automation Letters, 4*(2), 1684–1690. IEEE Robotics and Automation Letters. https://doi.org/10.1109/LRA.2019.2897372

Lin, J.-Y., Yoshida, K., Matsuki, K., Takikawa, K., Cosentino, S., Sessa, S., & Takanishi, A. (2016). The development of intraoral pressure control system on humanoid saxophone playing robot. *2016 IEEE International Conference on Robotics and Biomimetics (ROBIO)*, 1649–1654.

Lopes, N., Hélie, T., & Caussé, R. (2013). Control of an artificial mouth playing a trombone and analysis of sound descriptors on experimental data. *Stockholm Music Acoustics Conference 2013*, 521–528.

Murphy, J., Kapur, A., & Carnegie, D. A. (2014). Mechatronic keyboard music: Design, evaluation, and use of a new mechatronic harmonium. *Proceedings of the 2014 International Computer Music Conference*, 678–685.

NICTA-UNSW Robot Clarinet. (2008). UNSW Music Acoustics. http://newt.phys.unsw.edu.au/jw/clarinetrobot.html

Raes, G.-W. (n.d.). *Logos Foundation—Automatons and robots*. Logos Foundation. Retrieved May 26, 2022, from https://www.logosfoundation.org/instrum_gwr/automatons.html

Raes, G.-W. (2007). *Puff: An automated quartertone percussive organ*. https://logosfoundation.org/instrum_gwr/puff.html

Richards, O. (2003). *Investigation of the lip reed using computational modelling and experimental studies with an artificial mouth*. University of Edinburgh.

Robotic Trombone: The RoboTrombo Music Machine. (n.d.). Retrieved June 2, 2022, from https://hackaday.io/project/176463-robotic-trombone-the-robotrombo-music-machine

Rogers, T., Kemper, S., & Barton, S. (2015). MARIE: Monochord-aerophone robotic instrument ensemble. *Proceedings of the International Conference on New Interfaces for Musical Expression*, 408–411. https://pdfs.semanticscholar.org/de7a/ff6fdb9ac3e2cc0dfa5fd269368780944f27.pdf

Sidler, M., Bisson, M., Grotz, J., & Barton, S. (2020). Parthenope: A robotic musical siren. *Proceedings of the International Conference on New Interfaces for Musical Expression*, 297–300.

Solis, J., Chida, K., Taniguchi, K., Hashimoto, S. M., Suefuji, K., & Takanishi, A. (2006). The Waseda flutist robot WF-4RII in comparison with a professional flutist. *Computer Music Journal, 30*(4), 12–27.

Solis, J., Petersen, K., Ninomiya, T., Takeuchi, M., & Takanishi, A. (2009). Development of anthropomorphic musical performance robots: From understanding the nature of music performance to its application to entertainment robotics. *2009 IEEE/RSJ International Conference on Intelligent Robots and Systems*, 2309–2314. https://doi.org/10.1109/IROS.2009.5354547

Solis, J., & Takanishi, A. (2007). An overview of the research approaches on musical performance robots. *International Computer Music Conference Proceedings, 2007.* http://hdl.handle.net/2027/spo.bbp2372.2007.179

Solis, J., & Takanishi, A. (2011). Wind instrument playing humanoid robots. In J. Solis & K. Ng (Eds.), *Musical robots and interactive multimodal systems* (pp. 195–213). Springer.

Solis, J., Taniguchi, K., Ninomiya, T., Yamamoto, T., Sato, A., & Takanishi, A. (2007). Musical skills of the Waseda flutist robot WF-4RIV. *2007 IEEE/RSJ International Conference on Intelligent Robots and Systems*, 2570–2571.

Stevenson, S. D. (2009). *Experimental investigations of lip motion in brass instrument playing* (Dissertation, University of Edinburgh). https://era.ed.ac.uk/bitstream/handle/1842/4349/Stevenson2009.pdf?sequence=1&isAllowed=y

Sundberg, K., Barton, S., Walter, A., Sane, T., Baker, L., & O'Brien, A. (2018). A robotic percussive aerophone. *Proceedings of the 18th International Conference on New Interfaces for Musical Expression*, 409.

Takanishi, A., & Maeda, M. (1998). Development of an anthropomorphic flutist robot WF-3RIV. *International Computer Music Conference Proceedings, 1998.* http://hdl.handle.net/2027/spo.bbp2372.1998.267

Teissier, F., Gilbert, J., Campbell, M., & Petiot, J.-F. (2003). Comparative analysis of brass wind instruments with an artificial mouth: First results. *Acta Acustica United with Acustica, 89*(6), 974–979.

Uchiyama, J., Hashimoto, T., Ohta, H., Lin, J.-Y., Cosentino, S., & Takanishi, A. (2023). *A humanoid saxophone-playing robot based on instrument-centered design: Sound improvement with fingers using hybrid soft materials* (pp. 295–306). https://doi.org/10.1007/978-3-031-35634-6_21

Vergez, C., & Rodet, X. (1997). Comparison of real trumpet playing, latex model of lips and computer model. *Proceedings of the 1997 International Computer Music Conference*, 1–8.

Weinzierl, S., Lepa, S., Schultz, F., Detzner, E., von Coler, H., & Behler, G. (2018). Sound power and timbre as cues for the dynamic strength of orchestral instruments. *The Journal of the Acoustical Society of America, 144*(3), 1347–1355.

Zhao, F., Li, M., Zhou, C., Chen, G., & Lou, Y. (2020). Music melody extraction algorithm for flute robot. *2020 IEEE International Conference on Real-Time Computing and Robotics (RCAR)*, 38–43. https://ieeexplore.ieee.org/abstract/document/9303271?casa_token=m9zLDkkHsLEAAAAA:HKgrOHTlnCaMgnNPGDDYQz7rVVhIEB4HjvtBQ-54rXtk_en-hOSoLNp3GFqPuDd4Hj-Tjhpu

CHAPTER 7

AEROPHONES

Part 2: Design

Sonic Objects

With knowledge of aerophonic sound production and the history of machinic instruments that have put such ideas into practice, we start the design process with the sonic object. Aerophonic sonic objects can be acquired *off-the-shelf*, acquired *through rummage*, or *made from scratch*. The path chosen will depend on the complexity of the mechanisms and the resonator's materials. For example, a simple flute can be made *from scratch* by drilling holes in a PVC pipe. Making a metal trumpet with intricate valve mechanisms is another story entirely; far less resistance would be encountered by venturing to a music store. *Rummaging* presents interesting possibilities by providing access to objects whose aerophonic potential is latent: that ceramic jug in the basement is just waiting to sing. For those interested in making or modifying an aerophonic instrument, head to the Instrument Design section of the Companion Website ▶.

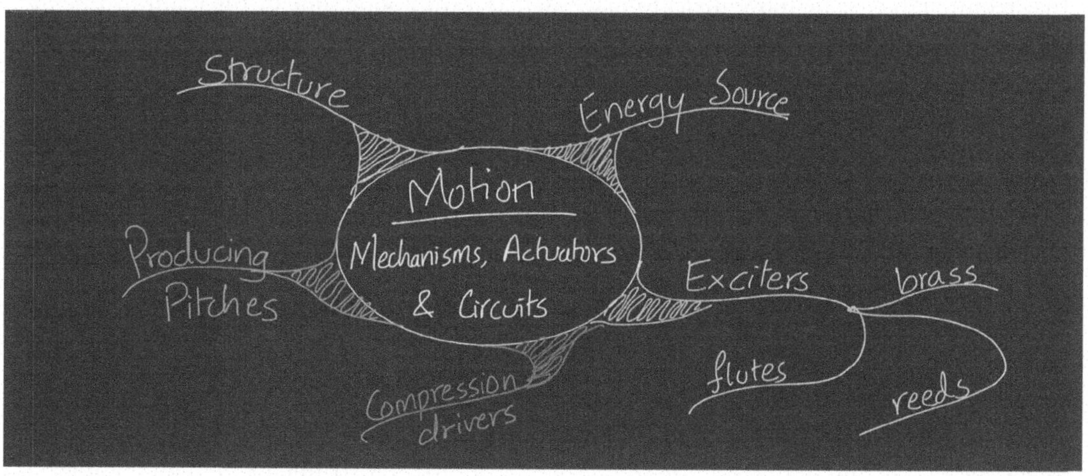

FIGURE 7.2.1. **Mindmap by Yash Garje.**

Motion

Designing a machine to play an aerophonic sonic object involves figuring out how the system will move. The following relation, introduced in Chapter 4: The Process of Making a Musical Machine, applies here:

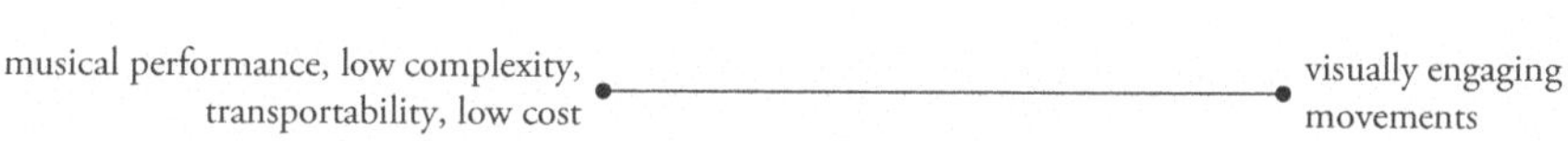

FIGURE 7.2.1A. **MUSICAL PERFORMANCE—VISIBILITY DIMENSION.**

For those prioritizing sound production in a studio environment, microscopic movements are not a problem. For those interested in gestures that are more visually engaging, the choice is either to design separate systems that differ in their visibility or to compromise with a single system situated between the poles of the above dimension.

The first option makes sound-producing and visually engaging movements independent. This strategy is employed in the Chinese bamboo flute-playing robot (J. Li et al., 2019). The *instrument play* system contains an air compressor and a solenoid-driven array of artificial fingers. The *body movement* system was inspired by observations of expert human musicians who use their heads and bodies to convey emotions and emphasize musical statements. To emulate such actions, they implemented a servo array to control the head, arms, waist, and costume. The authors realized the difficulty in imitating such human movements exactly; thus, they chose to interpret them in ways idiomatic to the machine. The body movements are aligned with specific musical scenes, purport to establish the machine's personality, and do not affect the sound-production movements.

The Logos Foundation also enables visually engaging gestures through motion. <*Klar*> (clarinet) moves along a pendular path. The instrument is suspended in a cradle connected to a wheel-based, motor-driven mechanism. Tilt sensors prevent positions that would potentially harm the system (Raes, 2012). <*Korn*> (cornet) and <*Bug*> (flugelhorn) move along two DOFs: vertically and horizontally. This movement creates visual interest, but it is also vital for sonic expressivity. High brass instruments are strongly directional, so human performers regularly adjust their position (Raes, 2023).

Another possibility is to make sound-production movements more visually engaging through a grander scale and more complex motion paths. Channel the creativity of Animusic: instead of microscopic solenoids positioned over tone holes, a dynamic, multi-legged structure could wander between the keys of a clarinet. This is easier for instruments that require larger-scale movements to play, such as a trombone slide or the foam paddle of a percussive aerophone. Another possibility is to combine these two options: dramatic sound-production gestures coupled with a separate system dedicated to expressive movements. With greater and more complex possibilities come increasing technical demands. More involved structures can be created by combining simpler elements, such as those introduced in the Mechanisms section of the Companion Website ▶.

Mechanisms, Actuators, and Circuits

The mechanisms, actuators, and circuits required to produce sound on an aerophone can be grouped into three primary categories introduced at the beginning of the chapter: **energy source**, **exciter**, and **resonator**. This organization is needed because some components, such as energy sources and resonators, are common to a variety of instruments. It allows builders to think in terms of mixing and matching components, leading to the creation of new musical instruments and innovative music.

For those requiring a primer (or those that would benefit from a refresher), the Companion Website ▶ contains sections devoted to *The Mechanical*, covering physical principles, mechanisms, motion conversions, connections, and methods, and *Actuators, Circuits, and Code*, providing technical descriptions and projects to help you turn information into physical action.

ENERGY SOURCE

Making sound with a machinic aerophone requires a device that generates airflow. Mechanized **pumps** transport air from one location to another, while **compressors** change the volume and density of the gas. *Positive displacement* compressors draw air into a chamber, reduce the chamber's volume, transport the compressed air (e.g., to a storage tank), and expand the chamber. **Bellows** consist of two flat surfaces (often hinged at one side) connected by a concertinaed container (usually a bag with folds). As the surfaces close, the container contracts, forcing the air out of a nozzle. As the surfaces separate, air is admitted via a one-way valve, expanding the container and the enclosed volume of air.

To choose the appropriate air-modulating device, determine the required **pressure range, volume flow rate,** and **capacity**. Do you need to be able to vary air pressure or flow while the instrument is playing? Consider **air quality**: oil-powered compressors may leak lubrication, negatively affecting air quality, making oil-free options more attractive. Regarding **size**, how portable does the air compressor need to be? Will it be fixed in position in a room, or will you need to put it in your car when the instrument is transported? Other pros and cons of specific compressor designs are summarized in Table 7.2.1.

TABLE 7.2.1. **Comparison of compressor types.**

Compressor Type	Pros	Cons
reciprocating / piston[1] 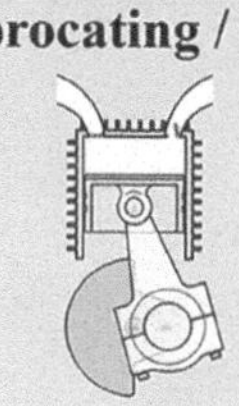	• compact portable; good for smaller applications • low energy consumption • low cost • easy to service	• not designed for continuous use • produce heat and noise • more moving parts that require lubrication • requires a reservoir • vibration
rotary screw[2] 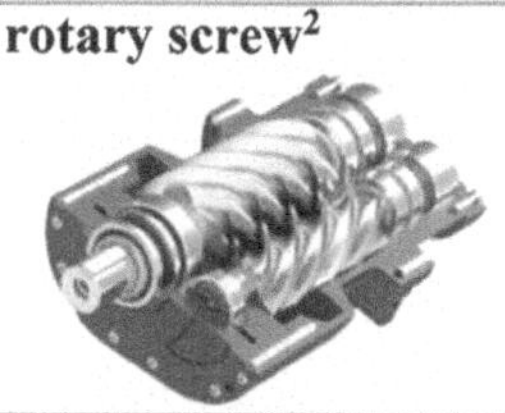	• produce significant air flow for large, industrial applications • continuous duty providing a constant flow of air • few moving parts; quiet and durable • produce cooler air that is easier to dry	• not as efficient • expensive • requires routine maintenance • small size or low rotor speeds are less practical
rotary vane[3] 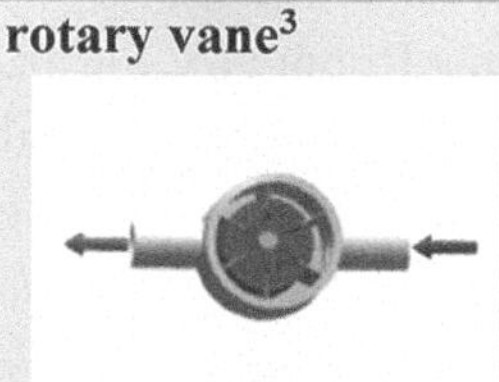	• continuous duty providing a constant flow of air • fewer moving parts: quiet • low air pulsation • compact • low cost	• Not as efficient as rotary screw • more difficult to service
rotary scroll[4] 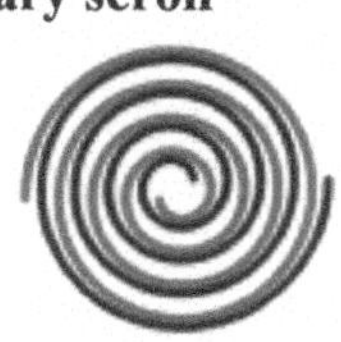	• few moving parts: reduced vibrations, durable, no lubrication needed, less maintenance • produce clean, low pulsation air flow • quiet	• produce relatively less power • smaller capacities • less efficient

[1] By Olivier Cleynen (this file); Original authors: User:Cdang, User:Zephyris—Own work based on File:Piston bielle vilebrequin coupe et schema cinematique.svg itself based on File:Piston bielle vilebrequin coupe et schema cinematique.svg, CC BY-SA 3.0, https://commons.wikimedia.org/w/index.php?curid=40046888

[2] By Airpol PPS—Own work, CC BY-SA 4.0, https://commons.wikimedia.org/w/index.php?curid=71251373

[3] By R. Castelnuovo, CC BY-SA 3.0, https://commons.wikimedia.org/w/index.php?curid=379575

[4] Image by Cacycle—Own work, Public Domain, https://commons.wikimedia.org/w/index.php?curid=784727

Bellows have been widely adopted in machinic aerophones. Vaucanson's *The Flute Player* featured nine bellows, each set of three attached to different weights to provide varying amounts of air (Wood, 2002). The Waseda Flutist Robot No. 4 Refined IV (WF-4RIV) features two bellows sealed in acrylic cases activated by motor-driven crank mechanisms. Varying the motor speed affects the airflow rate (Solis & Takanishi, 2011). The automated harmonium *Kritaanjli* (Murphy et al., 2014) features a DC motor-driven four-bar mechanism attached to the existing bellows of the instrument (traditionally hand-operated). Dimensioning the four-bar, which converts rotational into reciprocating motion, required determining the range of movement between the sound-producing threshold and that which would potentially damage the structure of the bellows. The rate of pumping (5–60 pumps / minute) was linearly mapped to MIDI velocities. The blowing module of the *Open Source Hardware Robotic Woodwind* (Clawson et al., 2022) consists of two circular end surfaces connected by a flexible concertina material. One end is fixed, while the other is moved by a stepper motor-driven lead screw, expanding and contracting the air reservoir to admit and expel air. The designs of some of these systems are represented in Figure 7.2.2.

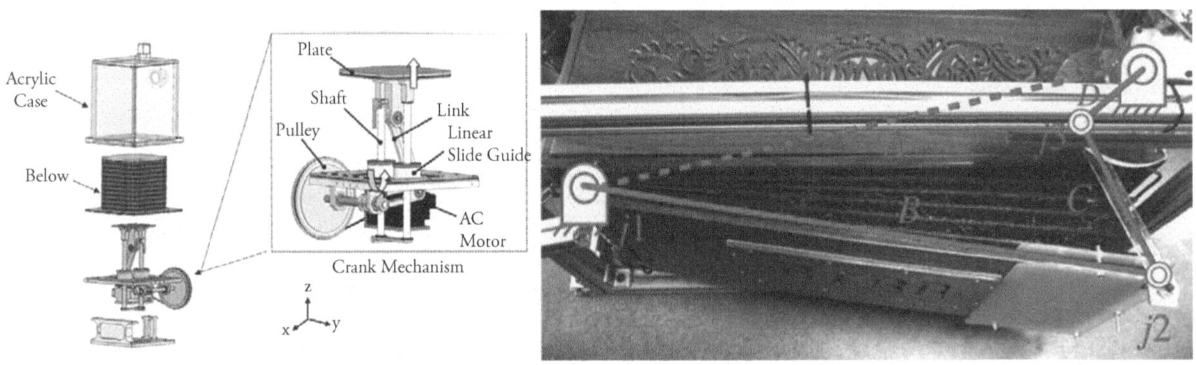

FIGURE 7.2.2. **Bellows pumping mechanisms. Left: WF-4RIV (Solis & Takanishi, 2011); Right: Kritaanjli (Murphy et al., 2014).**

Compressors and pumps are also featured in machinic aerophones, including the *mechatronic slide flute* (d'Andréa-Novel et al., 2016), the *Chinese bamboo flute* (J. Li et al., 2019), and the artificial mouth by Vergez & Rodet (1997). The Waseda Saxophonist Robot No. 2 utilized an eccentric crank connected to an air pump diaphragm (Figure 7.2.3, left), which was capable of producing a flow of 20 L/min and a pressure of 30 kPa (Solis & Takanishi, 2011). Logos used compressors in many of their organs, including <*Harma*>, <*HarmO*>, and <*Piperola*>, sometimes replacing the bellows mechanisms in the original instruments (Raes, n.d.). *McBlare* (Dannenberg et al., 2011) utilizes a custom-built pump that features a motor-driven gearbox connected to two reciprocating cylinders, which are visible and engaging to watch (Figure 7.2.3, right). This design was pursued because of the requirement for a high airflow rate and low pressure. The MPR Lab uses an industrial air compressor in the *Percussive Aerophone* and *Parthenope*.

Air compressors are convenient because they can continuously provide a wide range of pressures, but they are generally loud (particularly those designed for the construction industry). Some companies, such as Eagle, manufacture "quiet" compressors that are indeed less noisy than their industrial counterparts; however, they still require significant attenuation (e.g., placing them in a distant room) to be used in musical settings.

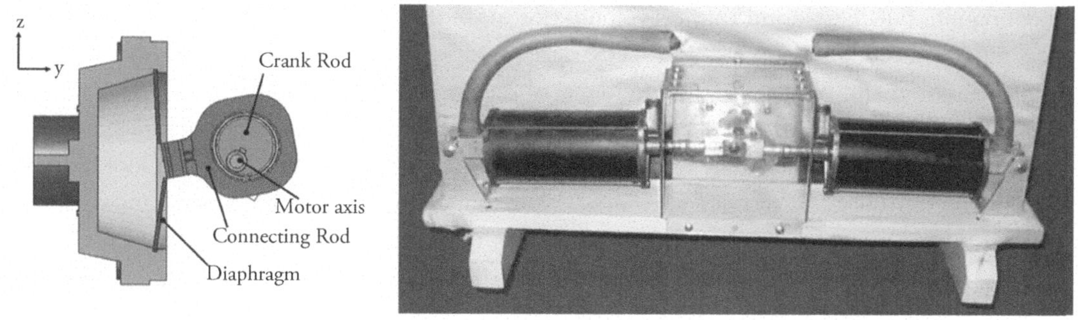

FIGURE 7.2.3. **Air supply mechanisms. Left: WAS-2 (Solis & Takanishi, 2011), Right: McBlare (Dannenberg et al., 2011).**

A quieter and more portable option is to use compressed air tanks, which have numerous benefits. The airflow output from pumps can be choppy, producing undesired sonic effects. Pumps generate pulsations of pressurized air rather than a smooth stream. When directly applied to an instrument, the result is similar to flutter-tonguing, rather than a fluid, sustained tone. To smooth this choppy air, a reservoir or tank acts as a **buffer**, collecting and calming the air before it is output to the next part of the system. Standard air compressors have reservoirs attached, but tanks without compressors can also be procured. Lin et al. (2016) experimented with reservoir sizes to mitigate air pulsations and found that a 2000 ml volume reduced the fluctuations to under 3%. Air can be compressed within a tank before a performance, sparing the audience from the distracting sound of the pump motors and obviating the need to transport the pump mechanism. Many builders integrate storage tanks for this reason (Dannenberg et al., 2011). The downside is that the source of compressed air is limited by the size of the tank: best for musical miniatures that need no rehearsal!

Airflow affects the pitch and dynamic of the tone produced. Depending on the instrument, producing a pitch requires intraoral pressure within a specific range. This has been described mathematically by Lin et al. (2016) in terms of pitch impedance:

$$P = Z_{note} \times U$$

where P is the intraoral pressure required to sound a particular pitch, Z_{note} is the impedance of that pitch, and U is the airflow rate input into the system. Different pitches have different impedances, so changing pitch requires accompanying changes in airflow (U) to continue producing sound (incredibly, humans learn to do this). Regarding dynamics, a linear relationship exists between airflow and output SPL (Lin, Kawai, Nishio, et al., 2019).

Producing specific pitches and dynamic levels thus requires a system where airflow can be measured, varied, and adjusted as necessary. Sensors such as the quick-response laminar flow sensor (QFS) used in the *Thai flute-playing robot* (Kato et al., 2018) or the Technoterm 5402 (Chatziioannou et al., 2019) measure airflow in the oral cavity (or other strategic locations). The rate of airflow is varied by changing the speed of the pump, which is achieved by modulating the electrical current delivered to the pump motor. Valves can also modify airflow. The UNSW-NICTA robotic clarinet regulates blowing pressure using a pump and a leak controlled by a PID loop (W. Li et al., 2014). Producing the range of pitches and articulations demanded by musical performance can be achieved by rapid and precise control over a valve. Lin et al. (2019) utilize a high-speed solenoid valve (K3-100HF-24) with a maximum frequency of 100 Hz in *WAS-5*. Kato et al. (2018) used a spool-type servo valve in the *Thai flute-playing robot*. A servo valve enables electrical control of air regulation, which can be varied from the PWM output of a microcontroller. The Logos Foundation utilized a servomotor-driven rotary valve with a source of compressed air for the artificial mouth of <*Bono*> (valve trombone). The sonic results were encouraging, but they found that they could not move the motor fast enough to generate musical pitches acceptably (Raes, 2010a). The right approach thus depends on both the technical implementation and the musical requirements of the builder.

With the energy source determined, the relationship between pitch, dynamics, and airflow over the instrument's range can be mapped. These values can be stored in a lookup table on a microcontroller and recalled as desired. Note that airflow affects both pitch and dynamic level simultaneously. Decoupling the two, for example, when trying to produce tremolo and vibrato, requires fingering techniques that will be discussed later in the chapter.

EXCITERS

We now have a stream of air, but something must set it into vibration to *excite* the volume of air in the attached vessel. For human players, the *exciter* includes the mouth and lips, which can be manipulated to produce diverse sounds. Some machinic examples emulate human physiology, attempting to produce sounds as humans do. Others, particularly those from acoustics research, focus on the physical variables involved in sound production. Those variables are manipulated by mechanical means that provide sufficient accuracy, resolution, and range, regardless of whether they are anthropomorphic or not. While such projects are situated more comfortably in the laboratory than the concert hall, the ideas are useful to one designing a machinic musical performer. The following sections will describe mechanistic approaches in key instrument categories, including flutes, reeds, and brass.

Flutes

Excitation of a flute involves a stream of air passing over an opening in the bore of the instrument. The *energy source* determines the rate of the input airflow, and the *exciter* directs that airflow in specific ways to produce sound. The *mechatronic slide flute* (d'Andréa-Novel et al., 2016) uses a plastic inlet that channels the air into a chamber where it is "stirred" and then passes to the "lips" and the resonator (p. 2). The Waseda Flutist contains 3-DOF artificial lips where the superior lip controls the thickness of the airstream, the inferior lip directs its angle, and the sideway lips its length (see Figure 7.2.4). The lips are made of thermoplastic rubber to provide the low hardness and high stiffness required (Solis & Takanishi, 2011).

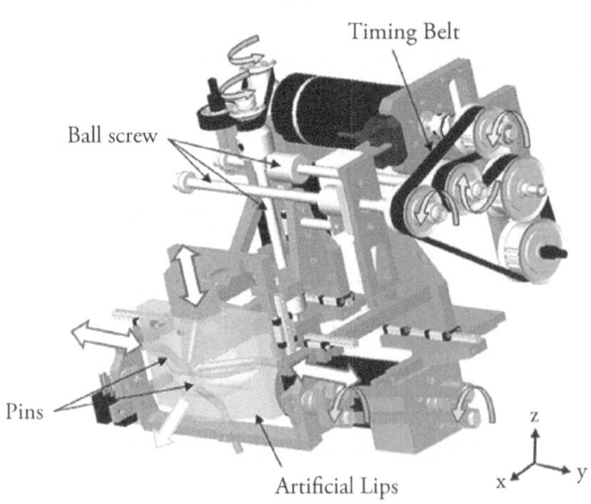

FIGURE 7.2.4. **Artificial lips of the Waseda Flutist (image from Solis & Takanishi, 2011).**

Vibrato is vital in wind instrument performance. The Waseda Flutist produces vibrato by varying the aperture of the glottis through motor-driven linkages connected to the vocal folds. Varying the aperture's amplitude and frequency produces vibrato effects (see Figure 7.2.5).

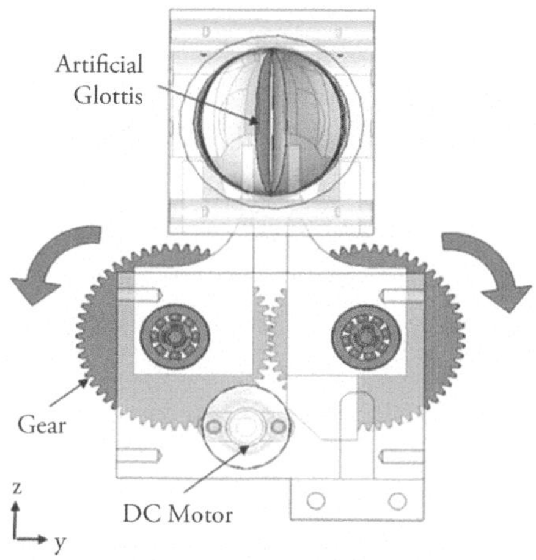

FIGURE 7.2.5. **Artificial glottis of the Waseda Flutist (Solis & Takanishi, 2011).**

Solis et al. (2006) compared the performance of the Waseda flutist (WF-4RII) to a professional human flutist. Similarities were found, though the authors acknowledged the need to improve sound clarity, brightness, and dynamics of the machine's performance.

The documentation of The Logos Foundation's *<Flut>* elucidates techniques for flute excitation (Raes 2020b). An initial experiment utilized a compressor-valve system to propel an airstream through an oval slit made in latex directed at a flute's opening. The jet angle and the distance from the lip plate (higher pitches require shorter distances) were varied to achieve proper intonation, requiring four DOFs. To avoid this complexity, they experimented with a membrane compression driver and an impedance converter (described in more detail in the Compression Drivers, p. 249, section) but found the sonic results to be "artificial and electronic" (Raes, 2020b). The design was modified by feeding the impedance converter's output into a tiny capillary inserted into the stop-end of the flute, improving sonic results (Figure 7.2.6). Further enhancements were realized by dynamically covering the lip plate orifice with a silicone membrane driven by a soft shift solenoid (Figure 7.2.6). Closing the flap completely extended the flute's lower pitch range by an octave. The ideal excitation waveform was determined by generating an acoustic response from the instrument using a recording of itself, recording the output, and then subtracting the two. The method is described in detail in Raes (2015).

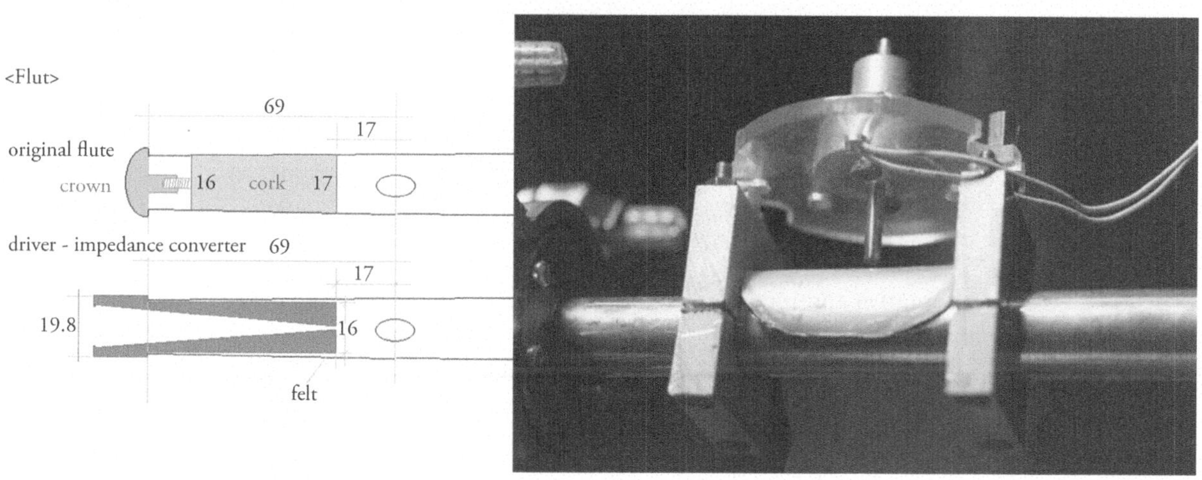

FIGURE 7.2.6. **<Flut> by The Logos Foundation. Left: impedance converter, Right: orifice-covering mechanism (Raes, 2020b).**

Reeds

Producing sound on a reed instrument requires balancing the air pressure in the mouth (intraoral pressure), the force applied by the lip to the reed (which is affected by the lip opening), and the position on the reed where that force is applied.

First, an oral cavity must connect to the air supply. Practically, this takes the form of a box with an opening connected to a proportional valve, which controls the input airflow. The volume of the cavity replicates that of human players, ranging around 170 cm^3 (Chatziioannou et al., 2019). The materials of the box can vary from plexiglass (Chatziioannou et al., 2019) to an elastomer such as SEPTON (Lin, Kawai, Matsuki, et al., 2019). The pressure required to produce sound depends on the characteristics of the physical system and the note desired, but devices have been built that achieve 10 kPa inside the cavity (Chatziioannou et al., 2019).

An interface between the oral cavity and the mouthpiece of the instrument is needed, often taking the form of artificial "lips." The lips of the Waseda Anthropomorphic Saxophone robot are made of SEPTON, an elastomer whose properties can be customized (Lin, Kawai, Matsuki, et al., 2019). The upper and lower lips move up and down via a mechanism that contains a motor-driven timing belt connected to a ball screw (Figure 7.2.7). As the lower lip moves, the pressure exerted on the reed is varied. An additional passive DOF allows the sideways configuration of the lips to be adjusted. The NICTA-UNSW clarinet's lip is made of flexible plastic, and its position and force (relative to and applied to the reed) are controlled by servo motors (W. Li et al., 2014). The lip force applied to the reed was varied by using a series of hanging masses incremented in 25 g quantities (Figure 7.2.10).

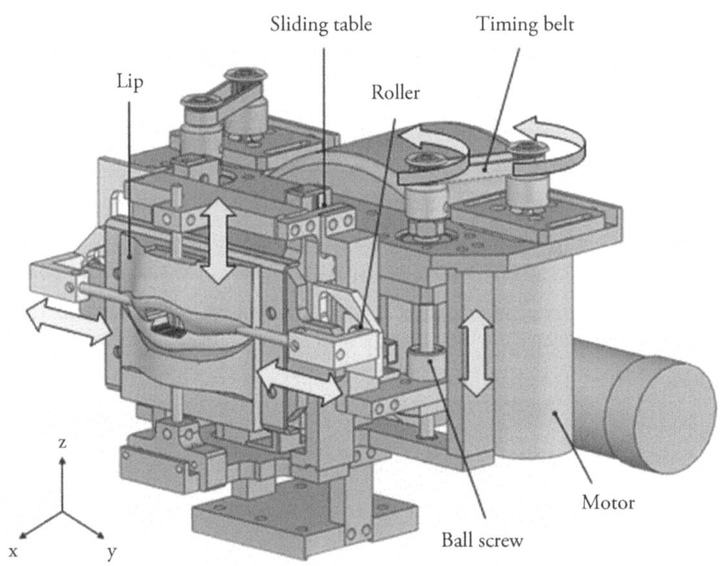

FIGURE 7.2.7. **Artificial mouth of WAS-2 Saxophone robot (Solis & Takanishi, 2011).**

An alternative and ingenious design appears in The Logos Foundation's <*Autosax*>. The third version of the instrument (2007) features an electromagnetically controllable reed-mouthpiece system (Figure 7.2.8). A solenoid applies force to the top of a traditional mouthpiece tip to control the aperture, while the position of a composite reed made of cane and spring steel is modulated using an electromagnet. The movement of the reed is controlled by an electrical current delivered to the electromagnet, which is driven by a waveform generated by a computer. The sound produced was "very good," and the system was capable of "a wealth of multiphonics, slaptongues, and other special effects" (Raes, 2022).

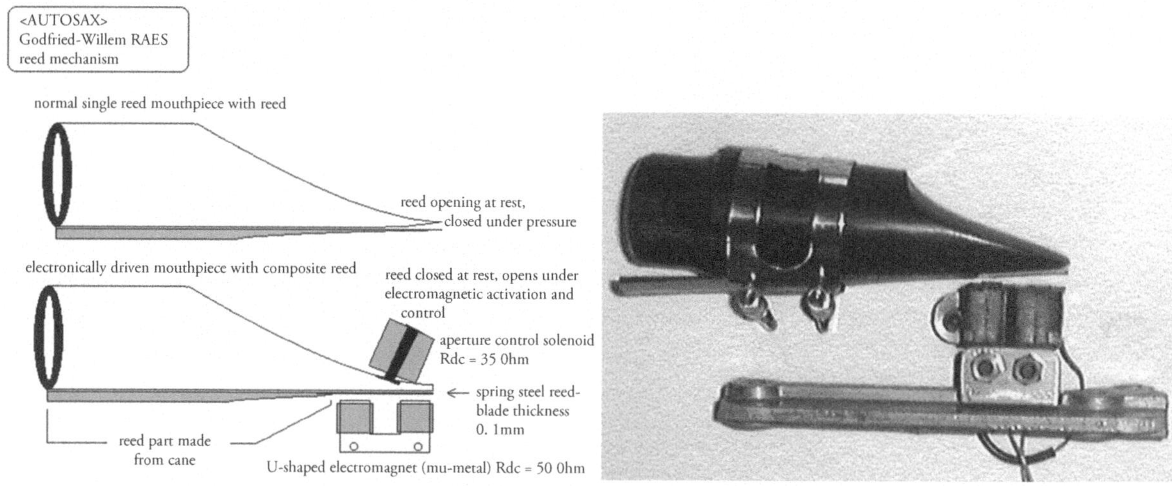

FIGURE 7.2.8. **Reed mechanism of <autosax> (Raes, 2022).**

A similar system was implemented in the *Post-Digital Sax* (Cybulski, 2022). The difference here is that a permanent magnet was affixed to the back of the reed. A custom mouthpiece was designed that allowed hinged movement (Figure 7.2.9).

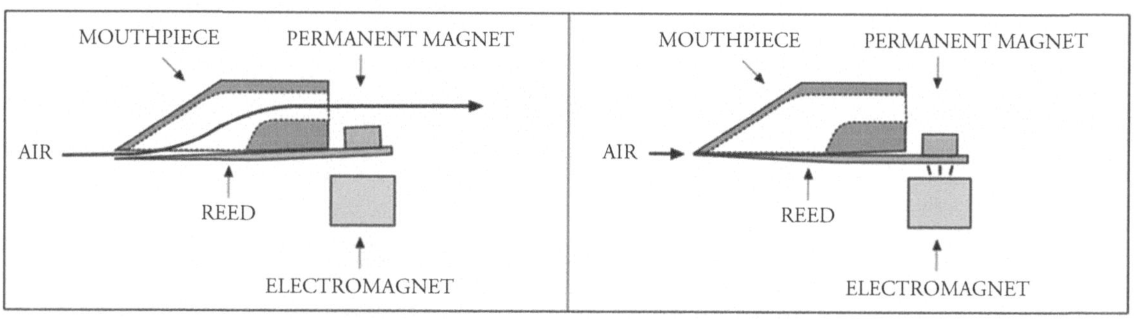

FIGURE 7.2.9. **Electromagnetic reed of the Post-Digital Sax (Image from Cybulski, 2022, CC BY 4.0).**

Cybulski notes that at 1 Hz, the motion of the reed resembles a pulse wave regardless of the waveform driving the electromagnet; the latter affects the duty cycle of the output. As the frequency of the input waveform increased, the output shape of the reed motion increasingly resembled a sawtooth pattern. This design depends on airflow from and the embouchure of a human player, thus enabling human–machine interaction and *cooperative* sound-making.

Tonguing is an integral part of articulating notes on a reed instrument. The physical factors involved in tonguing have been studied experimentally (Chatziioannou et al., 2019; W. Li et al., 2014, 2016; Lin, Kawai, Matsuki, et al., 2019). Many of these experiments utilized artificial systems to play reed instruments; designs are featured in Figure 7.2.10. The "tongue" of the NICTA-UNSW clarinet (W. Li et al., 2014; Figure 7.2.10, top left) is a rectangular prism made of polyurethane foam (the same as the lip). The tongue is connected to and controlled by a lever that is weighted with a 100 g mass at one end to apply force to the reed tip. Chatziioannou et al. (2019) made a tongue from a cylinder of cellular polyethylene attached to a linear actuator (SmartShaker) that can provide 31 N of force and can be positioned precisely. Using an artificial excitation system, the authors produced *portato* (smooth, pulsing notes) and *staccato* articulations depending on the contact time between the tongue and the reed (30 ms for *portato*; shorter durations for *staccato*).

The "tongue" of the Waseda Anthropomorphic Saxophone robot (Chatziioannou et al., 2019; Lin, Kawai, Matsuki, et al., 2019) consists of a configurable metal board positioned under the oral cavity (Figure 7.2.10). It can be moved so that it presses the reed against the mouthpiece, preventing it from vibrating. The torque required to rotate the "tongue" to this angle is a function of the board's moment of inertia, gravity, the pressure in the oral cavity, and the SEPTON material (see Lin et al., 2019 for specifics). The torques required for ascending and descending motion differ, which must be accounted for in the motor control algorithm. Evaluation of the system revealed that the tonguing mechanism, in conjunction with airflow control, yielded superior results compared to the tonguing mechanism alone. The authors concluded that while the products of the articulation system were promising, there were still qualitative differences between the sonic output of the machine and the human performer; thus, further dynamic variation was needed (Lin, Kawai, Matsuki, et al., 2019).

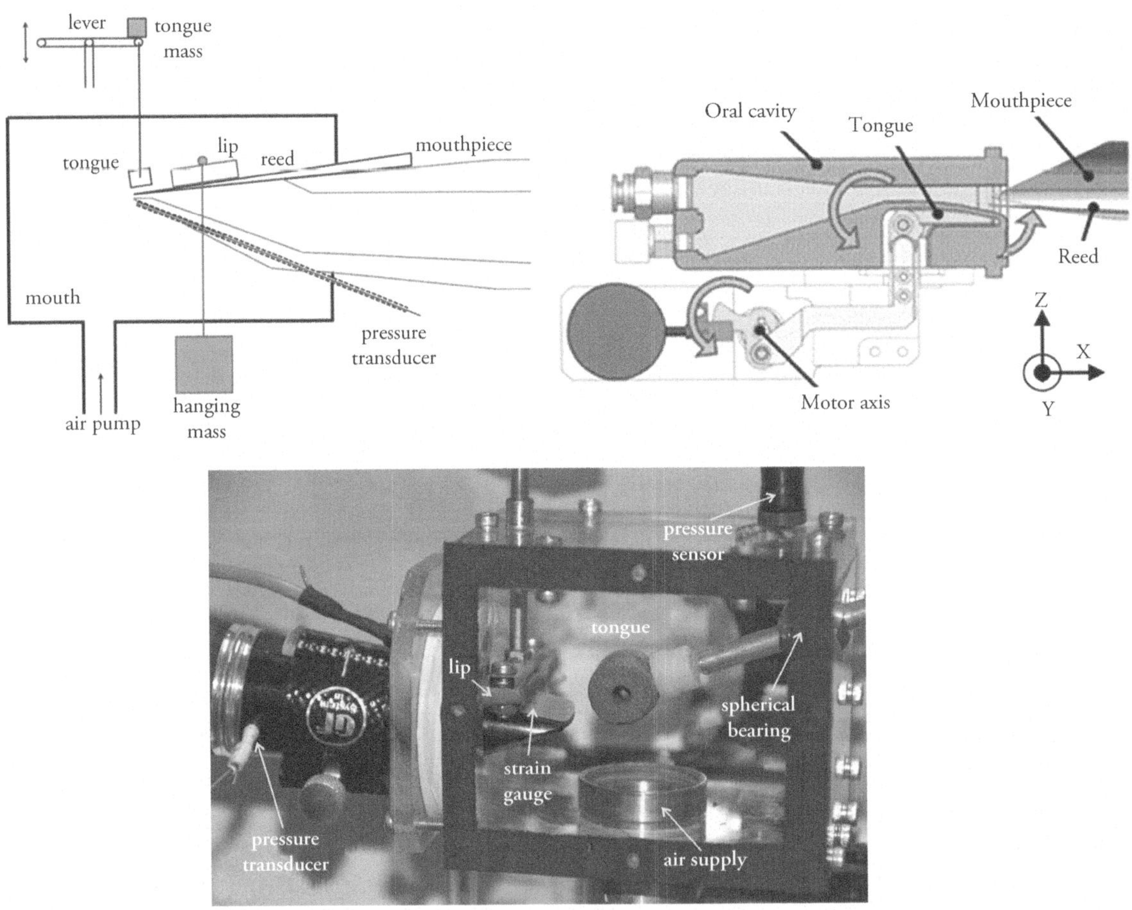

FIGURE 7.2.10. **Artificial mouths. Top Left: NICTA-UNSW Clarinet (image from W. Li et al., 2014); top right: WAS-5 (image from Lin et al., 2019); bottom: artificial mouth for clarinet (Image from Chatziioannou et al., 2019, CC By 4.0).**

Some reed-based exciters are part of *cooperative* systems that depend on human–machine interaction. The robot-assisted saxophone developed by Kurosawa and Suzuki (2010) features a "sensory mouthpiece" comprising a pressure sensor connected to a tube inserted into a standard mouthpiece. As a human player blows into the mouthpiece, changes in air pressure are used to determine fingerings corresponding to pitch sequences in pre-programmed scores. The fingerings are performed by mechatronic actuators, allowing a human and a machine to play the instrument collaboratively. In the *Post-Digital Sax* (Cybulski, 2022), a player provides airflow into the mouthpiece, affecting timbre through embouchure changes that coordinate with the electromagnetically vibrating reed. The sound from the instrument is projected acoustically, resulting in sonically unique and compelling effects (Media Example 7.2.1 ▶).

Brass

Machinic systems have been developed to study and play brass instruments, including the trumpet (Vergez & Rodet, 1997), trombone (Cullen et al., 2000; Gilbert et al., 1998), ophicleide (a keyed bugle), saxhorn, and bass trumpet (Bromage, 2007). When envisioning the design of artificial lips, fundamental principles apply. The width, height, and shape of the aperture between the "lips" (or their mechanical equivalent) affect sound production. As one would expect, sound frequency increases with lip frequency. Increasing pressure on the lips tends to increase the pitch of the produced note, while decreased pressure produces lower pitches (Lopes et al., 2013; Newton, 2009).

A prominent artificial mouth design features a cavity (e.g., a steel tube) that matches the volume of a human mouth. Artificial lips made from thin latex or polyurethane rubber tubes filled with water are positioned at the orifice. The stiffness of the lips can be varied either by the material chosen or by varying the volume of water that fills them. The lips rest on a channel (made of steel or plastic) that mimics human teeth (Figure 7.2.11). The trajectory of this research reaches back to Gilbert & Petiot (1997) and continues in the twenty-first century (Vergez & Rodet, 1997; Gilbert et al., 1998; Cullen et al., 2000; Teissier et al., 2003; Richards, 2003; Bromage, 2007; Lopes et al., 2013).

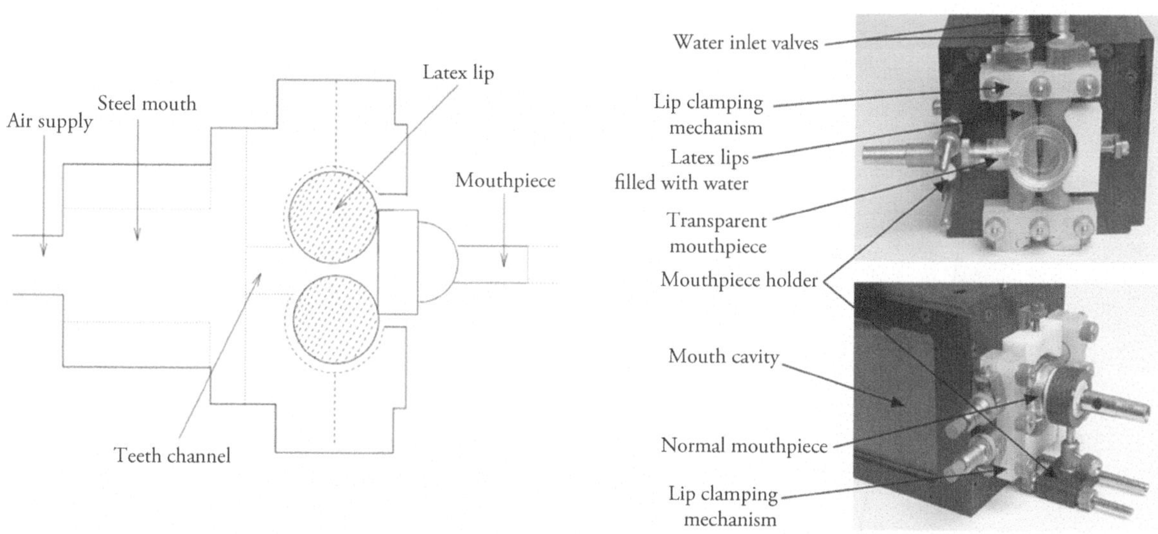

FIGURE 7.2.11. **Artificial mouth designs. Left: Vergez & Rodet (1997), Right: Richards (2003), image from Newton (2009); photo by Dr Mike Newton, Acoustics & Audio Group, University of Edinburgh).**

Adjustments in mouthpiece pressure, angle relative to the lips, and opening of the "jaw" affect sound production. Adjustments are made by changing the volume of water in the latex lips and mechanically moving the jaw, lips, teeth, and/or mouthpiece. Varying these parameters and the energy source (e.g., air pressure) yields an extensive range of pitches (Vergez & Rodet, 1997). Lopes et al. (2013) report that the artificial mouth could produce numerous stable notes using a basic mapping, but that higher-register notes were more challenging to realize. They recommended addressing the issue by varying the acoustic impedance of the mouth.

Sound quality and musical implications of artificial lip design are a priority for those who purport to use such systems for composition and performance. Gilbert et al. (1998) claimed their system was capable of "realistic playing performance" (p. 1628). Contrastingly, Campbell et al. (2021) report that the sounds produced by artificial lips are useful in the scientific study of brass instruments, but are likely of less interest to musicians due to their inability to replicate the complex movements of human embouchures. Within the domain of timbre, some researchers have focused on achieving a "brassy" sound with an artificial mouth. It was discovered that this timbre is not just a matter of lip motion or air pressure; a reconfiguration of the embouchure is necessary (Bromage, 2007; Stevenson, 2009). Changes in musical dynamics, such as a crescendo, require continual adjustments to the embouchure (Stevenson, 2009). Future designs capable of such dynamic changes may be more successful at producing musical gestures judged artistically satisfying.

An alternative artificial mouth model, identified as *Replica B*, was developed by Newton (2009) to minimize adjustment options while maintaining the necessary levels of control. The artificial lips were made from water-filled latex, as in previous versions, but here were glued to the lip blocks within the faceplate and decoupled from direct contact with the mouthpiece through a faceplate cover (Figure 7.2.12). The lip block profile could be changed; the only other control parameter was the internal water pressure of the lips. Stevenson (2009) used a similar design to study timbre, dynamics, and transients in brass instrument playing.

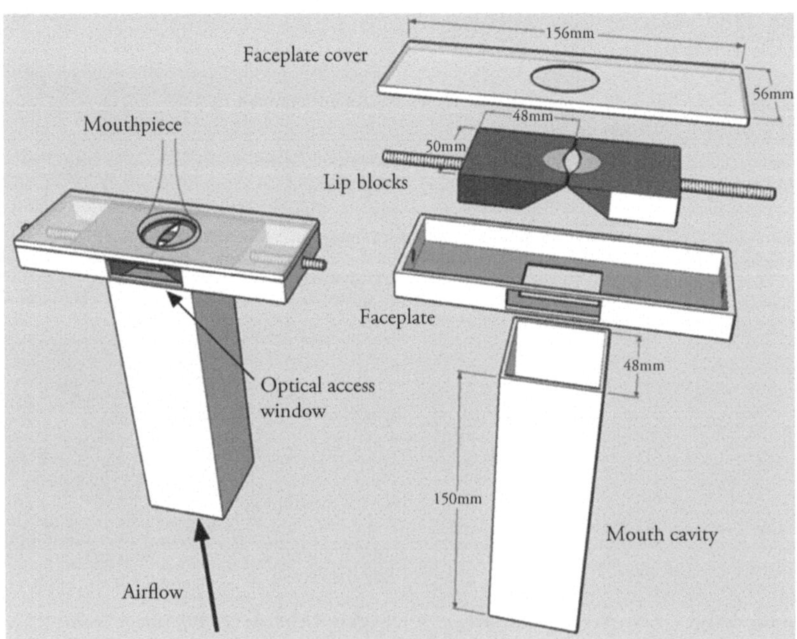

FIGURE 7.2.12. **Artificial mouth (Newton, 2009; photo by Dr Mike Newton, Acoustics & Audio Group, University of Edinburgh).**

Newton (2009) reported that after the device was assembled and configured, it remained stable and could be used over extended periods without requiring adjustments. The sonic results were "excellent" and "very close to human blown trumpet sounds" (p. 7). The design is capable of multiphonics and "chaotic" (p. 62) sounds, which may be of interest to musicians leaning toward the experimental. Stevenson (2009) concluded that the device produced sounds that were "consistent and repeatable" (p. 70) and "satisfactory" (p. 71). With that said, the lips were not sufficiently damped, which could be addressed by replacing the water with foam, gel, or a sponge. Stevenson (2009) produced transients using the artificial mouth by leaving a hole in the mouth box open while the pump was on, preventing a buildup of pressure. When the hole was manually plugged, the pressure increased to the level required to produce a transient. This method was simple and repeatable and could serve as a "tongue." Automating this process would be an opportunity for future improvement. Regarding dynamics, the artificial mouth could not play as loudly as human players (Stevenson, 2009).

The Logos Foundation's variation on the artificial mouth theme is evident in <*So*> (sousaphone) and the first version of <*Bono*> (trombone). The mouth contains silicon "lips" that are driven by a moving coil valve, which controls the aperture through which air passes to the mouthpiece from a small axial compressor. Dynamic control of the system is a function of the interaction between wind pressure and the amplitude of the signal that drives the moving coil valve for the lips. Higher frequencies result in smaller movements, requiring greater air pressure to compensate (Raes, 2021b).

COMPRESSION DRIVERS

Instead of providing an airstream from a pump or reservoir into an artificial mouth coupled to a mouthpiece, some builders have used linear electromagnetic drivers, diaphragms, and impedance converters as the input to machinic aerophones. A **compression driver** is a specialized type of loudspeaker used in this application, consisting of a moving coil-diaphragm system coupled to a waveguide or horn (Figure 7.2.13). The diaphragm diameter is significantly larger than the aperture of the horn that it is connected to, so the air is compressed as it is moved to the more confined space of the horn opening. The high pressure at the horn's input matches the lower pressure at the wider mouth area, resulting in efficient energy transfer (10 times that of direct-cone radiating loudspeakers; "Compression Driver," 2022). Compression drivers are advantageous because they can produce a wide range of frequencies and waveforms that elicit a range of sonic responses from an instrument. These driver systems are unique in that they combine **energy source** and **exciter** (which is why they are featured here in a separate section).

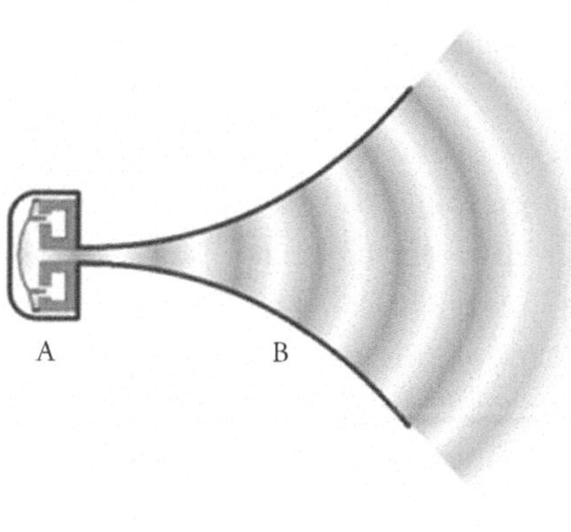

FIGURE 7.2.13. **Compression driver.**

In developing the reed instrument *<Ob>* (oboe), the Logos Foundation encountered difficulties achieving satisfactory sonic results using an air compressor coupled to the instrument's mouthpiece. As an experiment, they used a compression driver, but instead of coupling it to an exponential horn, they connected it to a custom-made impedance converter. The converter, modeled after a reed in a human mouth cavity, was fit into the oboe (Raes, 2010b). The sonic results were promising, leading to further implementations in *<Fa>* (bassoon), *<Asa>* (alto saxophone), and several brass instruments, including *<Trumpeter>* (Bb trumpet), *<Horny>* (French horn), and *<Bug>* (flugelhorn) (Raes, n.d.).

A model for the *compression driver–acoustic impedance* system was thus established. Its key features include a compression driver coupled to a membrane that forces air into an acoustic impedance converter. Within the converter, air flows through a capillary and then to the mouthpiece of the instrument (which sometimes needs to be redesigned), as shown in Figure 7.2.14.

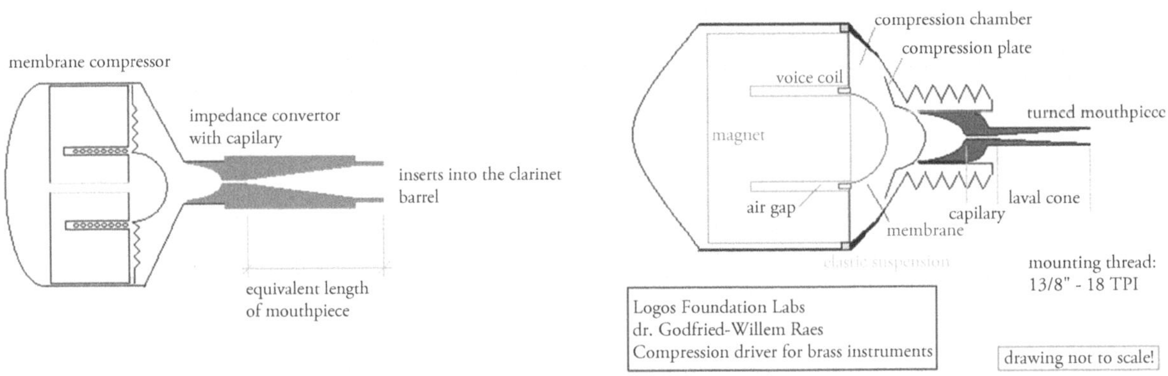

FIGURE 7.2.14. **Compression driver—Impedance Converters of The Logos Foundation's <klar> (Raes, 2012) (left) and <Trumpeter> (Raes, 2021c) (right).**

The dimensions of the impedance converter affect sound quality and vary depending on the instrument. When a note is requested, the appropriate tone hole, key, or valve combination is selected, and the driver generates a waveform (shaped according to a physical model of human players) at a frequency corresponding to the desired pitch. The resonances in the tubing of the instrument are excited, although air does not flow through it as in a conventional aerophone.

The sound quality is affected by the driver waveform. For reed instruments, trapezoidal waveforms work but with some modifications (Raes, 2010b). In *<Cornalto>* (saxhorn), an improvement in sound quality resulted when the attack portion of the waveform was modeled on vibrations of the lips and mouth cavity of a human player, followed by a sustain portion based on samples gathered from playing the instrument in different registers (Raes, 2021a). The resulting sound quality is somewhere between synthetic and natural. The system enables a range of microtonal temperaments and precise control over sonic parameters, including driver waveshape, noise, vibrato (frequency modulation speed and depth), tremolo (amplitude modulation speed and depth), volume, and amplitude envelope (ADSR), as illustrated in Figure 7.2.15 (Raes, 2020a).

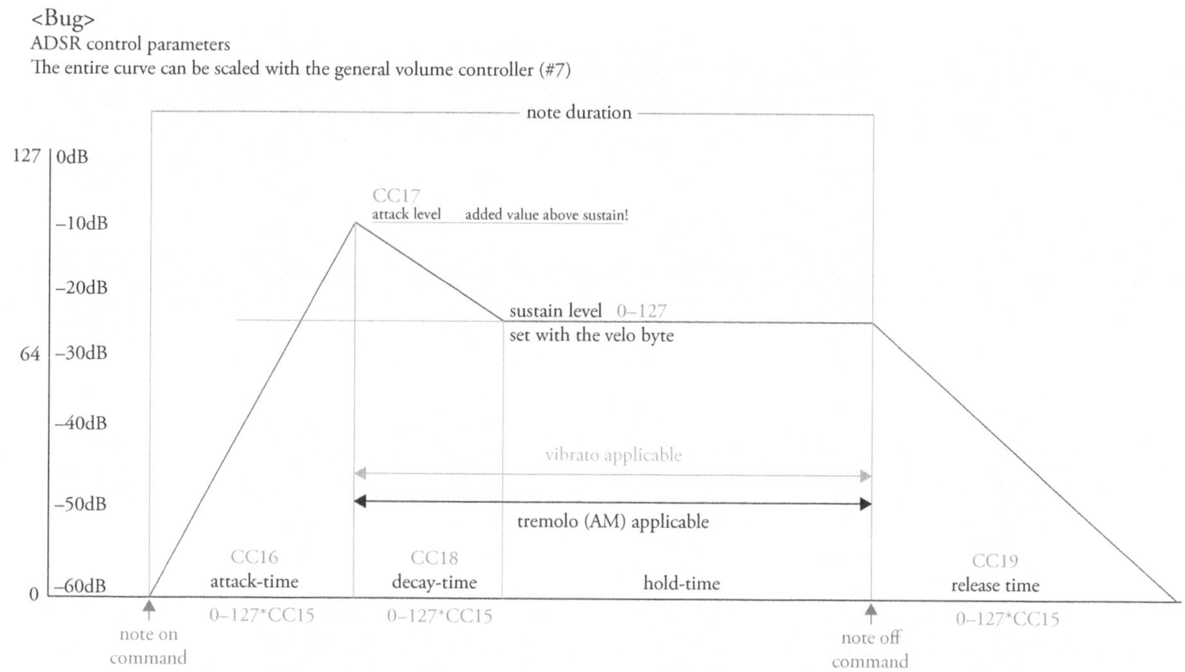

FIGURE 7.2.15. **Control parameters of The Logos Foundation's <Bug> (Raes, 2020a).**

In a variation on this theme, The Logos Foundation connected a small motor driver directly to a capillary and then to a cornet in *<Korn>*. The driver causes resonance in the tubing of the cornet, but there is no substantive airflow through the instrument. The resultant sound is described as "between synthetic / electronic and natural / acoustic" (Raes, 2023). Logos has also coupled impedance-matched loudspeakers to tuned PVC tubes and organ pipes, such as in *<Hybr>*, *<HybrHi>*, *<HybrLo>*, and the *<Pi>* group (Raes, n.d.).

Loudspeakers have also been utilized by acoustic researchers to gain a deeper understanding of the physical actions and responses involved in playing wind instruments (Newton, 2009; Richards, 2003). They offer designers and builders precise and wide-ranging control over the system's output via computers or other electronic devices. Modifying the input waveforms is easy, and the changes made can be subtle, drastic, or anything in-between. Depending on what the loudspeaker is coupled to, a wide range of acoustic responses can be heard and measured.

Producing Pitches

Producing different pitches on an aerophone can be accomplished by varying the effective length of the enclosed air column. One way to do this is by opening and closing configurations of **tone holes**. Human players control tone holes directly (i.e., by using their fingers) or through mechanisms (e.g., the keys on a clarinet). Musical machines require arrays of actuators, such as solenoids, to achieve similar control. These can be mounted directly over a tone hole or key mechanism or can be used to effect another structure such as an artificial finger (Chinese bamboo flute, Figure 7.2.17), acrylic hinge plate (Thai flute-playing robot; Kato et al., 2018), lever (e.g., The Logos Foundation's *<Ob>*, Figure 7.2.16), or rubber strip (MPR Lab's *Recorder Robot*).

FIGURE 7.2.16. **Fingering mechanisms of <ob> by the Logos Foundation (Raes, 2010b).**

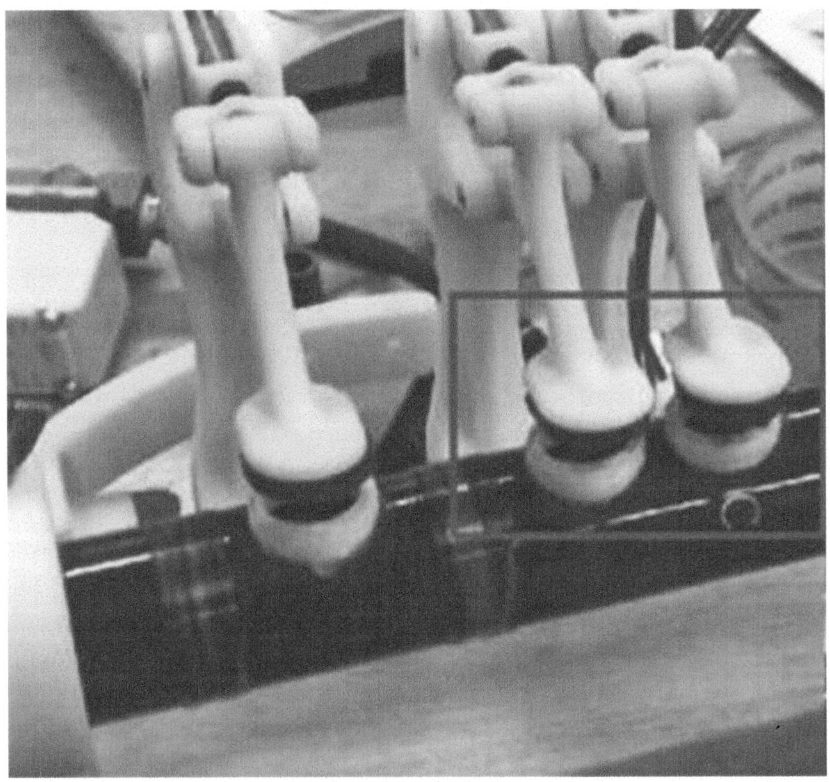

FIGURE 7.2.17. **Fingering mechanisms (J. Li et al., 2019).**

Closing a tone hole completely can be challenging. It is unlikely that the armature of a solenoid is going to create a seal when pressed against the instrument (it would also produce audible clicking on contact and could damage the instrument's body). Instead, a compliant material is needed that can deform and create a seal around the tone hole, thereby attenuating the sound produced by impact (as shown in Figure 7.2.17). At the same time, it should be robust enough to maintain its shape and withstand thousands of actuation cycles. Rubber and foam can work well; experiment with materials and thicknesses to see what yields the best results.

When depressing a key, the primary concerns are audible clicking on contact (as above), force, and speed to ensure the key is adequately closed. Uchiyama et al. (2023) experimented with soft (silicone rubber) pads on the fingers of the saxophone robot WAS-X, purporting to improve the sound quality. The amount of output force required by the actuators depends on the instrument; for example, Kurosawa and Suzuki (2010) found that 1.5–2.0 N were needed to close the keys of a saxophone. Key-pressing actuators require rapid movement over short distances, making electric solenoids or pneumatic cylinders suitable choices for this application.

Fingering can be thought of as a binary action that produces discrete pitches, but subtle variations can produce musical gestures such as glissando, portamento, tremolo, and vibrato. Kato et al. (2018) developed a fingering mechanism comprising an air cylinder and a soft stopper made from silicone and a urethane sponge (Figure 7.2.18). Varying the pressure in the cylinder affects the force exerted by the soft tip, modulating the aperture of the tone hole. The fine control required by the system precipitated the development of a high-precision quick-response pressure regulator (HPQR) (see Kato et al., 2018, for details). Using an oscillating pressure wave as input, the fingering control system produced a glissando between pitches a whole tone apart. Combining the fingering-force and airflow control allowed the authors to produce an independent tremolo.

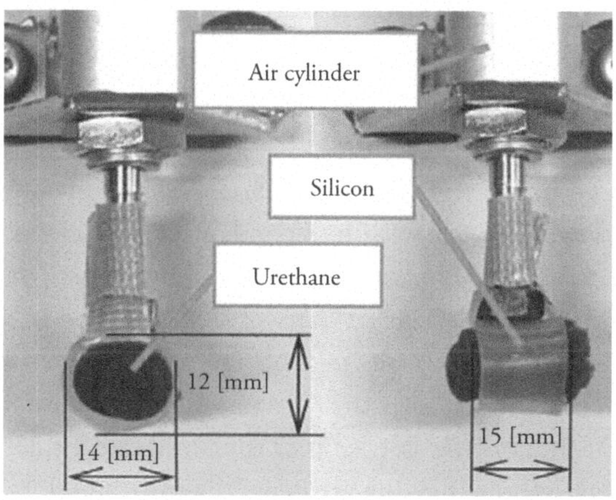

FIGURE 7.2.18. **Fingering mechanism of the Thai flute-playing robot (Kato et al., 2018).**

Sounding different pitches with a brass instrument requires coordinating the excitation and valve systems. The valve combination required to produce a specific pitch can be stored in the memory of a microcontroller and recalled as needed. The Logos Foundation's <*Trumpeter*> works in this way, making conventional and microtonal temperaments possible. Solenoids are typically used to actuate valves, with the amount of force needed depending on the instrument. The Logos Foundation's <*Horny*> (F-horn) required 2.5 N to depress the valves fully (Raes, 2013).

Fine control of tone hole, key, or valve actuators involves controlling parameters such as speed and stay (amount of time to remain in position after a note-off command) (Raes, 2020a). The timing of the excitation signal and the hole, key, or valve mechanisms must be coordinated. Typically, the faster parts of the system must be delayed to account for the latency in the slowest part. For example, in The Logos Foundation's <*Cornalto*>, a valve takes 10 ms to travel its length; thus, the driver signal (which has lower latency) is delayed by that duration to synchronize the two systems. The payoff of this system latency is more realistic sound (Raes, 2021a). The response time to operate the keys in <*Flut*> falls within a similar range of 5–10 ms. Articulation matters when calculating these delays: delay is not required for repeated notes using the same fingering or for legato playing (Raes, 2020b).

Slides are another way to vary pitches on an aerophone. The *mechatronic slide flute* utilized a linear motor with a goal acceleration of 8 m/s^2 and a maximum velocity of 1 m/s (d'Andréa-Novel et al., 2016). Slider positions were mapped to pitches by fully retracting the piston and then extending it in 3 mm increments, analyzing the sonic output via an FFT at each step to create a motor position–frequency curve. A precise position and, thus, pitch could then be obtained through linear interpolation between the points of the curve. Other mechatronic possibilities abound involving devices that transform rotary to translational motion. Rack and pinion, slider-crank, belt / chain drive, or even a cam and follower are all possibilities. In addition to force and speed requirements, visibility plays a role in this choice. For example, a rack and pinion system could be made *invisible*, while a slider-crank mechanism could make machinic identity *conspicuous*.

Structure

Beyond the aspects already mentioned in the *properties and relations* section, several structural questions must be addressed when designing a machinic aerophone. Will the air source be housed in a hidden enclosure, or will it be an integral part of the machine's identity? If the source is a large and noisy air compressor, it is best to separate and conceal it. The MPR Lab has utilized long tubing runs to locate air compressors far from the stage where the instruments are located, usually in adjacent rooms. If the air source is smaller and quieter, such as the pump of *McBlare*, then it becomes a choice of *form* and how *conspicuous* you want the mechanisms to be. All of this will affect *transportability*, with independent or peripheral elements being easier to detach and move. Regarding *mounting*, many aerophones change pitch using tone holes or keys, meaning small mechanical "fingers" are located near the instrument. Attaching the fingers directly to the instrument creates mounting complexities in most cases. It is usually easier to make a *frame* that holds both the "fingers" and the sonic object. The extent to which that structure is imposing or invisible is up to the builder. To that point, since the physical movements involved in aerophone playing are generally minimal, less structure is usually better, allowing the eye to focus on the action rather than its surroundings.

References

Bromage, S. (2007). *Visualisation of the lip motion of brass instrument players, and investigations of an artificial mouth as a tool for comparative studies of instruments* (Dissertation, University of Edinburgh). https://era.ed.ac.uk/bitstream/handle/1842/1966/Bromage%20S%20thesis%2007.pdf?sequence=1&isAllowed=y

Campbell, M., Gilbert, J., & Myers, A. (2021). Buzzing lips: Sound generation in brass instruments. In *The science of brass instruments* (pp. 63–100). Springer.

Chatziioannou, V., Schmutzhard, S., Pàmies-Vilà, M., & Hofmann, A. (2019). Investigating clarinet articulation using a physical model and an artificial blowing machine. *Acta Acustica United with Acustica*, *105*(4), 682–694.

Clawson, G., Moncur, B., Fogarty, K., Bennett, J., & Fox, C. (2022). Towards open source hardware robotic woodwind: An internal duct flute player. https://www.researchgate.net/profile/Garry-Clawson/publication/372288988_Towards_Open_Source_Hardware_Robotic_Woodwind_an_Internal_Duct_Flute_Player/links/64ae70deb9ed6874a51511c5/Towards-Open-Source-Hardware-Robotic-Woodwind-an-Internal-Duct-Flute-Player.pdf

Compression Driver. (2022). In Wikipedia. https://en.wikipedia.org/w/index.php?title=Compression_driver&oldid=1077348151

Cullen, J. S., Gilbert, J., & Campbell, D. M. (2000). Brass instruments: Linear stability analysis and experiments with an artificial mouth. *Acta Acustica United with Acustica*, *86*(4), 704–724.

Cybulski, K. (2022, June 16). Post-digital sax—A digitally controlled acoustic single-reed woodwind instrument. *International Conference on New Interfaces for Musical Expression. NIME 2022.* https://doi.org/10.21428/92fbeb44.756616d4

d'Andréa-Novel, B., Joly, C., Noël, T., & Senpauroca, J. (2016). Control of a slide flute: A mechatronic project. *Proceedings of the 3rd International Symposium on Movement and Computing*, 1–5.

Dannenberg, R. B., Brown, H. B., & Lupish, R. (2011). McBlare: A robotic bagpipe player. In J. Solis & K. Ng (Eds.), *Musical robots and interactive multimodal systems* (pp. 165–178). Springer.

Gilbert, J., & Petiot, J.-F. (1997). Brass instruments: Theoretical and experimental results. *Proceedings-Institute of Acoustics, 19*, 391–400.

Gilbert, J., Ponthus, S., & Petiot, J.-F. (1998). Artificial buzzing lips and brass instruments: Experimental results. *The Journal of the Acoustical Society of America, 104*(3), 1627–1632. https://doi.org/10.1121/1.424375

Kato, T., Shimazaki, K., Higashijima, K., Tokunaga, R., Ayuthaya, P. T. N., & Thumwarin, P. (2018). Refinement of a Thai flute-playing robot in Thai style. *2018 International Conference on Engineering, Applied Sciences, and Technology (ICEAST)*, 1–4.

Kurosawa, Y., & Suzuki, K. (2010). Robot-assisted playing with fingering support for a saxophone. *Proceedings of the International Computer Music Conference*, 2010, 1–4. https://quod.lib.umich.edu/i/icmc/bbp2372.2010.001/1/--robot-assisted-playing-with-fingering-support-for?rgn=full+text;view=image;q1=robot

Li, J., Hu, T., Zhang, S., & Mi, H. (2019). Designing a musical robot for Chinese bamboo flute performance. *Proceedings of the Seventh International Symposium of Chinese CHI*, 117–120. https://dl.acm.org/doi/abs/10.1145/3332169.3332264

Li, W., Almeida, A., Smith, J., & Wolfe, J. (2014). Tongue, lip and breath interactions in clarinet playing: A study using a playing machine. *Proceedings of the 21st International Congress on Sound and Vibration*, ICSV21, Beijing, China.

Li, W., Almeida, A., Smith, J., & Wolfe, J. (2016). How clarinettists articulate: The effect of blowing pressure and tonguing on initial and final transients. *The Journal of the Acoustical Society of America, 139*(2), 825–838. https://doi.org/10.1121/1.4941660

Lin, J.-Y., Kawai, M., Matsuki, K., Cosentino, S., Sessa, S., & Takanishi, A. (2019). Musical articulation system on humanoid saxophonist robot. In V. Arakelian & P. Wenger (Eds.), *ROMANSY 22–Robot design, dynamics and control* (pp. 392–399). Springer.

Lin, J.-Y., Kawai, M., Nishio, Y., Cosentino, S., & Takanishi, A. (2019). Development of performance system with musical dynamics expression on humanoid saxophonist robot. *IEEE Robotics and Automation Letters, 4*(2), 1684–1690. IEEE Robotics and Automation Letters. https://doi.org/10.1109/LRA.2019.2897372

Lin, J.-Y., Yoshida, K., Matsuki, K., Takikawa, K., Cosentino, S., Sessa, S., & Takanishi, A. (2016). The development of intraoral pressure control system on humanoid saxophone playing robot. *2016 IEEE International Conference on Robotics and Biomimetics (ROBIO)*, 1649–1654.

Lopes, N., Hélie, T., & Caussé, R. (2013). Control of an artificial mouth playing a trombone and analysis of sound descriptors on experimental data. *Stockholm Music Acoustics Conference* 2013.

Murphy, J., Kapur, A., & Carnegie, D. A. (2014). Mechatronic keyboard music: Design, evaluation, and use of a new mechatronic harmonium. *Proceedings of the 2014 International Computer Music Conference*, 678–685.

Newton, M. J. (2009). *Experimental mechanical and fluid mechanical investigations of the brass instrument lip-reed and the human vocal folds* (Dissertation, University of Edinburgh). https://era.ed.ac.uk/bitstream/handle/1842/3140/MichaelNewton_PhDThesis_May2009.pdf?sequence=1&isAllowed=y

Raes, G.-W. (n.d.). Logos Foundation—Automatons and robots. Logos Foundation. Retrieved May 26, 2022, from https://www.logosfoundation.org/instrum_gwr/automatons.html

Raes, G.-W. (2010a). <Bono> an automated rotary valve trombone. Logos Foundation. https://logosfoundation.org/instrum_gwr/bono.html

Raes, G.-W. (2010b). <Ob> an oboe robot. Logos Foundation. https://logosfoundation.org/instrum_gwr/ob.html

Raes, G.-W. (2012). <Klar> a robotic microtonal and moving alto clarinet. Logos Foundation. https://logosfoundation.org/instrum_gwr/klar.html

Raes, G.-W. (2013). <Horny> an automated French horn. Logos Foundation. https://logosfoundation.org/instrum_gwr/horny.html

Raes, G.-W. (2015). *Expression control in automated musical instruments* (p. 65). Ghent University College—School of Arts & Logos Foundation. https://logosfoundation.org/g_texts/expression-control.pdf

Raes, G.-W. (2020a). <Bug> a robotic and moving Bb flugelhorn. Logos Foundation. https://logosfoundation.org/instrum_gwr/bug.html

Raes, G.-W. (2020b). <Flut>: An automated flute. Logos Foundation. https://www.logosfoundation.org/instrum_gwr/flut/flut.html

Raes, G.-W. (2021a). <Cornalto> a robotic Eb alto saxhorn. Logos Foundation. https://www.logosfoundation.org/instrum_gwr/Cornalto/cornalto.html

Raes, G.-W. (2021b). <So>: An automated sousaphone. Logos Foundation. https://logosfoundation.org/instrum_gwr/so.html

Raes, G.-W. (2021c). <Trumpeter> a robotic Bb trumpet. Logos Foundation. https://logosfoundation.org/instrum_gwr/trumpet/trumpet.html

Raes, G.-W. (2022, August 15). <Autosax>an automated microtonal C-melody saxophone. Logos Foundation. https://logosfoundation.org/instrum_gwr/autosax.html#Autosax-english

Raes, G.-W. (2023, October 11). <Korn> a robotic and moving Bb cornet. Logos Foundation. https://logosfoundation.org/instrum_gwr/korn.html

Richards, O. (2003). *Investigation of the lip reed using computational modelling and experimental studies with an artificial mouth* (Dissertation, University of Edinburgh). Edinburgh Research Archive. http://hdl.handle.net/1842/11315

Solis, J., Chida, K., Taniguchi, K., Hashimoto, S. M., Suefuji, K., & Takanishi, A. (2006). The Waseda flutist robot WF-4RII in comparison with a professional flutist. *Computer Music Journal*, *30*(4), 12–27.

Solis, J., & Takanishi, A. (2011). Wind instrument playing humanoid robots. In J. Solis & K. Ng (Eds.), *Musical robots and interactive multimodal systems* (pp. 195–213). Springer.

Stevenson, S. D. (2009). Experimental investigations of lip motion in brass instrument playing (Dissertation, University of Edinburgh). https://era.ed.ac.uk/bitstream/handle/1842/4349/Stevenson2009.pdf?sequence=1&isAllowed=y

Teissier, F., Gilbert, J., Campbell, M., & Petiot, J.-F. (2003). Comparative analysis of brass wind instruments with an artificial mouth: First results. *Acta Acustica United with Acustica*, *89*(6), 974–979.

Uchiyama, J., Hashimoto, T., Ohta, H., Lin, J.-Y., Cosentino, S., & Takanishi, A. (2023). A humanoid saxophone-playing robot based on instrument-centered design: Sound improvement with fingers using hybrid soft materials (pp. 295–306). https://doi.org/10.1007/978-3-031-35634-6_21

Vergez, C., & Rodet, X. (1997). Comparison of real trumpet playing, latex model of lips and computer model. *The Proceedings of the 1997 International Computer Music Conference*, 1–8.

Wood, G. (2002, February 16). Living dolls: A magical history of the quest for mechanical life by Gaby Wood. *The Guardian*. https://www.theguardian.com/books/2002/feb/16/extract.gabywood

CHAPTER 7

AEROPHONES

Part 3: Making

Realization

BOM, FABRICATION, AND COST

The BOM and fabrication considerations for making a machinic aerophone are similar to those for other projects, so revisit the general descriptions of those topics in Chapter 4: The Process of Making a Musical Machine. I have learned over the years that for pneumatic components, quality and cost are related. Quieter, higher-capacity air compressors are generally more expensive. Many aerophonic machines use solenoid valves to control the release of pressurized air; less expensive options are sometimes susceptible to leaks and can become stuck. Buying the cheapest valves may end up being more expensive when you eventually need to replace them. At the same time, starting with a single, less expensive test component can help clarify the specifications needed. Deficiencies can be addressed through modifications (e.g., building silencing boxes for valves, as we did in the MPR Lab for the Percussive Aerophone), higher-quality components (e.g., solenoid valves), or by living with imperfections (the cheapest option and a skill that we should all work on developing).

Control and Feedback

For those requiring a primer (or who would benefit from a refresher), the Companion Website ⓟ contains sections devoted to *Sensing, Communications, and Control*, which feature technical descriptions and projects that enable you to collect and interpret information, and connect and control system components.

Control of a machinic aerophone requires coordination between systems that generate, regulate, and direct airflow, configure an artificial mouth, and modulate the effective length of an air column. An exemplary control scheme is used in the NICTA-UNSW robot clarinet (Figure 7.3.1). It includes a pump and a leak valve to adjust pressure inside the artificial mouth. The desired pitch is output as a MIDI note, which is associated with an air pressure level and lip, tongue, and finger positions stored in a lookup table on a microcontroller. A sensor measures the current air pressure in the chamber, and the value is fed into a PID loop whose output is used to adjust the pump and leak valve to achieve the desired air pressure level. The lip and tongue servos move to the appropriate positions, and the corresponding fingering solenoids are activated.

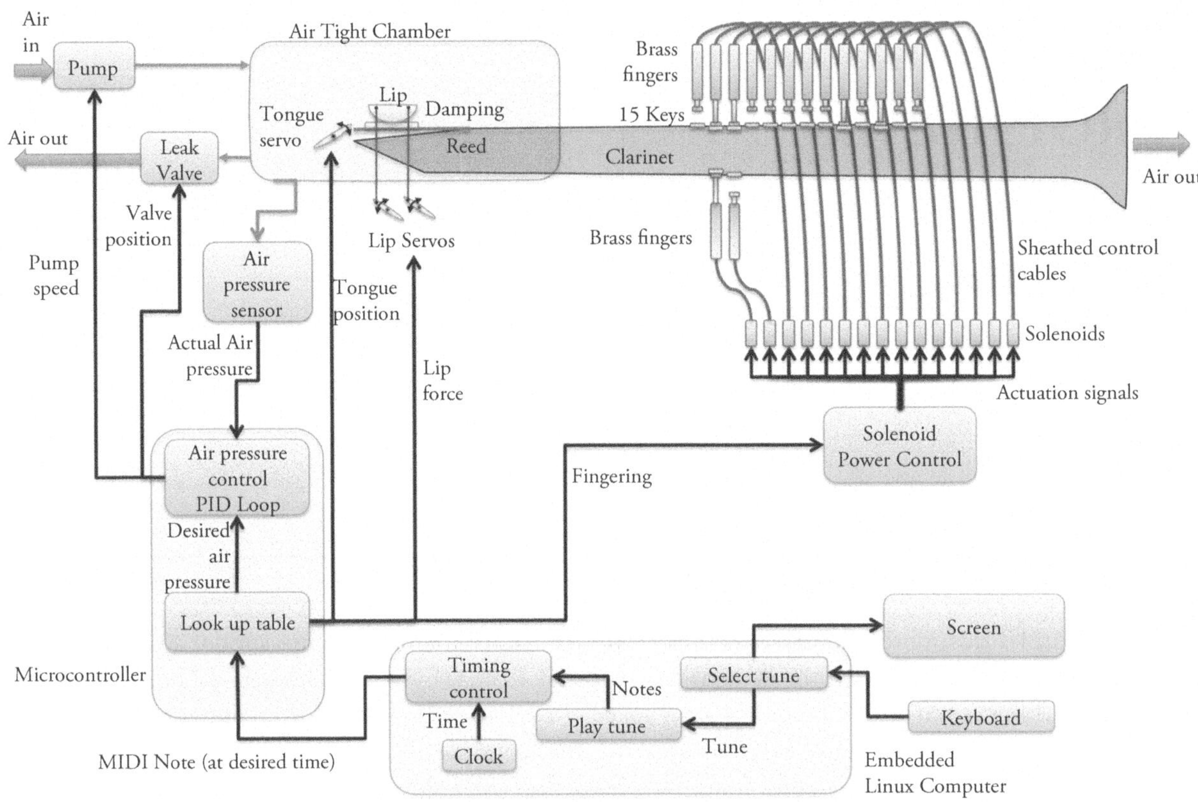

FIGURE 7.3.1. **System diagram of the NICTA-UNSW Robot Clarinet (image from Almeida et al., 2010).**

WAS-4 (Lin et al., 2016) employs a similar control system with feedback, which varies both airflow and lip opening width, known as the Music Pattern Generator (MPG). Using this control scheme resulted in shorter calibration times and musical performances with continuous sound production as compared to performances without the system.

Compelling musical performances contain variations in pitch, rhythm, and dynamics. The control scheme mentioned above can provide the pressure and fingering conditions required to produce a particular pitch at a specific time; however, the method of dynamic variation must also be specified. Expanding dynamic range is accomplished by modulating the airflow into the system. To translate this control into musically relevant segments, Lin et al. (2019) determined the minimum and maximum flow that would produce sound. They divided that range into eight segments corresponding to dynamics markings *ppp* – *fff*. The control system was then separated into two parts: the first generated the SPL corresponding to the desired dynamic marking, while the second stabilized the pressure in the oral cavity to ensure that sound was made. Different SPLs require different oral pressure ranges, which were achieved by changing the position of the lower lip.

Further musical nuance becomes possible through the coordination of control and actuation systems connected through feedback loops. The Auditory Feedback Control System of the Waseda Flutist (WF-4RIV), as described in Solis and Takanishi (2011), consists of an Expressive Music Generator (*ExMG*), a Feed Forward Air Pressure Control System (*FFAiPC*), and a Pitch Evaluation System (*PiES*). The *ExMG* outputs musical instructions according to rules defined by analyzing the performances of professional human flutists using artificial neural networks (ANN), with a focus on duration, vibrato frequency, and vibrato amplitude. The *FFAiPC* controls the air pressure from the artificial lungs during the attack of a note to articulate multi-note phrases. The *PiES* estimates the fundamental frequency of the note using the Cepstrum method and quantitatively evaluates its quality. When a note is played incorrectly, it is "marked," indicating to the control system to adjust required parameters, such as lip shape and air velocity.

Aerophonic sound is highly variable; a fundamental question in setting up a control scheme for a machinic aerophone is which parameters can be modified. The instruments of The Logos Foundation provide composers control over pitch (including bends and microtonal temperaments), noisiness, vibrato, tremolo, volume, envelope, and movement of the robot (e.g., vertical and horizontal rotation). Such parametric variety offers means to creatively sculpt musical phrases, opening the door to uniquely machinic territory. Conversely, variability increases the number of choices a composer or performer faces, which can have a paralyzing effect, leaving the full potential of the instrument undiscovered. It is an age-old question of interface design without a "right" answer; it depends on the user's skills and appetite. For aerophones, a sensible approach is to present configurations that allow basic musical functionality, such as intoning to familiar temperaments and scales, and a moderate dynamic range. More advanced articulations and manipulation of individual parameters (such as mouth configuration) can be available without confusing control of basic functions.

Regarding communications, many builders of machinic aerophones have used the MIDI protocol, given its ubiquitous adoption. Murphy et al. (2014) used control change (CC) messages to determine the speed of the motor driving the pump of the harmonium *Kritaanjli*. The Logos Foundation also uses the MIDI protocol to send messages to its machines. Given the potentially unique parameters of machinic aerophones, a customizable protocol such as OSC is also promising.

Evaluation

The methods used to evaluate aerophonic machines depend on the design, but three areas must be addressed in most cases. The first is dynamic range. Testing dynamic range requires a calibrated SPL meter or an RTA. The microphone gain (when adjustable) and position should be consistent across trials. The intensity of the sound and the perceived loudness that results are a function of the distance between the source and the receiver. What distance is relevant in the usage you anticipate? If the aerophone is to play in a small ensemble, you want to know the loudness at the positions of other ensemble members. Alternatively, you may choose a closer position to minimize environmental effects. Whatever method you choose, document your choices and be consistent across experiments. Loudness measurements of a clarinet from one foot in a small room are going to be difficult to reconcile with measurements from 10 feet in a concert hall.

The second area is time, specifically speed and precision. Regarding speed, it is desirable to know how fast an instrument can play when repeating a single pitch and when playing pitch intervals of various sizes. Test these rates at different dynamic levels. Regarding precision, determine the accuracy of a range of temporal intervals and the system's latency. These aspects of temporal performance can be evaluated using audio recordings. Programs such as Audacity allow you to record audio and precisely measure the time intervals between events at the millisecond level.

The third area that needs to be addressed is pitch range and accuracy. Determine the range of pitches that can be produced at different dynamic levels. Unlike a piano or a fretted string instrument, most aerophones can produce minute pitch variations, which can make intonation an issue. Test pitch accuracy across the instrument's range at various dynamic levels. As we have learned, the relationship between inputs and aerophonic sound production is more complicated than those of percussion or string instruments, so this testing will likely involve the manipulation of several parameters at once.

The other aspects of your machine that must be evaluated depend on the requirements you specified. Follow the guidelines given in Chapter 4 to ensure that the methods you use are appropriate, rigorous, and repeatable.

Example: Pneumatic Slide Whistle

In this project, you will learn the fundamentals of **pneumatic** systems, which create force through air pressure, and apply this knowledge to build a simple aerophone that plays a musical sequence.

GOALS

1. Build a pneumatic system that integrates an air pump, solenoid valve, electrical circuits, and code, utilizing variables and arrays, to produce the following musical sequence (Figure 7.3.2) on a slide whistle:

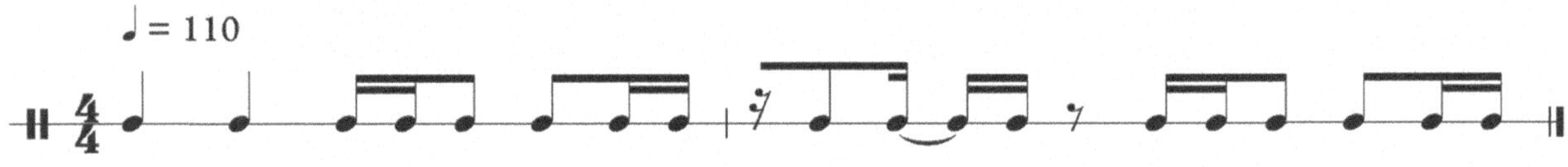

FIGURE 7.3.2. **Goal musical sequence.**

2. Implement a slide control system to enable pitch variation of the sequence.

PARTS

The Companion Website ▶ contains a parts list with updated vendor links.

- Arduino Uno
- USB cable
- Breadboard
- Jumper wires
- N-Channel MOSFET
- 10k ohm resistor
- Diode—1N4001
- 12V 5A switching power supply or 9V battery clips
- 2.1-mm barrel jack
- Micro air pump
- Solenoid valve
- Air reservoir
- Tubing

CONCEPTS

Pneumatic Systems

Essential components of a pneumatic system include

1. a source of pressurized air.
2. a reservoir to store and regulate the pressurized air.
3. a way to release the pressurized air automatically.

A common source of pressurized air is an industrial air compressor, which uses a motor or engine to force increasing amounts of air into a closed container, pressurizing it and creating potential energy. A pressure regulator converts the high-pressure air in the tank to a consistent, lower pressure needed for the application, as specified by a knob and gauge on the machine. Sensors measure the tank and regulator pressures. A pressure switch turns the motor or engine off when the maximum pressure is reached. The downside of industrial compressors is that their motors are loud (they are meant for job sites, not concert halls), large, and relatively expensive. Pumps are another option to create airflow, which come in a variety of sizes and can be quieter than industrial compressors. In this project, we will use an air pump powered by a DC motor, a reservoir (or container) that serves as a buffer to regulate and store pressurized air, and a solenoid valve to release the pressurized air automatically. Components will be connected by flexible tubing.

Pneumatic systems have unique requirements. A reservoir is needed as a supply of pressurized air. Suppose there is no reservoir and the pump is connected directly to the solenoid valve. The only volume of pressurized air would be in the tubing, which would be quickly depleted whenever a note is played. The pump must run continuously to maintain adequate pressure while the valve is in operation. Activating a pump for extended periods can be problematic if it is not rated for continuous use (check the pump's specifications). DC motors in air pumps also make undesirable noise when trying to make music. This noise intensifies when the pneumatic system is fully pressurized, which is wasteful as running the pump is unnecessary if the system is at full capacity. The preferred method is to fill a reservoir and then turn the pump off, leaving a (quiet) supply of pressurized air that can be released by the valve when sound is desired. Bigger reservoirs hold more air, which can support extended periods of music-making, but are also larger, heavier, and take longer to fill.

CIRCUIT

1. Lessons about how to control solenoids and DC motors (needed for the solenoid valve and pump in this project) are found in the examples in the Actuators, Circuits, and Code section of the Companion Website ⊙. Start with the specifications: What is the rated voltage for the pump and the solenoid valve? How much current does each motor draw? Use this information, along with the specifications of the electrical components, to select the circuit for each motor.
2. Connect the pump to the chosen circuit.

BUILD AND EXPERIMENT

Part A

- Write a program to turn the pump on for 1 second and then off for 1 second.
- While the pump is on, determine the intake and outlet (you can use a piece of paper or simply your finger). Connect the tubing to the outlet.
- Run the motor at half speed and press the tubing against the mouthpiece of the whistle. Run it at full speed and do the same. What differences do you hear? What explains these differences?

Part B

- Connect the solenoid valve to the circuit (do not connect the pump to the solenoid valve yet).
- Write a program to turn the solenoid on for 1 second and then off for 1 second. You should hear the solenoid click each time it actuates.
- Connect the pump to the solenoid valve using the tubing.
- Create a program where the pump runs, the solenoid valve turns on for 500 ms and off for 500 ms, and then the pump turns off. **Perform this experiment for only a few seconds.** What happens to the motor when the solenoid valve actuates? What might explain this phenomenon?

Part C

- Procure (see links on the Companion Website ▶) or make a reservoir. If building your own, find a container (e.g., a can) that is **strong enough to contain pressurized air safely and is free of parts that could dislodge under pressure, such as a lid or cork**. Make two small holes (slightly smaller than the outer diameter of the tubing) in the container. Connect the pump to one hole and the solenoid to the other using tubing. Make the connections airtight (use putty, glue, or silicone around the tubing-container junctions).
- Manually couple the tubing from the solenoid valve's output to the aerophone to produce sound. Rerun the program. What happens to the motor? What happens to the airflow? Does it sound different than in Part A? If so, why?
- Why is a reservoir a good idea in a machinic aerophone? What kinds of materials or objects make good reservoirs? What is the volume of your reservoir, and how much sound (in seconds) can it enable?

Part D

- Move the tube to different positions relative to the fipple of the slide flute: what are the effects on the sound produced? Is there an ideal position?
- If variability produces sonically desirable results, how can such adjustments be achieved mechanically? Which degrees of freedom are required? What actuators could enable these DOFs?
- Guided by your experiments, make a coupler that holds the tube at one end and the fipple of the flute at the other (this could be 3D printed or manually crafted, depending on your facility as a woodworker, potter, or machinist).

Part E

- Refer to the Mechanisms and Actuators, Circuits, and Code sections of the Companion Website ⓟ to identify systems that could be coupled to the slide to enable pitch variation. Possibilities include a linear actuator, a belt and pulley, or even a servomotor. Make a structure that joins the slide flute and the actuating mechanism.

CODE

With the system constructed, the next step is to write code that will play the musical sequence notated at the beginning of the project. Keep the following in mind as you write the program:

1. Run the pump until the reservoir is full and then run the rhythmic sequence using the solenoid.
2. The pump and solenoid should run independently, not in sequence. As stated previously, the pump needs to periodically pressurize the air in the reservoir and then turn off. This process should depend on the state of the pneumatic system rather than the rhythm being articulated by the solenoid. Implement code that will allow you to run processes in parallel.
3. The rhythm, as indicated in the goals, should be encoded using variables and arrays, as described in Project 2: Solenoid Percussion with Variables and Arrays on the Companion Website ⓟ.
4. Precise pitch control requires feedback about the position of the slide. Map slide positions (e.g., from an encoder or proximity sensor) to specific pitches and store the correspondences in a lookup table. When a pitch is desired, the associated slide position can be reached by running the motor. A control algorithm such as PID can also be used (Project 16: Motor Control with PID on the Companion Website ⓟ).

MUSICAL CONSIDERATIONS

Moving the slide is an obvious way to create different pitches, but can variations in the air supply also contribute to pitch gestures? Can such variations be combined with slide movements to create unique effects? Are there ways that you can modulate the timbre of the instrument? How can you extend dynamic range? What sorts of articulations are possible? What kinds of musical effects are possible with a mechatronic aerophone that would be difficult or impossible for a human to play?

Summary: Aerophones

Aerophones are instruments whose primary sound is generated by a volume of air. Most aerophonic instruments contain an **energy source** that generates airflow, an **exciter** that periodically interrupts that flow, and a **resonator** whose enclosed air is energized, producing sound. Categories of aerophones differ in how sound is made, including **edge-blown, reed, lip-buzzed, percussive, corrugaphones, sirens,** and **free**. Aerophones have been played by machines to expand expressive possibilities, entertain, educate, and facilitate research of human performance and acoustics. Examples can be categorized by excitation, pitch production, interface, and the *Properties and Relations* introduced in Chapter 4 (*player* ⇔ *instrument*, *form* ⇔ *function*, and *conspicuous* ⇔ *invisible* are particularly useful).

When setting the requirements for a machinic aerophone, standards introduced in Chapter 4 apply here, such as the range of IOIs or acceptable pitch accuracy. The dynamic range of aerophones varies by instrument. When establishing visibility requirements, most designers will grapple with the fact that the movements involved in aerophonic sound production are minute. The augmentation of these movements or the inclusion of complementary ones depends on the purpose of the instrument and the desire of the builder. *Internal* interaction is essential for machinic aerophones as subsystems need to be coordinated to produce sound. *External* interactions are possible, as for any other type of musical machine.

Sonic aerophonic objects can be acquired *off-the-shelf*, through *rummage*, or built *from scratch*. Which path you take depends on the instrument you are trying to make: a PVC pipe with tone holes is within reach for most, while intricate brass tubing with valves is not. Sonic objects can be made from wood, metal, ceramics, plastic, and glass. When making an *edge-blown aerophone*, manipulatable factors include blowhole size and shape, edge angle, and fipple design. Reeds are best acquired *off-the-shelf* or through *rummage*, but can be made if need be. The same can be said of mouthpieces. Other categories, such as *membrane* aerophones, *corrugaphones*, and *percussive aerophones*, are easily and cheaply made from materials found at a hardware store. Musical sirens must be made *from scratch*, given that the disc holes need to be configured precisely and the system requires a motor with speed control. Globular and tubular resonators can be made from scratch, though tubes are attractive as they are readily available. Tubes can be tuned using the calculations found in the Instrument Design section of the Companion Website ⓟ followed by iterative modification and evaluation. Pitch changes can be realized by varying input airflow, using multiple vessels, or integrating *tone holes*, *valves*, *slides*, or *magstrips*, which vary the effective vibrating length of the chamber.

When designing the movements of a machinic aerophone, consider what is needed to produce sound and what is visually expressive. Sound production typically involves small movements, such as mouth configuration or opening and closing tone holes. Visual interest could involve moving parts of the machinic player's body or the instrument itself. Combinations of the two—sound-producing movements that are visually expressive—are rarer but possible.

The mechanisms, actuators, and circuits of a machinic aerophone generate airflow, excite the adjacent air column, and produce pitch changes. Generating airflow is typically accomplished with a bellows or a pump that feeds into a tank, which serves as a reservoir of smoothed, compressed air. The tank is connected to a proportional valve and a leak, allowing the system to receive the appropriate pressure. Sometimes, the air is routed to an artificial mouth that contains "lips" that direct the airstream to the blowhole, reed, or mouthpiece of the instrument. In some cases, such as with reed instruments, an artificial "tongue" can create musical articulations. A different model of excitation using membrane compression drivers coupled to acoustic impedance converters was developed by the Logos Foundation and integrated into numerous machinic aerophones. Producing pitch changes on an aerophone involves pressing a tone hole, key, or valve, which can be accomplished with banks of linear actuators (such as solenoids). These can interact with the instrument directly or could drive intermediary "finger" mechanisms (consisting of levers, for example). Slides can be moved with linear motors, tracks, or belt and pulley systems.

When designing the structure of an aerophone, the location of the energy source, particularly if it is an industrial air compressor, must be addressed. A small pump or bellows system can be integrated into the instrument, while a noisy compressor with a large tank is best separated by long tubing runs. Many tubular aerophones contain many tone holes that require many solenoids to be positioned precisely around the instrument. The key in the design is then to consider how obtrusive that structure will be.

Control of machinic aerophones requires feedback and coordination between multiple systems. Typically, each note is mapped to a specific air pressure, mouth position, and finger configuration. Reaching the appropriate air pressure requires sensing the conditions in the system, which are then communicated to a PID loop that controls the energy source and leak valve. The motors in the mouth and pitch-changing solenoids move according to configurations specified in the microcontroller. Changes in airflow can be further modulated to expand dynamic range. More sophisticated control schemes are possible that integrate data from human performance and feedback about the quality of produced notes. Choosing which parameters are variable and available to instrument users reflects the machine's purpose and the builder's desires. MIDI has been used to communicate musical instructions in many examples, but more flexible protocols such as OSC are also possible. The chapter concludes with an example of a mechatronic slide whistle.

References

Lin, J.-Y., Kawai, M., Nishio, Y., Cosentino, S., & Takanishi, A. (2019). Development of performance system with musical dynamics expression on humanoid saxophonist robot. *IEEE Robotics and Automation Letters*, *4*(2), 1684–1690. IEEE Robotics and Automation Letters. https://doi.org/10.1109/LRA.2019.2897372

Lin, J.-Y., Yoshida, K., Matsuki, K., Takikawa, K., Cosentino, S., Sessa, S., & Takanishi, A. (2016). The development of intraoral pressure control system on humanoid saxophone playing robot. *2016 IEEE International Conference on Robotics and Biomimetics (ROBIO)*, 1649–1654.

Murphy, J., Kapur, A., & Carnegie, D. A. (2014). Mechatronic keyboard music: Design, evaluation, and use of a new mechatronic harmonium. *The Proceedings of the 2014 International Computer Music Conference*, 678–685.

Solis, J., & Takanishi, A. (2011). Wind instrument playing humanoid robots. In J. Solis & K. Ng (Eds.), *Musical robots and interactive multimodal systems* (pp. 195–213). Springer.

CHAPTER 8

INVENTOR PERSPECTIVES

In this chapter, we hear from a diverse group of designers, builders, and practitioners who have helped shape the world of musical machines. They were each invited to submit a reflection in response to the following questions:

- Why do you build musical machines? What inspires you to work on a new project?
- What is your ideation and realization process?
- What successes are you happiest with? What are the most important lessons that you have learned?
- What promise do these machines hold for making music?

These prompts were intended to be minimally restrictive, motivating subjective reflection. These artists, designers, and builders were invited to respond in whatever ways best captured the essence of their practice. The responses illustrate the varied perspectives and approaches within the field.

NICOLAS ANATOL BAGINSKY

Artist, instrument builder, programmer
http://www.baginsky.de

On Making Machines Make Music

In Switzerland in 1992, I was working on my first art installation that employed unsupervised learning ANNs when the idea hit me to combine these networks with a guitar robot. This installation I was working on is called "Transmigration—Educate your local Gallery," and it had developed some surprising responses to acoustic peculiarities of the exhibition space. Transmigration was an attempt to capture audience behavior in the gallery space during the last exhibition before the building was to be demolished. The Kohonen network controlling the installation was stimulated by motion detectors and microphones located in and around the showroom, and it responded to the sensory input by activating various actuators: gallery lights, opening and closing of doors and windows, and by swinging a small wrecking ball against a gallery wall. After a few days of audience participation, the system had chosen the very dim squeaking of a floorboard near the entrance as a signal to launch a recognizable sequence of activity whenever people would enter or leave the exhibition space. Closing my eyes and concentrating on the audible aspects only I could hear melodies in these sequences.

Musical Machines. Scott Barton, Oxford University Press. © Oxford University Press 2026. DOI: 10.1093/oso/9780197599891.003.0008

It became clear to me that I had to try giving a neural network control over a contemporary musical instrument. I chose the electric slide guitar because it is not chromatic, can play any note and interval, and because it's loud. After some extensive training of the system, it became obvious that this experiment had a tendency to play around blues harmonies (one, four, and five). After being puzzled by this observation for a while, I came up with an explanation. These harmonies are prevalent in the sound spectrum of an oscillating string, and the self-organizing maps that I had employed to control the robot must have detected the omnipresence of these harmonies that most of our modern Western music is based upon. What a humbling revelation. Even a small ANN consisting of only a few thousand simple neurons can develop a musical context similar to that of mankind. After all, it cannot have been so difficult to come up with pop music. The reason it is so catchy is that we already know the rules by instinct, even if we have not listened to a single pop song before. They are given by the physics of our planet, not invented by us.

It would now be very interesting to dig into other bodies of music in the quest to find other physical phenomena that may have been the origin of, let's say, Kabuki opera or Mongolian throat singing.

Although the slide guitar is very versatile, its polyphonic nature blurs the melodic results and makes things quite complicated. So I experimented with other instruments as well: a single-string bass robot, two vocal robots, and several others. Work on the vocal systems is particularly interesting, as it appears to me that many musical instruments mimic aspects of the human voice. The voice is probably the origin of all our music. My attempts to build singing machines are rudimentary, and only simple sounds can be produced; still, the melodic outcome is interesting and surprising and can clearly be identified as singing. Making a robot learn to sing is worth further research, and deep learning systems appear to be a new and promising approach to tackling this problem again.

Over the years of experimenting with ANNs and musical robots, I have listened to hundreds of hours of strange inhumane music. And even if my self-learning systems do recognize the importance of certain intervals, the musical outcome is outworldish and challenging to listen to. This practice has clearly broadened my musical horizon and has changed the way I listen to and play music.

Nicolas Baginsky, August 15, 2022

ANDY CAVATORTA

Artist
https://andycavatorta.com

There are three types of first-person stories: one in which you are the hero, one in which you are the victim, and one in which you learn something. The only type anyone wants to read is the third. So I will share what I think I've learned about the creative process in the past 12 years of being a full-time artist working with sound, motion, and machinery.

The Creative Lifecycle of a Commission

1. The Things We Carry

Before there is a project, before there is even an idea, there is the collection of all of your unexecuted partial ideas, vague yearnings, atomized excitements. The longer you work in a field, the more of these you possess to bring to each new project. Write them down. Take photos. Or just remember them if you can.

2. The Constraints

All commissions come with constraints. The budget, the schedule, the audience, the period and environment, the abilities of your studio, and the size of the smallest doorway between your studio and the venue. And the client often has specific requests about the subject or the medium. Everything commonly said about the importance of constraints is true. Even if you make them yourself. I have been asked to describe my "dream project, a project without any constraints." But a project with no constraints is a nightmare, not a dream. Much of the creative process is about problem-solving and discovery. Constraints create a direction and a framework in which new things can be discovered.

3. Irresponsible Fantasizing

I start ideating with something I call Irresponsible Fantasizing. I imagine that I am someone else encountering this future piece. I don't care who the artist is, and I'm not interested in being impressed by them. I fantasize about what I might experience there that would be meaningful and memorable. What I imagine is not limited to things that could actually be delivered.

4. The Dark Energy and The Rubik's Cube

The Dark Energy is what we think about when we don't even know we are thinking. It's invaluable and can have a major influence on the course of one's life. Behind the brain that I think of as me, there is another brain. Like my heartbeat or the complex algorithm by which we place adjectives in order, it belongs to nature and not to my conscious self. This is where the ideas and the excitement about ideas come from. It's ADHD like a team of 12 caffeinated horses, and it is always entertained. This other brain doesn't seem to believe it has a job. But you can give it one.

The Rubik's Cube is what I call the confluence of all of the project's dimensions. The form, the way it makes meaning with sound, the budget, the technical challenges, the type of audience, the essence of the irresponsible fantasies, everything. The goal is to spin and recombine this overwhelming number of variables to find the best cohesive vision that can be delivered within the constraints. Rather than reducing the options, this process usually produces whole new ideas that now must be included as facets of the puzzle.

There will not be a perfect solution. I don't fight the sprawling complexity. I mostly hand it to my dark energy to pick at obsessively while I do practical things like keeping myself fed and washed. The dark energy will report back occasionally and excitedly with partial solutions while I'm showering, biking in traffic, or trying to fall asleep. At some point, one of these exciting and imperfect recombinations really clicks. And the journey begins.

5. The Dark Forest of Possibilities

It's time to start building. I have an approximate goal, but I don't know exactly how to get there. It's like I know there is a clearing on the far side of a forest, but I don't know which combination of the endless forking paths leads to it. I call this The Dark Forest of Possibilities. The project and vision continue to evolve on this path. And we navigate it by making prototypes.

- If you know what you are doing, start building prototypes right away.
- If you don't know what you are doing, start building prototypes right away.
- Speculation is stasis. My projects only move forward when I am making informed decisions and moving to the next step.

All early prototypes should look like they were assembled at gunpoint. Some parts might need to be precise. But everything else can be made with scrap wood, cardboard, tape, and so on. We are not making early versions of a final product. We are just getting answers. We want to know how something looks, sounds, or feels—or if it's even possible. The more generations of prototypes you can build with your finite time and budget, the more mature your project will be. Speed is everything.

6. *Fake Rules*
As a project progresses through the paths of the dark forest, the number and opacity of the branching decisions can be overwhelming. I use some fake rules to help guide me and prevent overthinking. These are fake rules because I can think of exceptions to each. But they have served me well nonetheless.

- Duration: The success of a piece is proportional to the duration for which it is compelling. Always choose the path that makes the piece rewarding for longer.
- On Sound Alone: Any piece that incorporates sound should be able to work on sound alone. It should be meaningful and rich to an audience that knows nothing about it other than its sound.
- Sincerity: Only choose directions that are meaningful to you as an artist and that you believe will be meaningful to the audience. It's not hard to do just one of these. But both are needed. If you don't believe in it but think the audience will, it's disrespectful pandering. And if only you believe it, then it's pretentious wankery. You can fool some people either way. But most people can feel the insincerity.

7. *It's Done When You're Out of Time*
The piece is the sum of all of your implemented decisions at the point where you had to stop. You might feel that it could have been better if you had had more time and more budget. That might be true; it might not. It doesn't matter. Getting better at making the work includes getting everything important in place within constraints. I choose to believe that it's important to try to make every piece the best you've ever made. If you do this, too, the work will never get easier. But work will get better and better.

KOJI SHIBUYA

Professor, Dr. Eng., Mechanical Engineering and Robotics Course
Faculty of Advanced Science and Technology
Ryukoku University
Kyoto, Japan

Featured Work: *violin-playing robot* https://ryukoku-mechrobo.sakura.ne.jp/shibuya/research/research.html
Musical performances provide us with a variety of research topics in broader research fields, including psychology, biomechanics, musicology, AI, human–machine interface, robotics, and so on. My background is in mechanical engineering, and building robots itself is an interesting activity for me. However, only building robots is not enough for scientific and engineering research. So, musical performance offers a wide range of research themes that are beneficial for me.

When I was an undergraduate student, I wanted to choose a theme on the organ-playing robot. However, before my choice, my professor said that the project had just ended. Therefore,

I chose the next new theme: a violin-playing robot, specifically, building an algorithm for determining the bowing and fingering movements from a musical score. When I was a child and student, I learned to play an electronic organ and joined several wind ensemble bands, playing the Euphonium (brass wind instrument). Those performance experiences were probably the reason for the choice.

What inspires me for a new project is human performance. I believe that human performance is the best for a specific musical instrument. Thus, I will study human performance before beginning a new project if I use acoustic instruments.

The difficulty of building a violin-playing robot and researching music inspires and motivates me, too. Many people, including robotics researchers, told me that I should change the theme because it was challenging or impossible to build a violin-playing robot. However, human performance is incredibly sophisticated and fascinates me. Despite the difficulty, I strongly want to extract the essence of performance and apply it to a real robot.

Another reason is that robots performing with musical instruments attract not only researchers but also many people who are not familiar with music, including children and students, which I believe is very important. Some of the children and students who are fascinated by those robots will study science and engineering in the future when they become adults, which is crucially important for the enhancement of science and technology all over the world.

In the first year of my research as a master's course student, about 30 years ago, I did not know how to achieve my goal, and I devoted a year to elaborating a research concept without any experiments or building a robot. The research concept was constructed based on my view of a human performer's information processing from a musical score to real performance. In retrospect, the concept was very primitive, affected by my naive and personal experiences described above. However, the concept has been underlying my research for a long time with gradual modifications. We conducted experiments based on the concept in which human violinists' performances were recorded to reveal how they determine their performance from scores. The experimental results were abstracted and then applied to the robot's performance.

I believe that simulating the sound production of the violin is impossible using the present simulation technologies. Thus, producing real sounds is a must in my research. However, I am aware of the importance of simulation. Our group recently introduced reinforce[ment] learning to determine the bowing motion from a score. We want to use the real robot in the learning process, but it is unrealistic, as you know, because that will cause mechanical failure.

The goal of my violin-playing robot project is to build a system that can automatically determine the robot's body motion only from information obtained from scores. Of course, achieving the goal would make me happiest, but the goal is far from the ideal human performance in my present research state. The happiest time in the project was when we achieved the performance of a small musical piece by the real robot, which pleased me very much, even though it was small progress. In addition, we recently introduced reinforcement learning as described above, which is a very challenging theme for me, a mechanical engineer, to enter the new research field. We succeeded in constructing a basic framework and in determining the bowing motion for a very short four-bar phrase, which makes me happy, too.

From the experiments with a human violinist, I have learned that human performance varies even if the same violinist is studied. So, I concluded that it is impossible to draw performance trends based on averaged human data; instead, we should design a performance based on data from one violinist. I believe that we can easily design another performance if we identify the parameters strongly related to the performer's characteristics.

My view might be conservative. Musical performances using mechatronic technologies may exceed human performance with respect to some points, such as speed and accuracy. The produced sounds may be what humans have never experienced. However, it is questionable if the produced sounds fascinate audiences. Composers of Western classical music are composing contemporary music, but not all of them are accepted by audiences. Instead, music by Mozart and Beethoven is still performed because those pieces have been accepted by audiences. I admire the efforts for constructing new music performances, but produced sounds must be evaluated by humans. Only machines producing the accepted sounds will survive for a long period.

SERGI JORDÀ

Associate Professor, Music Technology Group
Dept. of Information and Communication Technologies
Universitat Pompeu Fabra
Barcelona, Spain
https://www.upf.edu/web/sergi-jorda

Sensu stricto, my work with mechatronic and robotic musical instruments consists of a robotic quartet (electric guitar, electric monochord violin, drum kit, and bagpipes) we developed in 1997 together with mechatronic artist Roland Olbeter[1] for Catalan artist Marcel·lí Antúnez's show *Afasia* (Jordà, 2002a). That said, "digital luthier" (a term I coined for my PhD thesis—Jordà, 2005a) is probably the denomination that I find most suitable for describing several decades of my professional activities. Since this personal quest started in the early '80s, I have been trying to persuade computers for helping us perform and create music in real time.

In 1978, I started studying physics and playing the saxophone. Although none of these activities are totally related to my present professional situation, both were in fact quite determinant for my current activities as a digital luthier. With the saxophone, I began playing free jazz, which turned practicing scales into a quite boring and dissatisfying activity, and while studying physics, I discovered computer programming. In 1983, already knowing some basic (and BASIC) programming, I was shocked by two albums (Laurie Anderson's "Big Science" and Herbie Hancock's "Future Shock") that made me realize that computers could be used for making music, "even free jazz", I thought. Convinced that computers were far better suited than me to repetitive and unexciting tasks (they learn anything at the first take, and they can repeat it forever without a glitch, I thought), I finally dropped practicing scales on the saxophone, bought my first computer (an Oric Atmos, soon to be followed by a Commodore Amiga), and started programming real-time MIDI applications, hoping that, in a not too distant future, I would be able to improvise "my" own music with those machines. This is what, to some extent and with different degrees of success, I have been doing since then. It started as a very personal and idiosyncratic trip, but in the mid-90s, I was already working on wider projects.

My ideation and realization process varies a lot depending on different types of projects, whether it's a strictly personal idea, a proof of concept, a commission, a new product, something more research-oriented, and so on. In that sense, Reactable was unique in many aspects. It started in 2003, as a tabula rasa, free from any technical constraints, trying to conceive the best music instrument I could imagine, for both novices and professionals alike, based on my past 15 years of experience as a digital luthier and a computer music performer. Of particular importance were also the very unlikely circumstances and resources that happened to be available for the occasion: the most brilliant possible team of collaborators someone could dream of

(Günter Geiger, Martin Kaltenbrunner, Marcos Alonso, and Ross Bencina) and an "infinite" 2- to 3-year horizon in front of us (Jordà et al. 2005b; Jordà et al. 2007). Its conceptual precursor was the GUI-based online software synthesizer, FMOL, which I developed between 1997 and 2002 (Jordà 1999, 2002b). FMOL had proved to be a good musical instrument used in live performance by many musicians including myself, and it was with this know-how and with the idea of surpassing FMOL's mice limitations that the Reactable project started in 2003. Reactable and FMOL are certainly the creations I'm happiest with. I'm also very proud of ROTOR, the iPad app we launched in 2017, which constituted the ultimate incarnation of the Reactable's lineage.

For the last years, I've been mostly working with AI and deep learning models for music generation (Haki et al. 2022). I am very excited about what AI might bring to music creation. I would like to see AI taking the form of creativity crutches, inspiring, bringing new ideas, and cheering experimentation to practicing musicians. More importantly, I would also like AI lowering the entry barrier and encouraging and bringing new creators to the game. Being creative is good for your mind and your mental health; it makes people happier and more critical. I would thus like to imagine a world in which "popular music" progressively recovered the status it had before being kidnapped by the record industry in the mid-twentieth century. I can imagine popular music as being again the music "of and by the people" rather than the music "for the people." I could imagine indeed all these lovely things, but I'm more worried that the tendency might rather be the opposite. I'm scared by the industry taking music AI, cheapening production, homogenizing contents, and, ultimately, rendering music creators obsolete.

Complementary Material

The ultimate goal for any new instrument could arguably be the potential to create a new kind of music. In that sense, baroque music cannot be imagined without the advances of sixteenth- and seventeenth-century luthiers, rock could not exist without the electric guitar, and jazz or hip-hop, without the redefinitions of the saxophone and the turntable. Yet, this extremely ambitious objective is often beyond the reach of its creator (80 years separate Adolphe Sax from Coleman Hawkins, and no less than 30 go by between Les Paul and Jimi Hendrix).

Being a bit more pragmatic, as a performer, I want to create instruments that are enjoyable to play and that mutually enhance the experience when playing with other musicians. Thereby allowing me to create or co-create music that will surprise me as much as possible, that will keep revealing little hidden secrets at every new performance. Music not necessarily better, nor worse, than a piece that I could compose in a studio, but music, in essence, that could not have been created in any other possible way.

Making instruments for oneself is not the same as for "everyone"; other considerations need to be taken into account, although for me the overall goals remain unchanged. My aim is to create instruments that people can enjoy playing; instruments that will be able to enrich and mature the performers' experiences in any imaginable way; instruments that allow scope for the performer (particularly in the case of a non-expert user) to be proud of the music created.

In order to survive in the extremely demanding instrumental ecosystem, any new instrument should clearly excel in something. It should either be able to do one thing that no other instrument could or, at least, it should do it better (whatever this can be and whatever "better" may mean). One should be especially careful about the instruments' conceptual capabilities and consider how new instruments impose or suggest new ways of thinking to the player, as well as new ways of establishing relationships, new ways of interacting, new ways of organizing time and textures, and new ways of playing, in short.

References

Jordà, S. (1999). Faust music online: An approach to real-time collective composition on the internet. *Leonardo Music Journal, 9*, 5–12.

Jordà, S. (2002a). Afasia: The ultimate Homeric one-man-multimedia-band. *NIME*, pp. 132–137.

Jordà, S. (2002b). FMOL: Toward user-friendly, sophisticated new musical instruments. *Computer Music Journal, 26*(3), 23–39.

Jordà, S. (2005a). *Digital lutherie: Crafting musical computers for new musics' performance and improvisation* (PhD Thesis, Department of Information and Communication Technologies).

Jordà, S., Kaltenbrunner, M., Geiger, G., & Bencina, R. (2005b). The reactable. *The Proceedings of the 2005 International Computer Music Conference,* 1–4.

Jordà, S., Geiger, G., Alonso, M., & Kaltenbrunner, M. (2007). The reacTable: Exploring the synergy between live music performance and tabletop tangible interfaces. *Proceedings of the 1st international conference on Tangible and embedded interaction* (pp. 139–146).

Haki, B., Nieto, M., Pelinski, T. & Jordà, S. (2022). Real-time drum accompaniment using transformer architecture. *3rd Conference on AI Music Creativity (AIMC 2022)*, 1–10.

AJAY KAPUR

Associate Provost for Creative Technologies
Director, Music Technology: Interaction, Intelligence and Design (MTIID)
California Institute of the Arts
http://www.ajaykapur.com

I was a jazz drummer starting when I was young, playing in school and jazz bands. During that period, it bothered me that the sound of my instrument was always the same. This pushed me into the realm of electronics with a question in mind: How can I use the skills that I have developed to change the sound produced? I started building things that allowed conventional gestures to produce a range of different sounds. I noticed that many systems were built for Western music, but there was nothing for my culture, which is Indian. I wanted to transfer some of this knowledge to my culture. I built a number of systems specific to sitar or tabla players who could use familiar gestures to create different sounds, which bridged Indian practices to things we are doing in the West.

I got sick of the speaker (I hate to say). I wanted music to be more immersive. I was jamming on my sitar, and someone at NIME had an automated rhythm machine, which I thought was so cool, but I didn't want the sound to come from a speaker. I went to India again and saw this incredible Saraswati in a temple that had multiple arms playing instruments, and I thought I want to build this; I want to make this real. That is when I did my PhD, with the goal of building a system that could make that Saraswati real, to make a robot that I could play with. It was all about the interaction. I am interested in human extension and human–machine symbiosis. You need sensors so that the machines know what to do. This raised questions: How do you perform? How does the machine know what you are doing? How can we use what we have been doing in the NIME (New Interfaces for Musical Expression) and ISMIR (International Society for Music Information Retrieval) communities to give systems intelligence and the ability to react to humans? How do you get instruments out in space and part of the audience? How do you make them immersive? I've evolved from thinking about musical robotics in terms of conventional performance and am thinking about it more in terms of installations or immersive experiences.

When I got to CalArts, I made an Orchestra of instruments. I made these practices part of our curriculum, training students in programming, signal processing, and robotics. I have been trained in engineering programs, which is why I am equipped to do what I do. The parallel challenge is I never felt like I belonged in an engineering environment. I definitely belonged at an Art school, but the question is, how do I get students to develop engineering chops? I added engineering courses to our curriculum so that students are ready to do things that are significant and cool. One of my top students is Meason Wiley, who is an executive at Meow Wolf and was the art director at Gremlin. Gremlin is the biggest robotic music sculpture in the world, now a permanent installation in Denver. Students in New Zealand are running a robotics program and making interesting instruments. It has been an incredible journey.

Ideation

In terms of the ideation process, I think of what I want the audience to experience first. How is it unique and new? What emotions do I want them to feel? I think about the musical composition. I then build instruments that can affect those experiences. Sometimes there is a story arc, and then it is a matter of building instruments that realize that story. It is art first, and then we use whatever tech is available. We are always looking at papers and what the community is doing to learn and apply knowledge to what we are trying to do. We are not trying to make technology for technology's sake. We are trying to make art that tells a story.

The Sonic and the Visual

If it is a mechatronic instrument that is moving, then you have to show it to the audience. Either it has to be immersive and people can come up to it and look at it or if it is on stage then you have to put cameras on it and project it. At one concert, there were nine screens so that the audience could experience the movement of the machines. I always put things in the audience now, such as in the seats, which helps fill the audience's imagination. Visual and audio parts are equal: it has to look good. Digital fabrication is so easy at this point: laser-cut materials, 3D-printed parts, 80/20; it is amazing! You can make things look good and professional really quickly.

Challenges

When I started, I was in a band and eventually I wanted to do my own thing. I was so excited that robots were going to be my band and that I was going to start traveling. The first question that then came up was, how do you move these things? That is a key aspect of design. Everything has to be able to fit in a suitcase that conforms to weight limits so that it can actually travel from country to country. You have to think about having parts that might be needed, putting them in multiple different suitcases so things would arrive at your destination. These travel aspects were a major learning experience. I've had parts of spinning robots fly off during a concert. If something goes wrong, you have to make sure it doesn't hurt anyone. We do safety checks, especially with machines in the audience, because we don't want anyone to get hurt. It took me years to find the right MOSFET that had the magic to do what I wanted, and now I can't get them. Sourcing parts is always an issue. Now I have storage bins of components for every project in case something needs to be replaced. I have instruments that I have to redesign because I don't have the parts.

Accomplishments

We have had a strong focus on human–machine performance. Subtlety and quietness is a key thing. We have worked with Indian and Korean dancers: as they move their hands, the instruments also move in subtle ways. I call it a *butterfly* when you take a solenoid and move it really fast but softly. It flutters, producing this beautiful sound when you multiply it by however many you have: it sounds like cicadas that are surrounding you. You can't experience that unless you are in a room. You can't experience that with speakers (unless you have a lot of them). I love watching things move. My biggest accomplishment is creating a curriculum and seeing what my students have done: it makes me so proud. I think this is an art form, and it is very exciting. I studied with Trimpin—he trained all of us hard—we owe a lot to him. It is in his name and his legacy that we have made this curriculum, and it continues.

The Future

I think the music performance space still needs to be figured out. For me, I can't get the business model to work. Every time I've tried, I set up a venue, it takes three days requiring 40 people, I do the show for two or three nights, and you barely make enough to pay the people. For experiences, you set it up once, and it can work for a long time. You have to do some minimal maintenance, but all these people can come and see it. It is a different type of experience that can be sustainable financially. We never talk about that because you are not supposed to talk about money with art, but it is really important. You have to think like an entrepreneur. It has to make sense. The way in which artists do this full-time, they have to think about this. It is why the greats have evolved to experiences and kinetic sculptures.

I'm really excited about venues like ARTTECHHOUSE (https://www.artechouse.com) or Lighthouse Immersive (https://lighthouseimmersive.com) that produced Immersive Van Gogh. Those are venues for multichannel audio and floor-to-ceiling projection, which is becoming a standard format in this country, which is cool and exciting. Places like teamLab (https://www.teamlab.art) and Meow Wolf (https://meowwolf.com) are bringing all the stuff that we have been building in the art and technology space and putting it on display with a throughline story. That is the future. That is what people want. We can make experiences where families pay $40 a ticket that provide 4–5 hours of fun, education, and a story. That is super exciting. That is where I am putting my attention as I go forward. Over the next 20 years, we are going to be building these immersive experiences that not only need projection mapping but also mechatronics and sound design that are part of the story. That is the key thing; it is all about the story. The things that have been successful are story first, technology second. As celebrities get involved in these spaces, it will bring the public to expect these kinds of experiences. There are so many exciting movements in the space; it is inspiring. It can be simple, and it doesn't have to be crazy. When you take kids to immersive experiences, they are engaged. The public has to be involved; we can't be so snooty where only academics or art people can understand what we are doing, or we will only play for ourselves. The question is, how do we bring these things into a place where the public can see? I'm not using the word "mainstream"; it's just "public." It doesn't have to be pop, but it has to be accessible. Artists and students can learn the tech and figure out how to bring it to the art forms they are creating: mix it with rap, mix it with techno. We need to get out and do it in public and get reactions from the real world.

STEVEN KEMPER

Associate Professor of Technology in Music and Related Arts
Oberlin Conservatory
Oberlin, Ohio, USA
https://stevenkemper.com/wp/

When I first started creating robotic instruments as part of Expressive Machines Musical Instruments (EMMI) in 2007, I was excited at the prospect of working collaboratively with a team, which isn't something that composers are typically encouraged to do. I also enjoy building things and was happy to combine my background in this area with my musical endeavors. At the time, I was skeptical that mechatronic musical instruments would be capable of producing meaningful music. I suppose I was still attached to the idea that human performers who had spent years training on their instruments would be much more "musical" or "expressive" than the machines we were building.

One of the most formative experiences of my career came late at night while EMMI was developing our first instrument, the Poly-tangent Automatic multi-Monochord (PAM). After a few months of development, we turned the instrument on (or likely just plugged it in since there was no power switch) and fed it some data to play. Suddenly, the instrument came alive—it was playing by itself. We had given life to an inanimate object, one that was physically moving in space and producing sounds through its actions. I came to understand the wonder of seeing an inanimate object produce musical sound as anthropomorphism. My mind was imbuing this machine with human-like qualities since it was engaging in the quintessentially "human" activity of performing music. However, I quickly discovered that audiences react differently to humans and machines even if they are performing similar types of music. For example, when human performers play music that is very fast or is extremely rhythmically complex, audiences embody the struggle to perform that music. Machines, on the other hand, do not necessarily struggle with speed and rhythmic complexity and thus don't necessarily generate the same level of excitement.

This realization led me to explore the idea of mechanical "struggle" and to think about ways that machines can fail. Mechanical failure, or the threat thereof, can evoke a powerful response in an audience. One mode of simulated failure that I've explored is the stuck loop. Machines are typically designed to execute repetitive tasks, but they can get stuck in a loop where they will repeat a function until they break, or the power runs out. Simulating a stuck loop causes the audience to wonder if the instrument is functioning properly and produces a sense of relief and resolution once the machine emerges from the loop.

The instruments that I have designed in recent years all actively embrace their mechanical nature. I enjoy exploring the novel sonic and gestural possibilities of a given set of actuators and resonant objects. Based on EMMI's approach, I design all of my robotic musical instruments to be highly portable so they can perform in a variety of different spaces. I have recently moved away from presenting robotic instruments in traditional musical contexts where there is a divide between the performer and the audience. I prefer allowing audiences to freely walk around and observe these instruments from a variety of different vantage points.

I have also recently shifted to exploring HRI in ways that allow human bodies to engage with robotic systems. My personal interest in this area is less focused on artificial musical generation or accompaniment, and more on the idea of the cyborg—creating human/machine hybrids. To date, I have constructed three systems where sensors detect bodily movement and translate that into control data for mechanical actuation. These include (1) the *Remote electro-Acoustic Kinesthetic Sensing (RAKS)* system co-developed with Aurie Hsu, a sensor-interface for dance that interfaces with robotic instruments, (2) the *Manus Tremens system*, where a performer directly controls the intensity of vibration motors through accelerometers worn on the wrists, and (3) the *Tremolo-Chimes kinetic sculpture*, which tracks the movement of participants in a space, allowing them to interact with robotically controlled chimes.

While musicians and audiences since the eighteenth century have critiqued mechanical music for not being "human enough," I would argue that such critiques miss the point. My practice seeks to let machines be themselves and encourages audiences to understand them on their own terms, filtered through our own anthropomorphic lens. There are certain musical gestures that humans perform well and places where we struggle. Likewise, machines are similarly more naturally suited to certain types of music. Exploring this continuum not only allows us to develop new musical ideas based on machines' capabilities, but it also enhances our capacity to relate to the machines around us as they become increasingly integrated into our lives.

JIM MURPHY

Lecturer in Sonic Arts
Victoria University of Wellington
Wellington, New Zealand
http://jim-murphy.com

Mechatronic and mechanical sound art has been the focus of my practice for more than a decade. During this time, my interest has shifted to different niches within the field: I first explored human–machine performance ensembles, then engaged in the creation of new parametrically rich mechatronic instruments, and (most recently) have undertaken a series of installation-specific works that try to cut to the core of what makes mechanical sound artworks compelling to audiences.

The ability of a mechanical sound artwork to convey the complete chain of events resulting in a sonic output is what makes them most interesting to me. Synthesized or recorded sounds necessarily exist within the black box of an electronics/loudspeaker system, with the process by which they have been realized being concealed from the audience. Mechanical means, however, allow an artist to convey the physical gestures that lead to or result from a particular sound-making event while still utilizing compositional practices that have emerged from computer music paradigms. I feel that this physicality, wherein sounds are produced in a causally visible manner in a space shared by sound-maker and audience, is at the heart of what makes mechanical sound art intriguing and worthy of exploration.

I think that we are at an interesting moment in the history of mechanical sound artworks. For centuries, the trend has been to create clockwork mechanisms able to play back, in a highly deterministic manner, pre-programmed compositions. The player pianos and water organs of earlier centuries are not so different from the MIDI-controlled solenoid drum sets of more recent years. However, I notice an emerging rejection of such determinism in mechanical music: starting perhaps with the works of Fluxus artist Joe Jones and continuing to grow in momentum in recent years, artists begin to create mechanical music systems that embrace unpredictability: sound sculptor Zimoun's works of sublime indeterminacy are a key example of this trend, showcasing the artist's role as a sort of "world builder" who creates an environment from which audiovisual events may arise in an emergent manner rather than as a clockmaker whose decisions inform every aspect of a machine's performance.

My own recent works take a cue from such chance-informed sound sculptures. In my ongoing "Machine Gesture" series, I attempt to produce works that emphasize the mechanical gestures that are produced in anticipation of (or in response to) the creation of a sound event. These works explore the tension between the deterministic and the unpredictable, containing elements both inspired by clockwork automatic music systems and those of more indeterminate intent.

I live in Wellington, New Zealand. Getting parts, materials, and access to tools can be trickier here than in places with a heavier industrial base. This relative shortage of surplus materials, tools, and components has deeply informed the practical aspects of my work: where practitioners in countries with more heavy industry often make heavy use of repurposed electromechanical hardware and a quick-turnaround prototyping cycle afforded by extensive industrial infrastructure, I often opt to build componentry from scratch, starting with raw stock. On a practical level, this paradigm results in smaller bespoke works whose from-scratch creation is more evocative of an artisanal process than of the automation engineering seen in some other artists' works. Conceptually, my use of raw materials leads me to embrace the physical materiality of the component materials in a manner that would likely be de-emphasized if I was making use of prefabricated surplus components.

Mechanical sound sculpture fascinates me more and more as time passes: I truly see this practice as a life-long exploration of art and craft. The act of rationalizing an artwork from a sonic/visual idea into a sculpturally effective working sound machine is something that I find no less rewarding now than when I first began exploring this field well over a decade ago. As technically informed art continues to embrace ever-deeper levels of abstraction (in the form of virtual and augmented reality, among other things), I believe that artworks that are real and artworks that contain physical motions with their inherent unpredictability and imperfections will be ever more valued by audiences.

TROY ROGERS

Composer, roboticist, creator, educator
http://www.troy82.com

the sounds of steel making steel out of rock
and loons with their isolated solitary calls in the night
a garage band in the basement
yielding to chamber music

revisit sound as plastic moldable material
through DSP
loudspeakers as performers found to be lacking
blowing out the cheap plastic-housed solenoids
in a Yamaha Disklavier
pissing off the piano tech

a pencil, a rubber band, a paper clip
balloon membrane stretched over steel mixing bowl
a better little solenoid, a microcontroller, and Max/MSP

apparently, this makes a robot (PercusBot)
uncanny bridge vs. uncanny valley
turns out imperfect robots endear

failure remains the best teacher
chaos can be compelling
a little noise in the system
noise, the only source of new patterns

if buildings become kinetic music making machines
architecture thaws; no longer music frozen in time (h/t Future Forms Lab)

building automated string and percussion instruments in a black widow-ridden shed in central Virginia
we tap into a stream of human endeavor two millennia in the making (EMMI)

interlude: if you can take residence in a stainless steel tetrahedral concert hall
housing the world's largest robot orchestra
fifty-plus musical robots of all manner
controlled by humans dancing naked in a matrix of lidar sensors
why wouldn't you?
(Godfried-Willem Raes + Logos Foundation)

what was perceived as esoteric noise
becomes accessible when rendered through electromechanical musical machines

put them on a cart
take them for a walk
don a gas mask and a teddy bear
become an automated instance of T. Coffey's 'gentle guerrilla'
this is/was Robot Rickshaw

in an error of irrational fear, xenophobia, general dull-minded paranoia, and so forth
art that does *not* summon the authorities should be considered suspect

if you're not getting physically assaulted
and having your life's work torn apart by meth heads
on one block in high spring Santa Barbara
and having a group of nubile coeds flash you and your robots
are you even living?
mere minutes in the life of an early 21st century semi-nomadic robot herder
a blip in the vast 60,000 miles driven in a grafittied up former u-haul box truck

where does it all lead—
utopia or oblivion
both within our grasp
like Bucky said

we need more applied futurists
all artists are in some sense applied futurists
musical roboticists have the advantage
of tapping into those two wondrous things that delineate and characterize humans:
our tools and our music
not to mention the rich mythology of robots

the children of today will build the robots of tomorrow
third graders get it just as well as MIT grad students
call it STEAM or don't

so we formed the Institute for Psyborganology
and the Center for the Creation of Sustainable Belief Systems
(which explores love as the highest form of human technology,
the one most likely to assist us in surviving our technological adolescence)
wielding a backpack of robotic musical instruments
sonifying air quality data
gathered by group of citizens
formed after the latest local refinery explosion

and out of isolation desperation
we constructed a sousaphone boat
reminiscent of the royal drinking boat with musical automats
a millenium or so ago
the coast guard was unamused

in a recent act, we destroyed the physical manifestation of our life's work
twelve years of musical robots demolished
with a hatchet
not just any axe, but an electric hatchet wielded by the chief psyborganologist himself
after giving an electric hatchet haircut
at a local festival

this was an act of love
not carried out in a spirit of violence
but instead with Octavia's words ringing through
the ole gas mask-clad noggin
a split second human-machine improvisational decision

another Bucky-ism:
ingenious adaptable humans will make a life raft out of the piano lid from a ship wreck if necessary
the problems come when we cling to piano lids when better options are available

related to that maybe quasi-McLuhanism about the tools we shape shaping us
but more so to than Onyx Ashanti-ism
about building a chrysalis
over and over again
and needing to break out of each one to become the next ourself

and now everything is broken and beautiful again
like it always was

ERIC SINGER

Artist and Technologist
ericsinger.com

As an artist, I am interested in the intersection of disciplines. I believe some of the most creative work can occur at these junctions. Being both a musician and an engineer, this led me very naturally to the idea of electronic musical instruments. I built my first one as an undergrad at Carnegie Mellon University—a rather clunky MIDI saxophone, using an Apple II computer as the brain. I fabricated it in the woodshop of Dr. Roger Dannenberg, a genius of music software and my first computer music professor. From there, I moved on to Berklee College of Music, meeting my second influential teacher, Dr. Richard Boulanger. The odd professor out, Dr. B., was more about software and alternative instruments than synthesizer jingles. As his assistant, I wrote software for alternative instruments such as the Radio Baton and Power Glove. This turned me onto the idea that with electronics, anything could be made into a musical instrument.

Returning to intersections, I started casting about for objects to cross with electronics and create improbable instruments. The question "How can I turn this scrap rubber tube into an instrument?" yielded the Sonic Banana, with bend sensors providing MIDI control. Similarly, transformed slinkys, slime, and wind chimes became the SlinkOTron, SlimeOTron, and ChimeOTron, all human-playable controllers that send data into a computer, where software sonifies it into a wide spectrum of timbres, pitches, and textures via algorithmic music generators. Many other instruments followed. Then, as often happens, I arrived at the question, "OK, what next?" How about reversing the formula and sending musical data in the other direction, out to devices? Thus, musical robotics and the seeds of LEMUR.

LEMUR began as a spinoff of the Madagascar Institute, a Brooklyn Dadaist arts combine I co-founded in the late '90s. With 100 or so artists and technologists, the question, "Who's interested in making musical robots?" garnered a group of people with a wide range of skills. The only element missing was someone with experience in making robots. Determination plus a generous Rockefeller Foundation grant afforded us the ability to experiment and learn. Everything you know how to do, you have to learn for the first time once. The group was mainly interested in creating machines that were their own instruments rather than ones that played existing instruments. This added another dimension of creativity to our work: finding materials that made sounds we liked, then turning them into instruments. Later, we were called upon to roboticize existing instruments, though not for the purpose of replacing humans. In fact, they often performed alongside human instrumentalists.

I've never said that musical robots play better than humans—only differently. I've been asked, "What does a musical robot sound like?" This is akin to asking, "What does a piano sound like?" Composer Conlon Nancarrow, seeking to produce polyrhythms unplayable by humans, wrote for the player piano by cutting his own piano rolls. Whether this ability is "better" than that of Glenn Gould or Thelonious Monk is, to me, irrelevant.

Improvisation, whether in pitch, rhythm, or timbre, has always been an important musical element to me. LEMUR's machines, designed to respond in real time, enabled improvisation with acoustically based instruments. This figured into their use in interactive installations, algorithmic compositions, and live interactive performances, where a computer "listens" to players and responds based on a composer's software (another approach to composing).

Before undertaking a project on this scale, I had to have a good answer to the question, "Why?" Is there any point to this besides the "cool" factor? I believe there was and is. We already have electronic sound and music output devices—synthesizers, of course—and they are amazing machines. We literally have the ability to create any sound imaginable. A crucial limitation though is that their sound always comes out of a speaker. If you walk into a space with your eyes closed and hear music, you can tell immediately whether it's live or recorded. The acoustics of live instruments and the way they interact with a space are far more complex than anything played through speakers. So, right away, there's a reason for leveraging acoustic sounds. Physically produced sounds themselves are sonically more complex than electronic ones, whether synths or recorded instruments. Inevitably, recording and reproduction filters out sonic complexity at many stages of the process, then plays it from point sources, where the resonance of physical instruments is lost. Even physical modeling synthesis can't approach the complexity, nuance, and pleasing randomness of a live instrument. These are properties our human ears and brains enjoy very much. We prefer chaos over perfection. The visual elements can't be downplayed either. Besides being more interesting to watch than a glowing apple on a laptop, seeing sounds being played adds to the sensory experience. This is more than sound plus sight. The brain integrates visual and sonic experiences into more than the sum of their parts.

But really, we're not thinking about any of that when we create robotic instruments. We want to present a new musical experience, entertaining, exciting, and moving the audience. And adding to the panoply of musical, sonic, and performative possibilities is what we as instrument builders seek.

MO H. ZAREEI

Artist
https://millihertz.net/

Motivation

On one hand, I don't view composing with machines fundamentally different than writing for a string quartet, or making an indie rock tune. These are all different ways of organizing musical thoughts and expressing them using different mediums, tools, and technologies. On the other hand, I do enjoy the flexibility and freedom that comes with building your own musical tools and instruments. Since I subscribe to a very broad definition of music (i.e., any artistic activity that has sound at its core), I don't see why I should limit myself to a specific set of timbres, frequencies, and overtones. Of course, that isn't to say that I don't find artistic value in works that do follow the more conventional blueprints. I do, however, find the creative process more interesting when I get to have a say, not only in the input and the output, but also in the design and functionality of the very medium that turns the input into the output.

I find the exposure of the artistic process to be an incredibly powerful tool. I am very much interested in exposing the materiality of sound and the process of sound production, and I build musical machines to help me do so. Within laptop-based electronic music (a practice that I've engaged in for more than a decade), even though the machine, that is, the computer, has the most consequential role, the artistic process is tucked away inside the black box, at least to a large degree. When employing physical and visually present artifacts, however, the process can be revealed and appreciated in a much more accessible way.

Inspiration

I am not very good at improvising. I never build things for the sake of building things and hoping to end up with something interesting in the end. The "technology for technology's sake" attitude does not interest me. I always need a clear-cut concept for inspiration. The concept itself can range from some form of physical or mathematical principle to a natural or socio-political phenomenon that has caught my attention. Very rarely is my work based on purely abstract and "compositional" ideas. Of course, I do integrate, and borrow from, a wide range of musical and compositional techniques and strategies. But these techniques are always employed to serve the realization of a clearly defined concept.

Ideation

As mentioned, coming up with a clear concept is always key. Regardless of the concept or the medium that I'm working with, I am always particularly drawn to stripped-down, functional, and minimalist approaches. As such, once I have the concept, I then try to plan out the most efficient ways to realize it. Through an iterative design process, I then develop the work from a prototype into its final stage. Throughout this process, I pay close attention to the visual aspect of the work as well. In fact, in many instances, it is the visual or physical aesthetic qualities of the work that influence a number of key decisions. I always aim for a form of audiovisual symbiosis. If I use machines, tools, or objects that are visually present at the output of the work, the way in which they are presented and seen is as important to me as the sonic output of the work.

Successes

When working with DIY technologies, I am very conscious of the danger of falling for some form of tool fetishism or gimmickry. I'm happiest when every single element that I have integrated into a work plays an equally significant role in the artistic expression.

Lessons Learned

1. Sometimes you have to choose being practical over being purist, and that is fine.
2. Document your work.
3. Document your work.
4. Document your work.

The Promise of Musical Machines

I don't really see a fundamental difference between working with laptops, violins, or motors and microcontrollers. I don't view these as completely different musical species. Music is the art of sound organization, and sound is the vibration of air particles caused by some form of physical disturbance. So musical machines can be as interesting (or as uninteresting) as any conventional musical tool.

That said, thinking about the dividing line between fine arts and sound art, I think the practice of developing your own musical medium holds great potential for blurring this line. Integration of custom-designed physical objects, tools, and machines within sound art can also lead to less emphasis on the conventional notion of "musicality," and more emphasis on the "art." I find that valuable. Due to a set of complex reasons, sound art has always suffered from inattention of the established art institutions. This revitalized wave of technologically driven and physically realized sound-making practice is certainly instrumental in changing the status quo on that.

CHAPTER 9

COMPOSING WITH AND FOR MACHINES

Generating Musical Ideas with Machines

I assume readers have varying degrees of musical experience. For those requiring an introduction to or refresher on fundamental concepts of music theory and composition, the Music Theory and Composition section of the Companion Website ▶ provides a foundation upon which this chapter builds.

How can machines inspire musical ideas? For many, the composition process is *intuitive*: nebulous, ineffable, and magical. Somewhat contradictorily, the unfettering of *intuition's* flow only comes through regimented practice, making thoughts and actions *automatic* expressions of will. Learning to compose involves practicing and building *techniques*, which are selected and implemented, shaping musical works and defining musical voices. For those who seek to write *for* others (including machines), this involves learning what players and instruments are capable of and what domains suit them best. For those who seek to make machines that write *themselves*, this involves translating techniques into *algorithms*. These autonomous machines may yield exciting results when acting "on their own." They also may function as interactive collaborators, providing a spark when your inspiration is quiet or when you find yourself on the same well-traveled road (which is also a byproduct of dedicated practice). In all cases, the question becomes, what processes or techniques should I develop? This chapter introduces compositional domains within which to find answers to that question.

THE BIG PICTURE

One can approach a piece from the *top-down* by focusing on abstract, *higher-level* aspects such as aesthetic preferences, intellectual interests, novelty, a central idea, psychological associations, emotional states, or spiritual expression. Doing so can help spark inspiration for *lower-level* ideas (e.g., rhythms, melodies) and act as a cohering force for the piece.

Aesthetic preferences are immediate, visceral, and intuitive and are a vital part of the compositional process. As composers, we are guided by our personal aesthetic sensibilities, but we are also influenced by those of collaborators, audiences, and communities, expressed through stylistic conventions. You may attempt to compose beyond your aesthetic preferences, making choices that do not necessarily suit your tastes or that you are indifferent to. John Cage advised, "You must free yourself of your likes and dislikes" (Kostelanetz & Cage, 1987, p. 108). Cage's compositions posed questions and provided answers through chance operations rather than making deliberate choices and exerting control (Kostelanetz & Cage, 1987). Emulating a genre, artist, or convention is another way to step out of your aesthetic shoes. These ideas may sound strange, but they are ways

Musical Machines. Scott Barton, Oxford University Press. © Oxford University Press 2026. DOI: 10.1093/oso/9780197599891.003.0009

to learn about your preferences and your place in the world. Self-inquiry can bring understanding and inspire creativity.

You might compose according to your **intellectual** interests. **Sonifying** phenomena—a flow of gases, the configuration of stars, or the movement of people through a city—are but a few possibilities. Mathematics can generate musical materials: the Fibonacci sequence can determine rhythmic intervals, or the golden ratio can dictate the proportions of a work's form. The results of these processes may appear unedited, or you can use your aesthetic preferences to select, hone, and structure results. The *transparency* of the generative process or the correspondence between phenomena and their representation in the work is a fundamental consideration. Too direct, and the results are obvious or predictable; too abstract, and the connections between the music and its inspiration are missed. Computers are well equipped for such analytical efforts, which often require interpreting large data sets.

Novelty and **innovation** are core values in the development of art and technology. Musical novelty can occur in melody, harmony, rhythm, timbre, instrumentation, form, stylistic synthesis, or performance techniques. Innovation can be achieved through visual presentation, movement, or technological features. A machine's novelty can influence the novelty of the musical work: unique abilities engender unique musical ideas. With all of that said, novelty alone is not enough to make a piece compelling. Novelty is exciting when it serves the music's or artist's purpose rather than its own.

A **central idea**—a form, narrative, image, scene, or relationship between characters—can give a work shape. Crafting structure first can help you develop and connect individual ideas. Doing so offers a perspective on the proportions of a work's sections. It provides a way to cohere disparate statements, one of the more challenging aspects of composition. While in the thick of composing, you may find that a central idea must adapt to prevailing conditions. This is OK: flexibility is a part of the artistic process.

The identity of the piece may be **emotional** or **spiritual**. There may be an understanding or feeling that you want to express, something that you want to evoke in the listener, or both. You may be apprehensive that machinic intermediaries will dilute the power of such messages. Those fears are allayed when recognizing that musical and artistic expressions are regularly mediated by technological objects, physical contexts, and social conventions. Which ones facilitate, amplify, or dilute is a matter of familiarity and acceptance. These are matters that reside in the minds and lives of producers and receivers, not just in mediating artifacts. New kinds of tools may help you express yourself, communicate with others, heal, find joy, and connect to a higher power.

Your piece's identity may be gained through **association** and **context**. Modern works that utilize the sounds of a Yamaha DX7 or Linn LM-1 drum computer transport seasoned listeners back to the era of hairspray, shoulder pads, and *New Wave* music of the 1980s. Associations can be evoked through rhythm, instrumentation, harmony, musical form, physical form, materials, **gestures** (a metaphoric motion to help express a thought or feeling), sampling, dress, graphics, performance spaces, lighting, text, media, and color. Associations connect external expressions to our cultural and personal lives, synthesizing the familiar and the novel into a **context** that sets a frame for an audience to orient perspectives, make connections, and form interpretations. This context can be composed. The challenge is to allow listeners to gain some fluency in the piece's language, enabling them to understand what is being said so that the work can set its own agenda and criteria for value. The meaning of the work is situated in a complex web of associations and not just the materials of its conveyance.

GROUPING AND ORGANIZATION

Music can also be composed from the *bottom-up* by temporally organizing sonic elements on *lower levels*. The lowest level consists of individual elements (e.g., a note). These elements are concatenated to form a group (e.g., a rhythm). The character of a group depends on the number of elements and their configuration, which can be described in terms of distance (e.g., a semitone, a measure) or proportions (e.g., 1:2 for the rhythmic intervals 250 and 500 ms). That group becomes an element on a more abstract level that can be combined with others to form larger groups (e.g., a phrase). As this process continues, more expansive groups and higher levels (e.g., sections, movements) emerge, creating a **hierarchy**. The structure of a hierarchy defines a work's formal identity over longer periods of time.

Groups emerge according to the characteristics of our perceptual and cognitive systems. Music's identity is not wholly external or absolute; it comes to fruition in our awareness. Concepts from psychology, particularly those involving perceptual organization, are valuable for composers who negotiate such tendencies when crafting musical statements. These concepts can be applied to different levels of a hierarchy (e.g., individual notes, phrases, sections).

We distinguish, connect, and relate musical elements based on their **similarity**. The rhythmic similarity between a bass line and a kick drum can fuse the two in the foundation of a soul track. The timbral dissimilarity between a distorted electric guitar and a piano can help distinguish one as the lead and the other as the accompaniment. Dissimilarity can also create a boundary between groups; for example, a simultaneous change in key, instrumentation, and tempo can signal the beginning of a new section. Similarity is about balance. We typically favor some degree of similarity as it engenders feelings of cohesion, but too much similarity can become monochromatic. **Contrast** is a vital aesthetic and structural ingredient. Similarity is a defining force when applied to musical form. Often, composers write coherent individual sections but struggle to put them together. Considering the similarities between adjacent parts in terms of individual voices, parameters, and the holistic picture can help make connections.

Similarity is a fundamental ingredient in the perception of **continuity**. Continuity provides a feeling of movement and direction, tracing contours and vectors along musical parameters and the work as a whole. Continuity is paused or muted at group boundaries, such as at the end of a phrase or section. The extent of continuity's fracture is proportional to the magnitude of the boundary (e.g., breaks between movements are typically more dramatic than those between melodic segments). Repetition is a core part of continuity, but so is transformation, which stimulates the sense of forward movement and progression. Breaks in continuity can be dramatic and exciting, particularly when foreshadowed (e.g., risers and drops in EDM). The topic is fascinating and fundamental to the compositional process; it was the subject of my dissertation, *Understanding Musical (Dis)continuity* (Barton, 2012).

The **proximity** of elements affects how musical groups are formed. Notes that are temporally "close" to each other tend to form groups, while those that are "distant" are separated by a group boundary ("close" and "distant" are relative terms). Proximal grouping results in *runs* of notes and *gaps* in between them, affecting our sense of where rhythmic patterns start (see Garner & Gottwald, 1968). Proximity is a factor in pitch perception; notes that are "closer" in height tend to be integrated into the same auditory stream (Bregman, 2001). Our ability to form perceptual streams allows us to distinguish musical **voices**, even when those voices are defined by similar timbres (e.g., in Nancarrow's *Studies for Player Piano*).

Image schemas are concepts derived from fundamental physical experiences used metaphorically to understand complex ideas and organizations (Snyder, 2000). They have a kinesthetic component and are thought to develop early in life as we navigate the external world.

Examples include *up / down*, *centrality*, *motion-linkage–causation*, *paths–goals*, and *inside / outside*. Image schemas are useful in describing musical relationships and are inspirational in creating them. *Centrality* is pervasive in musical expression, exemplified in the tonic of a key or the downbeat of a measure. A *center* could be a specific pitch, registral band, harmonic or rhythmic configuration, texture, or timbral quality that sets a home base. A voice(s) can wander from this location but eventually returns home, bringing a sense of familiarity and resolution. Musical elements *link* (depending on aspects such as similarity) so that one element is understood to *cause* the next, imparting a sense of continuous *motion* along a *path* toward a *goal*. A core group of instruments (e.g., a rhythm section) may provide an *inside* foundation, while *outside* spaces are populated by sounds that are distant or fleeting (e.g., an atmospheric pad). A related concept is that of **figure / ground**, which refers to a type of perceptual organization where one element(s) appears front and center as the *figure* while supporting others appear in the back*ground*.

IDIOMATIC WRITING

Composing requires awareness of what is *idiomatic* and what is possible for players and instruments. Some chord **voicings** (the "vertical" positions of notes in a chord) stretch beyond the span of some pianists' hands. When writing a rapid upward arpeggio for a flute, some pitch sequences are easier to produce than others. By studying orchestrational practices and working with performers, composers gain insight into these affordances and constraints, thereby forming a fundamental base of knowledge. While "orchestration" is often taught in the context of European Art Music, this practice is not confined to the orchestra. Whether working with digital synthesizers, guitar amps, or musical robots, all composers must learn about the affordances and constraints of instruments and players.

If idiomatic writing is essential, *what kinds of musical gestures are idiomatic to machines?* Exploring this question is the project of designers, builders, and composers. Of particular interest is how machines can facilitate our expressivity in ways that human players cannot. Some answers derive from the affordances and constraints of specific machines, while a journey through *rhythm*, *pitch*, and *texture* can also be revelatory.

RHYTHM

For those requiring an introduction or refresher to rhythm basics, see the Rhythm and Time section of the Companion Website ▶.

The temporal precision of musical machines opens the door to a panoply of rhythmic possibilities. Musical machines can produce complex **rhythmic configurations**. A rhythmic configuration refers to the proportions of temporal intervals that constitute a rhythm. Consider the rhythm in Figure 9.1:

FIGURE 9.1. **A rhythmic configuration.**

If we assign the value 1 to the beat (a quarter note here), then the proportions of the rhythm are

1. .25 .5 .25 .33 .33 .33 1.

This type of representation is easier to work with when exploring more complex rhythmic configurations. Consider the following rhythm:

.777 .43 .128 .907 .325 .49 .777

Traditional notation, optimized for low-integer beat divisions, is of little use in representing or realizing this rhythm. Machines are uniquely capable of performing such *microtiming* nuances accurately and repeatably; thus, this is a **machine rhythm** (Barton, 2016). Machine rhythms are notated as a string of numbers, making their generation, manipulation, and performance in machinic domains idiomatic. *Machine rhythms* evoke a quality that is distinct from those made by humans. As they cycle, grooves emerge that are precisely lopsided, forged by **micro-syncopations**. The musical possibilities of *machine rhythms* serve as a motivation to design, build, and utilize musical machines.

Machinic microtiming capabilities can also be applied to **expressive timing**. Expressive timing refers to the subtle temporal variation from ideal notated proportions, defining musical styles from the **rubato** of a Chopin piano prelude to the **feel** of Elvin Jones's drumming in a jazz tune. Some researchers have emulated human expressive timing using computers (see Cancino-Chacón et al., 2018, for a review). Alternatively, machinic microtiming variation is an area within which composers and programmers can be creative and contribute to the development of new musical styles.

Composers can manipulate rhythm using mathematical operations. Rhythmic patterns, including configurations and meters, can be altered via addition (e.g., add a quarter note at the end), subtraction (e.g., remove an eighth note in the middle), multiplication (e.g., quarter notes => half notes), and division (e.g., quarter notes => eighth notes) to yield new musical ideas. A **metrical hierarchy** (Figure 9.2) provides a probability distribution that composers can use to determine the position of new notes. Weighting stronger positions provides clear structure and regularity, while weaker positions provide **syncopations** and ambiguous configurations. I incorporated a version of this method in *Experiment in Augmentation 2* (2018), where a human performer controls high-level aspects of the musical texture (e.g., *density*), while algorithms determine the pitch and temporal position of individual notes.

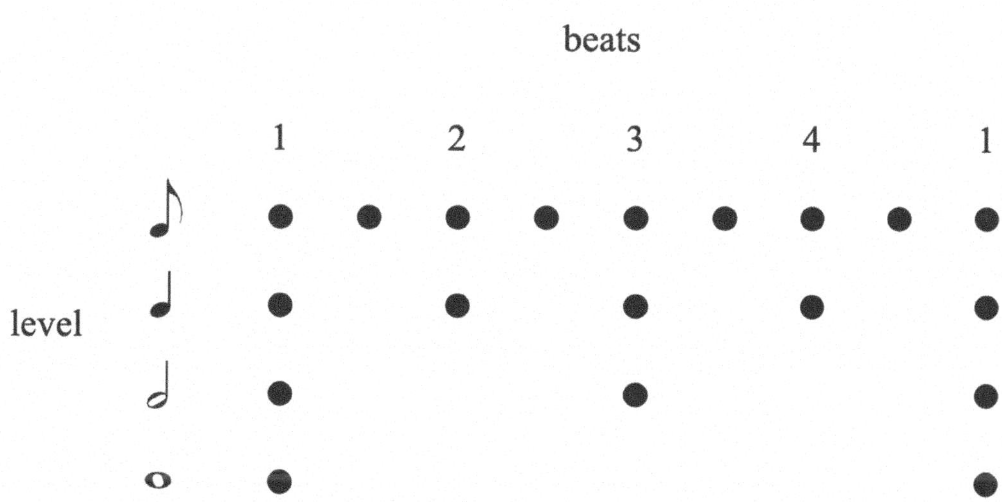

FIGURE 9.2. A metrical hierarchy for 4/4 meter.

The **rate** at which events occur is a fundamental aspect of rhythmic experience, affecting our perception of local patterns and global parameters such as **tempo**. Rate changes shape our sense of rhythmic identity (Barton, Getz, et al., 2017), pitch, and even **texture** when speed is taken to extremes. **Speed** has been a visible arena for machinic performance, given our fascination with virtuosity and the ease with which machines can play fast, slow, and everywhere in between. Rate changes may be terraced or continuous, following contours defined by a limitless number of mathematical functions (see Figure 9.3).

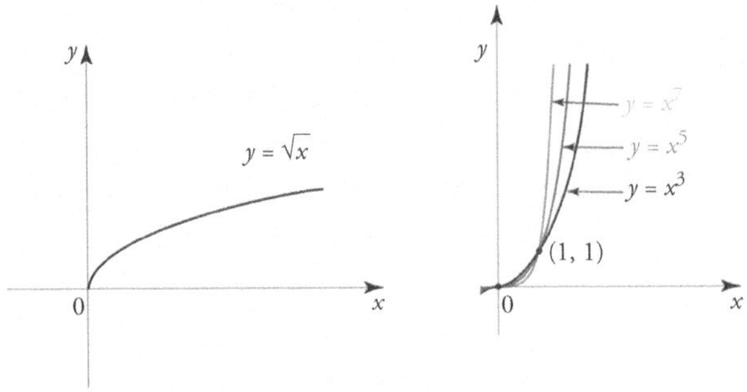

FIGURE 9.3. **Mathematical functions to apply to rate contours.**

Exploring tempo contours was a primary motivation for my composition *Tempo Mecho* (2019) (Media Example 4.1.1 ▶). The piece explores how the identity of a musical idea changes when heard at varying rates, both static and contoured (Figure 9.4).

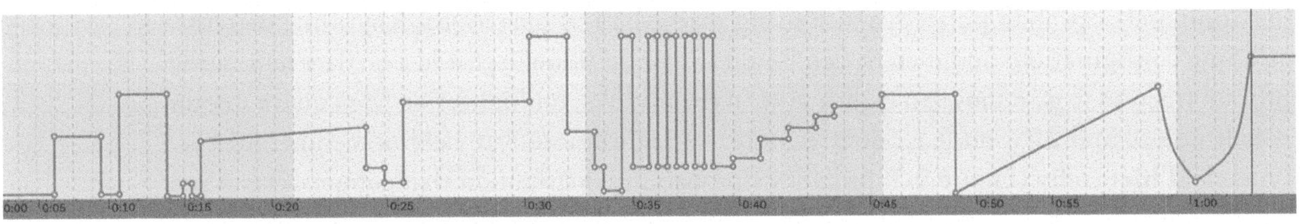

FIGURE 9.4. **Tempo curves of *Tempo mecho* (Barton, 2017).**

The faster and more complex the rhythmic expression (e.g., a terraced shift from half notes to 7:5♪, a meter change from 7/8 -> 17/32, an exponential tempo change from 88 to 110.7 bpm), the more difficult it is for humans to perform. Computer-based electromechanical systems perform mathematical calculations and physical actions with great speed and accuracy, making them well-suited for these applications.

Machines are adept at coordinating multiple voices, which can become challenging for human performers when complexity increases. Machines' temporal precision allows synchronization to be specified on the millisecond level. They can coordinate multitudinous voices in relationships that complex polyrhythms—cross-rhythms, polymeter, polytempo—require. Conlon Nancarrow explored these rhythmic possibilities in his *Studies for Player Piano*, setting tempo proportions from 12:15:17 (No. 17) to $\sqrt{2} : 2$ (No. 33) to $e : \pi$ (No. 40) in polytempic canons. The results illuminate rhythmic territory that remains fertile for future compositional exploration.

PITCH

The space of available pitches in any musical instrument or work is defined by a **tuning system** such as **just intonation** or the more familiar **equal temperament**. The less-explored world of **microtonality** exists between the steps of the chromatic scale. The precision and speed of musical machines make tuning options accessible in new ways. A machinic system with pitch feedback could ensure the stability and accuracy of a tuning layout. Alternate tunings could be selected in between and even during musical works, expanding the palette of pitch gestures and harmonic progressions. This was part of the motivation behind the MPR Lab's *Cyther* (Barton, Prihar, et al., 2017).

A composer can sculpt a pitch world using abstract, *top-down* ideas, including *centricity*, *function*, *implication*, and *progression*. A melody may deviate from and return to a *central* pitch. A type of chord (e.g., V7) may *imply* movement toward a goal (e.g., I) in a chord *progression*. These chords then *function* in specific ways: the V7 provides tension and anticipation of resolution in the I, which functions as a harmonic *center*. You can adhere to conventionalized harmonic systems (e.g., the major–minor tonal system) and their associated musical styles or create something new. While the latter may seem daunting, computers can help by allowing you to codify harmonic conventions algorithmically. For example, centricity can be achieved by ensuring that one pitch sounds more frequently than any other and that melodic lines that meander in a register return to a common starting point. Implication and progression can be established through rules or probabilistic systems such as Markov chains.

From a *bottom-up* perspective, musical machines allow pitch configurations that are not idiomatic to human instrumentalists. Non-anthropomorphic designs are exceedingly capable in **vertical** (chord voicings) and **sequential** pitch groups. Machine-produced chords can extend over wide pitch ranges, include many notes, and form complex shapes. Pitch glides (e.g., **portamento**) and sequences can comprise giant leaps and minute shifts, rapid changes of direction, and non-linear and asymmetric contours. These pitch shapes can define musical motives. The ability to explore pitch nuances motivated the MPR Lab's robotic musical siren *Parthenope* (Sidler et al., 2020), which generates pitch variations by adjusting the speed of rotating perforated discs.

Musical machines afford diverse **articulations** (e.g., staccato and tenuto) and **ornaments** (e.g., grace notes and trills). *Rise of a City* (2009) features a melodic line written for the robotic monochord *PAM* (by EMMI and the MPR Lab) that explores articulatory capabilities (Figure 9.5, Media Example 6.1.3 ▶). PAM's pitch-producing tangents produce rapid grace notes, while a solenoid-controlled damper creates durational articulations, including staccato and tenuto.

FIGURE 9.5. **Excerpt from the score of *Rise of a City* (Barton, 2009).**

The notation in Figure 9.5 gives a partial sense of sonic actualities in performance. The score was translated into a MIDI sequence on a DAW. The timing of elements, such as grace notes and damper engagements, was specified precisely in milliseconds (which traditional notation is not built for). Fine-tuning the score was critical in realizing my creative intentions. It involved discovering PAM's strengths and weaknesses, which shaped my perspective of the machine's performative *identity.* The process is similar to working with human musicians. A composer presents idealized notions in a notated score. Musicians interpret these ideas according to their abilities and those of their instruments, teaching the composer about what is possible and what works best. Musical machines and human performers are idiosyncratic, which requires an iterative process driven by curiosity and compromise.

Bottom-up harmonic possibilities are determined by the physical characteristics and capabilities of the musical machine. For example, a multi-string zither that tunes via tension could produce intonational idiosyncrasies depending on the individual and collective tension of the strings as well as the rigidity of the instrument's structure. The harmonies might be challenging to predict but also musically interesting, provided you adopt a flexible mindset.

I have seen many projects that did not afford the kind of control, particularly in pitch and rhythm, that we have become accustomed to in commercial instruments. Often, my advice is to stop trying to make the instrument something that it is not and instead embrace what it *can* do. This requires shifting how you think about music; it invites you to be curious about affordances and constraints. This is key in working with human performers and creative innovations generally.

TEXTURE

Musical **texture** emerges from the qualities of and relationships between voices. It is a **topography** whose surface is populated by variations in forms, features, relief, and contour. Some elements are small and distributed; others are salient and in focus. The organizing principle of *figure / ground* plays a role in texture as some elements appear front and center, while others color a complementary background. In the vernacular of music, these roles are described with terms such as *lead, soloist, featured, accompanist,* and *supporting.* The core idea is that musical aspects are prominent to varying degrees, with positions and roles determined by the qualities of elements relative to each other. The collective identity of these voices forms musical *texture.*

Music theoretic concepts such as **monophony** or **polyphony** can guide the creation of a musical texture, but they are not the only possibilities. Alternatively, one can use spatial or physical metaphors as a source of inspiration. For example, *thick* ⇔ *thin* could refer to the number of voices present or the "weight" of a sound that is a function of its spectral profile. *Wide* ⇔ *narrow* could refer to the "distance" of the pitch register over which voices are spread. Textures can be *static* or *dynamic* (more physical metaphors). They can start in a monochromatic, dense band and separate into striated lines marked by a cornucopia of colors. A single, cycled melody can initiate a section over which other voices are incrementally superimposed, creating a texture that is dense and full. Other dimensions are possible, including *rough* ⇔ *smooth,* and *dense* ⇔ *sparse,* which can be the inspiration and model for a section or work.

The quality of a texture is a function of the number of voices (or elements), how they are orchestrated, and their distributions in pitch, temporal, and dynamic space. These factors contribute to the *simplicity* <-> *complexity* of constituent elements and the relations between them. Measures have been developed to quantify musical complexity, which could be used in the compositional process (e.g., Essens, 1995; Shmulevich & Povel, 1998), but there is no obligation to be so specific. Consider the number of elements present and their alignment with each other or an external reference, such as a scale or temporal grid. Balance is usually good, for example, pairing a complex lead line with an unadorned accompaniment. Extremes may be enticing for the adventurous, such as tangling manifold elements in a web that obscures individual identities (e.g., the *micropolyphony* of György Ligeti's *Atmosphères*) or crafting minimalistic textures that meditate in unadorned simplicity.

The process by which such qualities are generated may be *top-down*, that is, having an image of the texture as a starting point (e.g., a flock of birds), or *bottom-up*, that is, enacting "local" rules that affect the relationships between individual elements out of which larger organizations emerge (e.g., the next pitch is more likely to move up than down). The nature of this process, which can be formalized or intuitive, will influence the simplicity or complexity of the resulting texture. Simple local rules can produce complex emergent structures; numerous, diverse voices can combine in singular gestures that are grand and powerful, yet simple.

Musical machines are uniquely capable of generating musical textures, as they can produce diverse voices simultaneously, independently, and with precise control. As any percussionist or pianist knows, getting two hands to perform separate actions simultaneously is a cognitive and motoric challenge. An 88-key piano- or percussion-playing machine with actuators over each key / object can produce 88 independent lines across a wide range of pitches and timbres. Actuation speed affords textural exploration. Machines can excite a drum or string at 20 Hz (50 ms IOI) or faster, melding individual articulations into a sonic blur. The temporal, pitch, and timbral effects possible with this kind of machinic actuation add colors to the textural palette. The synchronization of multiple machines with the capabilities described further expands the possibilities of new musical textures.

MUSICAL EXAMPLES

The preceding concepts are illustrated in compositions for musical machines, including Nancarrow's *Studies for Piano Player*, Ligeti's *Poème Symphonique* and *Continuum*, George Antheil's *Ballet Mechanique*, and Squarepusher's *Sad Robot Goes Funny*.

Conlon Nancarrow was a prolific and adventurous composer who voiced ideas almost exclusively through musical machines, specifically the player piano. His *Studies for Piano Player*, composed over the second half of the twentieth century, illustrate some of the myriad possibilities of machinic sonic textures. Compositionally, his constraints for the studies were rigid: he used only one instrument, which limited the available timbral palette. Timbre is one of the primary ways we distinguish voices, so this orchestrational choice also had implications on musical texture. Nancarrow responded to this restriction by working in register and rhythm to clarify and obscure voices and their interactions. Some studies begin with clear melodic lines that are transformed and woven into complex polyphony, culminating in massive, thick textures at the work's conclusion. For example, *Study No. 5* (Media Example 9.1 ▶) begins with hesitating, punctuated sonorities over which rapid melodic contours and chordal rhythmic patterns are superimposed. The texture evolves into a chaotic blur at the end of the piece, where semblances of previous themes percolate out of a sonic deluge only possible on a machinic piano.

The reverse of *Study No. 5*'s textural development is heard in György Ligeti's 1962 composition *Poème Symphonique*, for 100 metronomes (Media Example 9.2 ▶). The mechanical metronomes are set to different speeds, wound fully, and started simultaneously. A sonic cloud is unleashed, eventually dissipating to reveal individual rates that create a myriad of polyrhythms. The machines tire out one by one, leaving a single voice to conclude the piece. As there is little pitch, timbral, or dynamic variation between instruments, the work is fundamentally about rhythm and texture. In this case, both musical aspects are determined by mechanical capabilities, namely, the rates the metronomes can produce and the durations over which their spring-wound clockwork mechanisms can make sound. The nature of the piece is machinic, but a human element is also palpable. Ten performers start the metronomes manually, each responsible for ten metronomes. The devices are to start simultaneously, but exact synchronization is manually impossible, resulting in nuanced rhythmic phase relationships. The beginning could be synchronized more exactly through electromechanical means, but this would produce a different musical work.

George Antheil's *Ballet Mechanique* (1924) (Media Example 9.3 ▶) was one of the earliest and most ambitious works scored for an ensemble including musical machines, which expanded possibilities in musical texture. The original version was written for 16 player pianos, two pianos, three xylophones, seven electric bells, three propellers, a siren, four bass drums, and one tam-tam. Antheil sought to capture the essence of American industry and architecture through percussive material interactions between steel, wood, and rubber. Instead of attempting to represent the sounds of the industrial age directly, he commented on the simultaneous beauty and danger of an "unconscious mechanistic philosophy" (Schmidt-Pirro, 2006, p. 412). Synchronizing the instruments was not technically possible in Antheil's time. Such coordination became achievable later in the century, resulting in a fully machinic performance of the piece in 2006 at the National Gallery of Art in Washington, DC, realized by Paul Lehrman and LEMUR, Eric Singer, director.

The music is a torrent of activity. Small fragments repeat verbatim, bells sound alarm, swooping contours interject from player pianos, and low rumbling propels the herd forward, creating anticipation of impending action. Absent leaders and followers, the texture is an egalitarian plane formed by the collective. The components of the texture are fragments and gestures rather than melodies and themes. They momentarily pop out of the noisy barrage, grabbing attention before being subsumed by the crowd. The characters of these gestures are uniquely machinic in their precise timing, undifferentiated dynamics, and registral expanse. Their combination is an overwhelming confluence of noise and signal, expressing human feelings about a time and place.

György Ligeti's *Continuum* (1968) (Media Example 9.4 ▶) is a collage of rates and lengths in the form of superimposed ostinati. The texture is one of stasis and movement. The continual cycling of pitch patterns creates a feeling of rapid orbits of varying sizes. At the same time, movements are predictable in their repetition, giving the sense that gestures *stay in place*. Varying the length of pitch patterns affects their periodicity, which slows the sense of movement. The piece was originally composed for harpsichord but was arranged for and realized on machinic pianos and a barrel organ. The movement is consistent between the versions, but the differences in instrumentation have textural effects. The piano is cleaner and more discrete, while the organ produces something fuzzier, expansive, and amorphous.

Squarepusher's *Sad Robot Goes Funny* (2013) (Media Example 9.5 ▶) uses the considerable abilities of *Mach*, the robot guitarist of the band *Z-Machines*, to create musical textures that are uniquely machinic. The bridge section of the work (~2:25) consists of dense, extended chords and soloistic monophonic lines. It begins with chords segmented in time and connected by passagework. As the music develops, the space between chords shrinks, resulting in rapid, densely voiced harmonic shifts. At the same time, monophonic lines accelerate, lengthen, and are harmonized, creating undulating contours and scattered dispersions at incredible speeds. Unlike Nancarrow's *Studies for Player Piano*, the texture never becomes so dense that individual components are indistinguishable. Here, the listener can keep track of the action, though it occurs so quickly and contains so much information that it teeters on the brink of comprehensibility. The harmonies and soloistic lead lines are situated in jazz but have been transformed to create an expression that is characteristically machinic, defined by a texture that juxtaposes figure and ground in a single, coherent voice.

Human–Machine Interactions

The preceding musical examples were written exclusively for musical machines. Juxtaposing humans and machines in performance illuminates the expressive spaces unique to each and the similarities between the two. This was part of the motivation for *Rise of a City* (2009) (Media Example 6.1.3 ▶), for human-played electric guitar, the robotic monochord PAM, and robotic percussion. The first and final sections of the piece feature ideas sounded by both PAM and the guitar. Sequences are synchronized between the two instruments but also offset in time, providing different perspectives to view and compare the performative qualities of the player-instrument entities. The timbral processing of the two string instruments is similar but distinctive enough to match the character of the players and the parts. The middle sections of the work show the two voices in a figure-ground relationship: PAM assumes a lead voice that the guitar accompanies in the B section, with those roles reversed in section C. This structure illustrates how PAM's character differs from the guitar's when brought to the front of the stage. This difference emerges from exploring the idiosyncratic characteristics of each voice and writing lines that suit each of them uniquely.

COMPOSING CONTEXT—GUITARBOTANA AND MARBLE MACHINE

The context within which humans and machines interact is composable, as are the mediations that facilitate communications between them. Mari Kimura's *Guitarbotana* (2004) (Media Example 9.6 ▶) features a human violinist and LEMUR's robotic string instrument, *GuitarBot*. Philip Auslander (2017) argues that Kimura, who wrote, programmed, and played the piece, has a ventriloquial relationship with the musical machine. She assumes the role of a performer herself (violin) but displaces her agency as a composer and programmer onto the instrument. At the same time, because *GuitarBot* can respond to human input in various ways, Kimura acts as a mediating force of the machine's performance:

> In performing with a robot, I assume a dual responsibility—for my own musical trajectory as well as that of this metal partner that doesn't think . . . So my job as a performer is to merge with the robot in terms of musical thinking during performance . . . . (Kimura, quoted in Auslander, 2008, p. 113)

For her, this responsibility includes mediating in a way that creates the perception of the machine as endowed with human capabilities that it does not possess:

> My compensating for the robot's or computer's lack of musical "integrity" as the performance goes along should be hidden from, or unnoticeable to, the audience. In short, my aim is that the performance as a whole come across to the audience as if the robot or computer is thinking, feeling, and being sensitive; that it possesses the "rights and responsibilities" of a true musician. (Kimura, quoted in Auslander, 2008, p. 114)

Auslander (2008) argues that Kimura's mediations include the visual presentation of the performance. Kimura wears a black dress, a uniform of classical music performance, and physically interacts with the instrument as one would with another human, directing her gaze and physical gestures toward and in response to the machine. These aspects convey a performance of art music, evoking associations with concert halls, attentive listeners, and a musical canon that spans hundreds of years.

Compare *Guitarbotana*'s visual scene to that of Wintergatan's *Marble Machine* (2016) Media Example 3.1 (▶). Martin Molin of the band *Wintergatan* is also dressed in black, contrasted against a background bathed in white, facing a machine that is as intriguing as it is imposing. Molin's appearance and physical gestures signal musical performance but do not evoke associations with European Art or "classical" music. Molin wears a tight-fitting black jacket with decorative patterns on the collar and cuffs, buttoned in the front over a long-sleeved beige shirt. His black jeans are also tight, extending into black socks and black high-top shoes. His mid-length hair is disheveled in a stylized way, adorned by a pair of white headphones that bear a striking resemblance to a pair owned by one of my uncles in the 1970s. Molin taps, bops, stomps, and grooves to the beat of the music, throwing levers back and forth, fretting an electric bass, and entraining to the rotations of the crank mechanism. These gestures, in combination with the music itself (e.g., the production of the drum sounds), are spoken in the language of rock performance, inflected by a 70's vibe.

Like Kimura, Molin faces a similarly unfamiliar amalgam of mechanical parts, though the two machines differ in meaningful ways. *Marble Machine* is made primarily of wood, while *GuitarBot* is made of metal, evoking contraries such as warm versus cold, organic versus synthetic, and pliable versus unyielding. The means of action are transparent for both, but in varying degrees. *GuitarBot*'s moving bridge and picking mechanisms are prominently visible, although the control systems are not (e.g., circuit boards, computer code). *Marble Machine* is anomalous and even anachronistic in that it eschews computer control, instead relying purely on mechanical means, much like its golden-age predecessors. Unlike many of those predecessors, the mechanisms of *Marble Machine* are not hidden in an enclosure. Instead, they are visibly displayed, inviting the viewer's curiosity about how the machine works (illustrated in the video production, which traces the journey of a single black marble).

Kimura's and *Wintergatan*'s roles as composers in these cases are not limited to choosing which pitches occur at which times. They also compose an interactive context that shapes audience perspectives on the nature of the musical work and the identity, capabilities, and relationships of the performers. These performances have captured audiences' attention not just because of the technical ingenuity of the machines, harmonic complexity, or the catchiness of melodies. The contextual factors mentioned fit together "just right," creating performances that are grounded in cultural traditions, technically innovative, and aesthetically and intellectually compelling.

MACHINE AS INTERMEDIARY—CYTHER

Human–machine interaction can occur through a shared physical medium. The Music, Perception, and Robotics Lab's *Cyther* (Barton, Prihar, et al., 2017) is a human-playable robotic zither whose actuators are positioned underneath the 10 strings, allowing a human to play from the top of the instrument. This arrangement facilitates human–machine interaction that is not typically encountered in instrumental performance. As both the human and the machine can excite (pluck, pick, hammer) and damp each string, players assume varied roles. For example, when the instrument excites the strings rapidly, sounds dense chords, or some combination of the two, the strings' sustain can effuse harmonic mush. A human assumes the role of a filter, muting some strings completely, some partially, and allowing some to ring. Our physiological flexibility enables this to be accomplished with fingers, palms, hands, arms, or other objects at various angles and forces, creating different musical effects. The resultant **cooperation** between human and machine in creating music takes advantage of the unique physical capabilities of each performer.

In a *cooperative* interaction, machinic output can be predetermined, algorithmically generated, or produced by another person. The latter scenario facilitates human–human interactions through a machinic medium. Nate Tucker and I have explored these possibilities in improvisations where I would live-code *Cyther*'s output while Tucker would simultaneously pluck, hammer, drum, and mute the instrument's strings. One such improvisation, *Life's Node*, was recorded and featured on Tucker's 2015 album *Off Grid*. *Cyther* also functions robotically to facilitate *cooperative* human–machine interactions. In my 2017 work *Experiment in Augmentation 1* (Media Example 9.7 ⓑ), a human (me) plays the instrument, and the output is fed into a computer program that interprets what was played and generates musical responses. At times, I lead, with the robot imitating and transforming my ideas; at others, the robot generates materials that I filter and shape. In the middle of the piece, I use aluminum rods to mute and stop the strings. Toward the end of the section, I place the rods on the strings while the solenoid exciters hammer them from underneath. The rods are energized, bouncing unpredictably on the strings in terms of rate and location, creating a sonic texture driven by physical interactions that are only achievable with *Cyther*. I had not heard sonic textures like these before, which excited me. The moments were made possible through human–machine *cooperation* within a shared physical medium. They have convinced me of the potential for musical machines to open new doors of creative expression.

RESPONDING BY IMITATING AND TRANSFORMING

The idea of machines responding to human input begs the question of how such abilities are possible. How does a machine listen? What should it respond to? How will it generate a response? When will it respond? The following are compositional techniques that can help answer these questions.

One voice can **imitate** another. The first step is determining what will be **repeated**, often a pitch or rhythmic sequence, but could also include aspects such as dynamics or articulations. This requires time to interpret what was played, marking the start and end of statements. The respondent must then choose when to reply. For human performers, these requirements mean that such dialogue usually occurs in different sections, such as **call-and-response**. Machines can process information and execute actions so quickly that they can mimic another performer in *real time*. There will be some latency in the response, but it can be minimal enough so that the effect of the gesture is that the two performers appear to be playing together, particularly during slow passages.

From imitation comes **transformation**. Transformations can occur along any musical parameter and be accomplished algorithmically, making them particularly well-suited for a computer-controlled instrument. One method is to transform the **pitch** content of an input musical statement while maintaining its rhythmic and temporal structure. **Transpositions** shift pitches up or down by a specified interval. **Inversions** involve "flipping" a pitch contour about a defined axis, such as the first note in the sequence. The **retrograde** of a musical sequence is achieved by reversing the order of its elements. **Retrograde inversion** involves flipping and reversing a sequence of pitches. Each of these examples is shown in Figure 9.6:

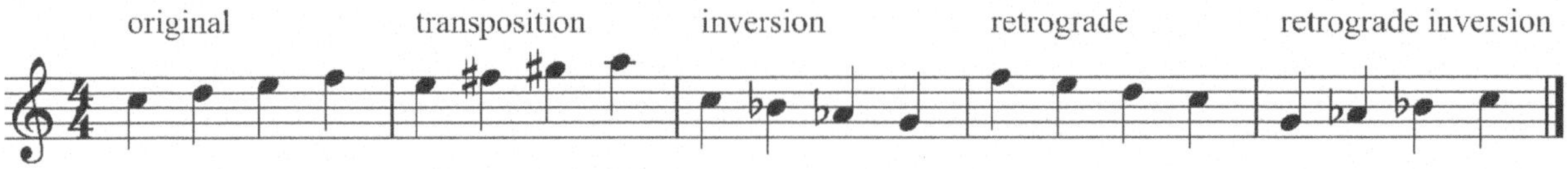

FIGURE 9.6. **Pitch transformations.**

The **rhythm** of musical sequences can also be transformed. Like pitch sequences, the order of the durations of a rhythm can be reversed. Durations can be expanded or contracted by a specified ratio, known as **augmentation** and **diminution**, respectively (Figure 9.7).

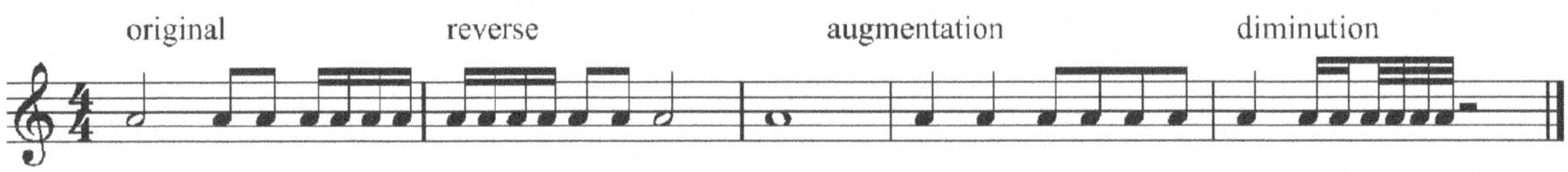

FIGURE 9.7. **Rhythmic transformations.**

Some transformations can be applied to both pitch and rhythmic sequences. Pitch and rhythmic sequences can be **rotated** so that elements appear in different positions. Rotations can affect pitch and rhythm together or independently (Figure 9.8).

FIGURE 9.8. **Pitch and rhythmic rotations.**

Elements can be **added** or **subtracted** from sequences to make textures more or less dense. The number and position of elements and the modification frequency can be fixed or variable, with variability determined by random or probabilistic processes. Subtractions are more straightforward as they involve determining the number and position of elements to be removed in a specified period. For example, if there are five notes in a phrase, the fifth note will be removed after 5 seconds.

Additions are more complicated. They also involve determining how many elements will be added, but each new element's pitch, duration, and velocity must also be specified. Pitches may be chosen from a set (such as a key or scale), a group (such as those present in the previous measure), or determined from a sequence of intervals (e.g., a Markov chain). Velocities are usually similar between proximate notes, with some variation being desirable. The position of new notes is perhaps the most complicated aspect. First, will simultaneities (i.e., chords) be allowed? If not, existing temporal positions of notes must be specified so they are removed as possibilities for added notes. The temporal positions of new notes are selected according to a time scale. Scales with small gradations (e.g., milliseconds) provide for rhythms that are effectively **free**. Beat-based rhythms use a temporal grid anchored by a note value (often a quarter note) of variable duration divided and concatenated by 2^n. Larger divisions (e.g., eighth notes) may (or may not) be more common than smaller ones (e.g., 32nd notes) when setting up a probability distribution for selecting the position of a new note. Once the position is selected, the duration of the new note must be adjusted according to the position of the subsequent note. The duration of the preceding note also needs to be adjusted to ensure it does not overlap with the onset of the added note (this step depends on the instrument: overlaps may be desirable with some instruments, such as pianos, while problematic with a flute or monochord). The results of these processes can be seen in Figure 9.9:

FIGURE 9.9. **Addition and subtraction of elements in musical sequences.**

Adding and subtracting notes can affect cross-rhythms, phrase length, and meter changes. The length of a rhythmic sequence is usually the same as the pitch sequence to which it is bound, but that does not have to be the case. **Isorhythm** is a compositional technique that emerged in the polyphonic vocal music of thirteenth-century Europe, applying a repeating rhythmic pattern called the *talea* to a pattern of pitches called the *color*, which may be of the same or different lengths. The idea of repeating patterns of various lengths is a powerful one that has inspired music worldwide. It remains vibrant today and is particularly suited to the mathematical capabilities of computers.

Examples of how transformations could be implemented algorithmically:

- Every measure there is a pitch rotation of two elements to the left while the rhythmic sequence remains the same.
- Every seven beats there is a 50% chance of a pitch rotation to the left with the amount of the rotation determined according to a Gaussian probability distribution with a mean of 3. Rhythmic rotations of three to the right occur every 10 beats.
- Every 5 seconds there is a 25% chance that elements are added, with the number of elements chosen randomly between 1 and 5.
- If there are more than 17 notes in a measure, there is an 85% chance that notes will be removed; if there are fewer than 5 notes in a measure, there is an 85% chance that notes will be added; the number of elements in both cases is determined by a probability distribution. If the number of notes is between 5 and 17, there is a 50% chance that one note will be added and a 50% chance that one note will be subtracted every measure.

The preceding examples are just a few among countless possibilities for musical transformations. These have been used by composers and improvisers for hundreds of years and are well-suited for algorithmic and computational implementations. Modifying the parameters of such rule-based systems can be laborious, but it is a process that reveals the character of your aesthetic preferences as a composer, musician, and listener and gives insights into the type of musical statements that are exciting and innovative (or not).

RESPONDING WITH SOMETHING NOVEL

There comes a time in any interaction, whether in conversation or musical improvisation, when it is time to change the subject and move to something new. The preceding descriptions of interaction techniques have involved machinic *responses* to human inputs, but if the machine is to be a true partner and collaborator, shouldn't it sometimes be able to direct the discourse? If your answer to this question is yes, then the *hows* and *whens* of such machinic interjections must be determined, which are complicated problems. The fundamental issue is that, as humans, we get tired or bored of situations, which prompts us to act in ways that reinvigorate our interests. This may involve introducing something **novel** that has not been experienced before or bringing back something familiar but not recent, which effects a *change of pace*. The problem is that these actions are motivated by *feelings*. Feelings of interest and boredom are alive; they grow and diminish according to personal and contextual factors across a range of time scales. These aspects are fundamentally human (or, more accurately, animalistic) and are outside the boundaries of machine capabilities.

The preceding does not mean that the introduction of "new" ideas by a machinic collaborator is impossible; instead, such actions must result from motivations that are not based on feelings. One approach is to use random or probabilistic processes to determine when a new idea is introduced. A simple rule might be that the more a statement is repeated, the more likely it is that a new one will be introduced. Another approach is to analyze the conventions of specific musical works or a collection of works that constitute a musical genre. How many individual elements are contained in rhythmic patterns, melodies, phrases, chord progressions, and sections, and what are their durations? How often do these elements change? Do some elements change more than others (e.g., a drum pattern vs. a lead vocal)? Such analyses, whether manually or computationally performed, can help make machinic responses contextually relevant. Feedback from collaborators (including *robots*) may influence whether or not to introduce a new idea. For example, a visual gesture can indicate that an idea should be sustained, which the machine considers when choosing how to act (developing such a system of signs is an integral part of collective human improvisation).

The *content* of new ideas needs to be forged. Designing an idea that will spark interest in collaborators and audiences is a delicate matter. If introduced at the beginning of a performance, expressive space abounds as boundaries and expectations have not yet been erected (though some do exist depending on factors such as the performance space, concert program, attire, and configuration of the ensemble and audience, etc.). If introduced in the middle of a performance, the degree of similarity to previous material is a factor. Too different, and the statement is contextually insensitive or even oblivious; too similar, and the thirst for change is unsatiated. There is no "right" answer to this balance; exploring possibilities is part of the fun in making art.

Generating a new idea involves establishing and choosing within musical parameters that provide shape and structure. These involve *low-level* parameters, such as pitch, time, dynamics, and timbre, as well as *higher-level* aspects, including rhythmic groupings, phrasing, density, and contour. The difference between generating new ideas in "isolation" and in an interactive exchange has to do with context. Interaction involves other agents with unique sets of knowledge, abilities, interests, and preferences, which must be considered when contributing new ideas to the dialogue. Figuring out the specific nature of this context and how it will affect machinic responses is a human problem; it is the question of how humans can understand the communications of others and be expressive themselves.

Summary

This chapter covers ideas and techniques relevant to those interested in composing for musical machines. One way to approach a piece is by focusing on abstract, *higher-level* aspects such as aesthetic preferences, intellectual interests, novelty, a central idea, psychological associations, emotional states, or spiritual expression. Music can also be composed from the *bottom-up* by organizing sonic elements in time on *lower levels.* Doing so results in linear and hierarchical groupings shaped by principles that describe how we perceive and organize information. These organizational principles include similarity, proximity, continuity, image schemas, and figure / ground. Groupings are the basis for musical structures that delineate a work's form and proportions. Structural elements include **motifs**, **sequences, phrases, melodies,** and **themes**, which combine, repeat, and transform into various musical **forms.**

Composing requires awareness of what is *idiomatic* and what is possible given the players' physical characteristics, whether human or machine, and their instruments. The musical gestures idiomatic to machines can be defined in the worlds of *rhythm* and *time*, *pitch*, and *texture*.

Rhythm involves grouping elements into temporal patterns. Meter is a form of temporal organization that is **hierarchical** and can be used to generate rhythms algorithmically. The temporal precision of musical machines enables complex **rhythmic configurations** involving **machine rhythms**, **micro-syncopations,** and interval relationships of greater complexity than the low-integer ratios typically encountered in human-made music. Machines are adept at **expanding** and **contracting** rhythmic sequences, in terms of both the number of elements, resulting in syncopation and meter changes, and rate, resulting in **augmentation** and **diminution.** Rate changes can occur according to complex ratios and contours, which can be derived from mathematical functions. Rates in human-played Western music are typically related mathematically by 2^n, though more complex ratios, traditionally notated as **tuplets**, are comfortably in the realm of machine performance. More complex rhythmic composites arise from **coordinating** diverging voices, such as **hocket**, **polyrhythm, cross-rhythm, polymeter, phase** relationships, and **polytempo.** Conlon Nancarrow's *Studies for Player Piano* explore these rhythmic possibilities in depth, using the compositional technique of **canon**. **Expressive timing** is an essential aspect of human performance that provides a model for machinic performance. Systematic approaches to machinic microtiming variation are an area for future exploration.

Musical organization involves combining pitches *vertically* (e.g., in a **chord**) and *horizontally* (e.g., in a melody); these groups form the basis for **harmony.** The space of available pitches is defined by a **tuning system** such as **just intonation** or **equal temperament**, which opens the doors to the less-explored world of **microtonality**. The precision and speed of musical machines make these worlds accessible in new ways. Harmonic relationships involve higher-level ideas, including *centricity*, *function*, *implication*, and *progression*, which are the ground upon which new conventions can be built. *Bottom-up* possibilities also abound where harmonic content is determined by the physical characteristics and capabilities of the musical machine. Musical machines allow pitch configurations that are not idiomatic to human instrumentalists in terms of **chord voicings, sequential movement, glissando** and **portamento, articulations**, and **ornaments**, which help to shape compositional and performance identities.

Musical **texture** emerges from the qualities of and relationships between voices; it is a **topography** whose surface is filled with variations in forms, features, relief, and contour. Texture can be described using formal concepts from music theory or spatial or physical metaphors, including *thick ⇔ thin*, *wide ⇔ narrow*, *rough ⇔ smooth*, *dense ⇔ sparse*, *stasis,* and *movement*. The quality of a texture is a function of the number of voices (or elements), how they are orchestrated, and their distributions in pitch, temporal, and dynamic space. These factors contribute to varying degrees of *simplicity <-> complexity* of constituent elements and the relations between them. These qualities can be generated through *top-down* or *bottom-up* processes that can be formalized or intuitive. Musical machines are uniquely capable of realizing novel musical textures, as they can generate multiple voices simultaneously, independently, and with precise control. Examples include Nancarrow's *Player Piano Studies*, Ligeti's *Poème Symphonique* and *Continuum*, George Antheil's *Ballet Mechanique,* and Squarepusher's *Sad Robot Goes Funny.*

Juxtaposing humans and machines in performance illuminates the expressive spaces unique to each and the similarities between the two. The context within which humans and machines interact can be composed, as can the mediations that facilitate communications between them. Mari Kimura's *Guitarbotana* and Wintergatan's *Marble Machine* are studies in such mediations involving attire, physical gesture, materiality, and the transparency of how sound is produced. Human–machine interaction can also occur through a shared physical medium, as evidenced in cooperative musical machines such as The MPR Lab's *Cyther* and musical works such as *Experiment in Augmentation 1.*

A machine can respond to others by **imitating** or **transforming** rhythmic and pitch sequences through **transposition, retrograde, rotation**, **inversion**, retrograde inversion, **subtraction**, and **addition**. Machines can also introduce new material, which requires crafting contextually relevant content and determining optimal timing. Generative models can be based on analysis of human performance, rule-based systems, or probabilistic algorithms. Interaction involves interpreting signals (e.g., cues) and considering the knowledge, abilities, interests, and preferences of external agents. Enabling such sensitivities in machines is one way that humans can understand the communications of others and be expressive themselves.

References

Auslander, P. (2008). Live and technologically mediated performance. In T. C. Davis (Ed.), *The Cambridge companion to performance studies* (pp. 107–119). Cambridge University Press.

Auslander, P. (2017). Lucille meets GuitarBot: Instrumentality, agency, and technology in musical performance. In T. Bovermann, A. de Campo, H. Egermann, S. -I. Hardjowirogo & S. Weinzierl (Eds.), *Musical instruments in the 21st century*: Identities, configurations, practices (pp. 297–314). Springer. https://doi.org/10.1007/978-981-10-2951-6_19

Barton, S. (2009). *Rise of a city* [unpublished score].

Barton, S. (2012). *Understanding musical discontinuity* (Doctoral dissertation, University of Virginia). ProQuest. https://doi.org/10.18130/V35C3S

Barton, S. (2016). Creativity in the generation of machine rhythms. *Proceedings of the 1st Conference on Computer Simulation of Musical Creativity*, 1.

Barton, S. (2017). *Tempo curves of tempo mecho.* Ableton Live [Computer Software]. https://www.ableton.com

Barton, S., Getz, L., & Kubovy, M. (2017). Systematic variation in rhythm production as tempo changes. *Music Perception: An Interdisciplinary Journal, 34*(3), 303–312.

Barton, S., Prihar, E., & Carvalho, P. (2017). Cyther: A human-playable, self-tuning robotic zither. *Proceedings of the International Conference on New Interfaces for Musical Expression*, 319–324. https://doi.org/10.5281/zenodo.1176266

Bregman, A. S. (2001). *Auditory scene analysis: The perceptual organization of sound: Vol. 2nd MIT Press paperback.* MIT Press, 1999. http://cognet.mit.edu/library/books/view?isbn=0262521954

Cancino-Chacón, C. E., Grachten, M., Goebl, W., & Widmer, G. (2018). Computational models of expressive music performance: A comprehensive and critical review. *Frontiers in Digital Humanities*, *5*, 25.

Essens, P. (1995). Structuring temporal sequences: Comparison of models and factors of complexity. *Attention, Perception, & Psychophysics*, *57*(4), 519–532.

Garner, W. R., & Gottwald, R. L. (1968). The perception and learning of temporal patterns. *The Quarterly Journal of Experimental Psychology*, *20*(2), 97–109.

Kostelanetz, R., & Cage, J. (1987). The aesthetics of John Cage: A composite interview. *The Kenyon Review*, *9*(4), 102–130.

Schmidt-Pirro, J. (2006). BETWEEN THE EUROPEAN AVANT-GARDE AND AMERICAN MODERNISM: George Antheil's "Ballet Mécanique." *Soundings: An Interdisciplinary Journal*, *89*(3/4), 405–429.

Shmulevich, I., & Povel, D. J. (1998). Rhythm complexity measures for music pattern recognition. *1998 IEEE Second Workshop on Multimedia Signal Processing*, 167–172.

Sidler, M., Bisson, M., Grotz, J., & Barton, S. (2020). Parthenope: A robotic musical siren. *Proceedings of the International Conference on New Interfaces for Musical Expression*, 297–300.

Snyder, B. (2000). *Music and memory: An introduction*. MIT Press.

Wintergatan (Director). (2016, March 1). *Wintergatan—Marble Machine* [Video]. https://www.youtube.com/watch?v=IvUU8joBb1Q

CHAPTER 10

REFLECTING INTO THE FUTURE

The previous chapters have revealed questions, knowledge, and artifacts that define the landscape of musical machines. With richer perspectives on modern robotic efforts, we see them less as disruptive innovations and more as continuations of older and storied traditions that had been lost in the sizable wake of the twentieth century's sonic reproduction technologies. The intrigue of eighteenth-century European viewers of musical automata is alive today. We marvel at complexity and ingenuity, are curious about how things work, and ponder philosophical questions about the nature of human experience. Artists and technologists remain fundamentally motivated by such matters and engage with them whether communicated explicitly or not. For any designer, builder, composer, or performer who works with musical machines, these big-picture ideas are in play. They provide connections with others across time and geography as well as inspiration to create novel works. This was true then, remains true now, and will continue to be true in the future. It is where the book started and where it ends.

What also persists is that the objects we use to make art, whether *materials*, *tools*, or *instruments*, both reflect and enable our expressivity. Our needs, desires, and ideas are *mediated* by a host of factors attached to the use of such objects, including the time and labor required to develop requisite knowledge and skills, the availability of resources, and the ability to connect to others who can help. These factors interact with our creative ideas in a symbiotic way; both shape the realization process *cooperatively*. Instrument builders and composers do not typically have ideas that materialize in pristine and complete forms. Creative vision is directed when hands get dirty: from working with strings, wooden resonators, motors, and computer code. The objects that result shape the expressions of their users in *message* and *identity*, both as seen internally and externally. They connect builders, composers, performers, and audiences through a shared physical medium.

Our fascination with musical machines and the inherent connection between expression and the objects we use to make art are ancient and persistent. What changes is the nature of those generative objects. The quest to find new ways to be expressive, or to reflect our current condition, is aided by the technologies that define the present-day (which could mean deliberately not using those objects, of course). At present, these are microcontroller platforms, motors, 3D printers, and neural net-crunching GPUs, which have expanded the scope of what is possible and provide us with a unique perspective on the big-picture questions posed earlier. Developing proficiency with these objects is a project for artists, engineers, and makers, enabling them to connect creative expression with technical knowledge and skills. The boundary between artist and engineer is porous, where engineers think about expression, creativity, and implications, and artists think about mathematical relationships, formalized processes, and design specifications. The interior chapters of this book are designed specifically to facilitate such disciplinary transience.

Musical Machines. Scott Barton, Oxford University Press. © Oxford University Press 2026. DOI: 10.1093/oso/9780197599891.003.0010

The presentation of such knowledge is not necessarily intended to make artist-engineers proficient in all areas, obviating the need for collaboration with others. Quite the opposite, one of the most important lessons that I have learned is when to stop and find someone who knows more about the topic than I do. To do so requires a foundation of shared knowledge in the form of concepts and terminology that allows for conversation, enabling you to convey ideas to someone else about what you want and understand their responses. These kinds of conversations are among the most fertile ground for creativity, for they reveal what is possible, sparking inspiration and dialogue about future possibilities not previously known. This is how a compositional desire for a timbre or articulation is translated into an electromechanical system that can produce such sound. As artists and engineers share knowledge, ideas are appropriated and remixed, and new forms of expression emerge. This is because expression does not occur in a vacuum; it is fed in an environment that is culturally and technically bountiful. It is facilitated by people who connect to others. The purpose of this book is to help build that shared knowledge and motivate those connections. It is up to you to employ those ideas in new tools, instruments, and musical experiences.

NOTES

CHAPTER 4

1 The creations of Animusic (*ANIMUSIC*, n.d.) are a wonderful source of creative inspiration for designing motion.

CHAPTER 8

1 https://olbeter.com/

INDEX

For the benefit of digital users, indexed terms that span two pages (e.g., 52–53) may, on occasion, appear on only one of those pages.

Tables and figures are indicated by an italic *t* and *f* following the page number.

The manufacturer's authorised representative in the EU for product safety is Oxford University Press España S.A. of El Parque Empresarial San Fernando de Henares, Avenida de Castilla, 2 - 28830 Madrid (www.oup.es/en or product.safety@oup.com). OUP España S.A. also acts as importer into Spain of products made by the manufacturer.
Printed and bound by CPI Group (UK) Ltd, Croydon, CR0 4YY
14/09/2026
02216311-0001